AND HIS CRITICS

David L. Hull

The Reception of
Darwin's Theory of
Evolution by the
Scientific Community

DARWIN AND HIS CRITICS

DARWIN

AND HIS CRITICS

The Reception of Darwin's Theory of Evolution

by the Scientific Community

DAVID L. HULL

The University of Chicago Press

Chicago and London

The University of Chicago Press, Chicago 60637
The University of Chicago Press, Ltd., London

© 1973 by David L. Hull
All rights reserved.   Published 1973
University of Chicago Press edition 1983
Printed in the United States of America

90 89 88 87 86 85 84 83   1 2 3 4 5

Library of Congress Cataloging in Publication Data

Hull, David L.
Darwin and his critics.

Reprint. Originally published: Cambridge, Mass.:
Harvard University Press, 1973.
Bibliography: p.
Includes index.
1. Darwin, Charles, 1809–1882. On the origin of
species.   2. Darwin, Charles, 1809–1882. The descent
of man.   3. Evolution.   I. Title.
QH365.08H84   1983   575.01′62   83–4855
ISBN 0–226–36046–6

DEDICATED IN APPRECIATION TO

Virginia and Glenn D. Bouseman

Iris and Robert H. Reid

Preface

Alvar Ellegård (1958) chronicled the reception of Darwin's theory of evolution by the general public as reflected by the reviews of Darwin's works in popular British journals. Ellegård also included an excellent chapter on the relation between the theory of evolution and nineteenth-century philosophies, but he intentionally excluded any extensive investigation of those reviews published outside Great Britain or the scientific objections raised to Darwin's theory by the scientists of his day. In a history of ideas, the effect which Darwin's theory had on the general public is very important, but from the point of view of understanding the etiology of scientific revolutions, the reception of evolutionary theory by the scientific community is of much greater importance. After all, evolutionary theory is first and foremost a scientific theory. Darwin himself claimed not to be especially interested in the reception of his theory by the general public and was a little embarrassed that it proved to be so popular in certain quarters. Rather, he was primarily concerned with the acceptance of his views by scientists, most especially by a few key biologists and geologists.

This book will deal with the reception of Darwin's theory of evolution by the scientific community as reflected in the reviews written by competent scientists. Whenever possible, these papers are presented in their entirety. On occasion, however, extensive quotation from the *Origin of Species* and other redundant passages have been omitted. Prefacing each review is a series of quotations from Darwin's correspondence, the purpose of which is to provide some suggestion as to the value which Darwin and his correspondents placed on the review, from both a scientific and a sociological point of view. Darwin was interested not only in the truth of his theory but also in its acceptance by the scientific community. Following each review is a short essay which presents a few pertinent facts about the author and which attempts to place the review in its conceptual and historical setting.

A second major concern of this book is the influence of philosophy of science on the reception of evolutionary theory. One would expect men like John Herschel, William Whewell, and John Stuart Mill, who had carefully investigated the nature of science as such, to be better able to evaluate a new and novel theory than ordinary scientists and philosophers. Such was not the case. All three men rejected the tenets of evolutionary theory either entirely or in part. As the numerous comments scattered throughout the reviews of the *Origin of Species* included in this anthology amply attest, one of the major objections to evolutionary theory was that it was not sufficiently "inductive." The reviews will speak for themselves as far as the scientific objections to evolutionary theory are concerned. The philosophical objections are another matter. The purpose of the Introduction to this volume is to set out Darwin's own ideas on the nature of science and to trace the interconnections between nineteenth century philosophies of science and the reception of evolutionary theory. In the Introduction I have tried to bring some order out of the chaos that confronts the modern reader attempting to understand why so many nineteenth-century philosophers and scientists found evolutionary theory as a theory so objectionable.

The three works which come closest to addressing themselves to the same subject that is treated in this volume are Gertrude Himmelfarb's *Darwin and the Darwinian Revolution* (1959), Gavin de Beer's *Charles Darwin, A Scientific Biography* (1964), and Michael Ghiselin's *The Triumph of the Darwinian Method* (1969). Himmelfarb approaches her subject as a historian and as such manages to transmute the usual plaster of Paris image of Darwin into flesh and blood, but the philosophic and scientific discussions in her chapter, "Analysis of the Theory," are extremely untrustworthy. One wonders why someone with such a marked distaste for science in general and evolutionary theory in particular should have chosen to write a biography of Darwin.[1] Both de Beer and Ghiselin are competent biologists. In their works they argue that Darwin's scientific method is roughly equivalent to the hypothetico-deductive model set out by Karl Popper (1959) and Carl Hempel (1965). According to these philosophers, the truth of a scientific hypothesis is independent of its source. No method of discovery can guarantee the truth of the resulting generalizations. What counts is verification. It does not matter how Darwin came to formulate his theory of evolution by chance variation and natural selection. What does matter is that his theory was empirically meaningful and

1. For reviews of Himmelfarb's book, see Mayr (1959b) and Appleman (1959).

could be confirmed or at least refuted by scientific investigation. Darwin conformed to this model of science because he spent most of his adult life trying to show that evolutionary theory was true—or at least not false. In the Introduction to this volume, Darwin's method and the appropriateness of the modern hypothetico-deductive model will be examined somewhat more critically. The primary purpose of the introduction, however, will be to trace the interrelations between the reception of Darwin's theory of evolution and the philosophies of science held in Darwin's day.[2]

Some explanation must be given of the rationale for the selection of the papers to be included in this volume. Two considerations were primary: the relative accessibility of the review to the modern reader and its scientific or philosophic interest. The reviews of T. H. Huxley and Asa Gray were among the most important defenses of evolutionary theory, but they have been omitted because they have been anthologized repeatedly and are easily available (see, for example, Huxley [1896] and Gray [1876]). Thus, the pro-evolutionary forces are represented by less notable authors like Henry Fawcett and Chauncey Wright. Since Ellegård limited himself to British journals in his discussion of the reception of evolutionary theory, I have included several reviews from foreign publications, three from American journals,[3] two translated from German and one from French. These translations appear here for the first time. The reviews of the *Origin* published by Marie Jean Pierre Flourens (1864) and R. A. von Kölliker (1864) are actually more typical of the response from continental scientists, but they have been excluded because they are too long and because they are still easily available. Huxley replied extensively to these two reviews in his "Criticisms on 'The Origin of Species'" (1864), reprinted in *Darwiniana* (1896).

I have had the benefit of advice and criticisms from numerous friends and associates in producing this volume. I wish to thank Michael Ghiselin, Alex Michalous, Susan Musante, Michael Ruse, and Mary B. Williams for suggesting improvements in the Introduction. Special thanks are due to John Greene, Ernst Mayr, and an anonymous reviewer who read and commented on the entire manuscript. The papers by Bronn and von Baer were translated in collaboration with Bella Sélan. Ernst Mayr helped to

---

2. Other books which are worth reading are C. C. Gillispie (1951), William Irvine (1955), Loren Eiseley (1958), and two by J. C. Greene (1959 and 1963). For summaries of the influence which Darwin had and continues to have on philosophy, see John Dewey (1910), R. Hofstadter (1944), P. P. Wiener (1949), John Passmore (1959), and J. H. Randall (1961).

3. See also Daniels, ed. (1968).

produce the finished version of the von Baer paper. Wayne Gailis offered numerous suggestions for improving the translation of Pictet's essay. Katherine Kirkish, Dawn Klemme and Dorothy Dietrich labored to type and retype the lengthy manuscript. This anthology was prepared under National Science Foundation Grants GS-1971 and GS-3102.

D.L.H.

Contents

I
Introductory

II
Reviews

I
Introductory

This leads me to remark that I have always been treated honestly by my reviewers, passing over those without scientific knowledge as not worthy of notice. My views have often been grossly misrepresented, bitterly opposed and ridiculed, but this has been generally done, as I believe, in good faith.— Charles Darwin, *Autobiography,* p. 125.

I have got fairly sick of hostile reviews. Nevertheless, they have been of use in showing me when to expatiate a little and to introduce a few new discussions.

I entirely agree with you, that the difficulties on my notions are' terrific, yet having seen what all the Reviews have said against me, I have far more confidence in the *general* truth of the doctrine than I formerly had.—Charles Darwin to T. H. Huxley, Down, December 2, 1860, *Life and Letters,* 2:147.

1 · Introduction

Darwin expected theologians, people untrained in scientific investigation, and even those scientists who were strongly religious to object violently to his theory of evolution. He had also anticipated the skepticism of even the most dispassionate scientists. He had not labored over twenty years for nothing gathering facts to support his theory and attempting to discount those that apparently conflicted with it. But he had not anticipated the vehemence with which even the most respected scientists and philosophers in his day would denounce his efforts as not being properly "scientific." To the extent that these latter reactions were genuine and not the result of religious bigotry, they can be explained by reference to the philosophies of science popular in Darwin's day. In this chapter Darwin's understanding of the philosophies of science of his day and his own views on science will be set out as fully as possible.

Darwin had both the good fortune and the misfortune to begin his scientific career at precisely that moment in history when philosophy of science came into its own in England. Of course, philosophers from Plato and Aristotle had always written on epistemology and, after the scientific revolution, they were presented with the added advantage and obligation of reconciling their philosophies with the current state of science. Some of these philosophers were also themselves scientists. But the works of Descartes, Locke, Hume, Berkeley, Leibnitz, and Kant do not exhibit the same concern with the accomplishments of science and the nature of the "scientific method" which has come to characterize philosophy of science.[1]

1. I do not want to exaggerate the difference between epistemology and philosophy of science. I have no serious quarrel with those who want to identify the two, as does Gerd Buchdahl (1969). However, I am primarily interested in the reception of evolutionary theory by scientists in the nineteenth century. For this purpose, I have chosen to narrow the focus of my discussion to those areas of epistemology which seem most closely connected to the scientific enterprise.

Commencing with John Herschel's *Preliminary Discourse on the Study of Natural Philosophy* (1830), English-speaking scientists became self-conscious about the proper method of doing science. During the years 1837–1842, when Darwin was residing in London and working on the species problem, the great debate on the philosophy of science erupted between William Whewell (1794–1866) and John Stuart Mill (1806–1873). In 1833 Whewell contributed his *Astronomy and General Physics Considered with reference to Natural Theology* to the Bridgewater Treatises. In 1837 he published his *History of the Inductive Sciences* and in 1840 *The Philosophy of Inductive Sciences, Founded upon their History.* Darwin was impressed by the breadth of knowledge exhibited in these five volumes and remembered Whewell as one of the best conversers on grave subjects to whom he had ever listened.[2]

In 1843 Mill brought out his influential *System of Logic, Ratiocinative and Inductive, Being a Connected View of the Principles of Evidence, and the Methods of Scientific Investigation.* Although Mill had gathered most of what he knew of science from reading Whewell's volumes, his *System of Logic* was largely an empiricist attack on Whewell's reworking of Kant's rationalist philosophy. The key word in this dispute and in the methodological objections raised to Darwin's theory was "induction." It would be nice to be able to set out at this point the meaning which the disputants attached to this word, but I cannot. Everyone meant something different by it, and in the works of a single man, one is likely to find many different uses of the word. Initially it was used by Francis Bacon (1561–1626) to contrast his abortive "inductive method" with the Aristotelian "deductive method." As popularly misconceived, the deductive method consisted in an irresponsible leap to a conclusion of high generality and the subsequent deduction of consequences of these generalizations regardless of observed facts. Perhaps the scholastics at their worst were guilty of such maneuvers, but the above characterization was not only a parody of Aristotle but also fails to identify the actual weaknesses in Aristotle's system. The inductive method, also as popularly misconceived, began with observation and proceeded by the cautious construction of generalizations of greater and greater generality. The deductive method proceeded from the peak of the pyramid of knowledge down to its base, whereas the inductive method started at the base and worked up. The superiority of a pyramid resting on its base rather than its peak was obvious.

Although Bacon came close to advocating the inductive method as set

2. Darwin (1958), p. 66.

out above, Herschel, Whewell, and Mill were all aware of its shortcomings. Science had not and could not proceed by the method set out by Bacon. Yet Bacon was the patron saint of the scientific revolution, and "inductive" was an honorific title not to be discarded lightly. All three men wanted to reserve the term "induction" for the process by which scientific knowledge is attained. Simultaneously they also wanted to use it to refer to the means by which such knowledge was proved factual. For Herschel and Mill induction was the discovery of empirical laws in the facts, reasoning from the known to the unknown. Concurrently, this inductive method insured the truth of these laws. For Whewell, induction was the superinducing of concepts on the facts by the mind. Experience might stimulate the mind to form a concept, but once the appropriate concept had been conceived, truth was guaranteed. For Herschel and Mill, both Kepler's laws and the parallel line postulate of Euclidean geometry were inductions from experience; for Whewell they were self-evident truths whose truth could be known *a priori*.

Whewell did not reply to Mill for six years, then published his *Of Induction, with Especial Reference to Mr. J. Stuart Mill's System of Logic* (1849), just in time for the third edition of Mill's *System of Logic*. The controversy which ensued was gradually decided in Mill's favor, not because Mill's position was especially superior to that of Whewell, but because Whewell's version of Kantian philosophy ran contrary to the rising empiricism of the time. It was in the midst of this controversy over scientific method that Darwin published his *On the Origin of Species by Means of Natural Selection, or the Preservation of Favoured Races in the Struggle for Life* (1859). For Darwin, this coincidence was, to say the least, a mixed blessing.

In his last year at Cambridge, Darwin read Herschel's *Discourse* which, along with Humboldt's *Personal Narrative* (1818), stirred in him "a burning zeal to add even the most humble contribution to the noble structure of Natural Science."[3] Herschel had written his book to elicit precisely this reaction in his readers. Newton had brought about *the* scientific revolution. Henceforth, all scientists had to do was to fill in the details of Newton's great structure, perhaps expanding upon it, but always within the Newtonian framework. Nor was there any conflict between science and religion. Geology had challenged the literal interpretation of the Bible, but scientists and theologians had made their peace. Scientists could concern themselves with the workings of the material world, once created, but the first cre-

3. Darwin (1958), p. 67.

ation, life, mind and soul were the province of the theologian. "To ascend to the origin of things, and to speculate on the creation, is not the business of the natural philosopher."[4] Not only were scientists making great contributions to the noble edifice of science and to mankind by applications of science in medicine and industry, but their discoveries also lent support to religion through natural theology. As they discovered more clearly how nature worked, they showed how great the creator's wisdom had been. In his youth Darwin had hoped to join in this great parade of scientists and men of God marching arm in arm to produce a better world. Instead he stopped it dead in its tracks.

In his early publications, Darwin gave every appearance of contributing his share to received science. His journals concerning the voyage of the *Beagle* were in the best tradition of such narratives. His monographs on living and fossil barnacles certainly were in no way controversial. Even the publication of his papers on the formation of the Parallel Roads of Glen Roy (1839) and of coral reefs (1842), though theoretical in nature, did not detract from his growing reputation as a true inductive scientist. In the first, he explained the appearance of a series of parallel shelves on the sides of a glen in Scotland in terms of the gradual elevation of the land. The parallel roads were actually the remnants of former beaches. In the second, he explained the formation of coral reefs in terms of the gradual subsidence of the ocean floor. New coral grew on the old as it fell beneath the surface of the water. But with the publication of the *Origin of Species,* large segments of the scientific and intellectual community, turned on him. Both Adam Sedgwick, the eminent geologist, and Richard Owen, the leading comparative anatomist of the day, had encouraged Darwin in his early work. After the *Origin,* their praise turned to ridicule. Sedgwick (1860) complained that Darwin had "departed from the true inductive track." Owen, while admitting that he himself had casually entertained the notion of natural selection, had judiciously refrained from enunciating it. It was "just one of those obvious possibilities that might float through the imagination of any speculative naturalist; only the sober searcher after truth would prefer a blameless silence to sending the proposition forth as explanatory of the origin of species, without its inductive foundations."[5]

Darwin was prepared for the abuse which the *content* of his theory, especially its implications for man, was to receive from certain quarters,

4. Herschel (1830), p. 38.
5. Owen (1860).

but he was not prepared for the criticism which his *methodology* was to receive from the most respected philosophers and scientists of his day. Most contemporary commentators tend to dismiss these criticisms as facile, disingenuous and superficial, suspecting that they stemmed more from a distaste of the content of Darwin's theory than from his methodology, but this dismissal is itself too facile. Certainly the repeated invocation of the Baconian method by many of Darwin's critics and even by Darwin himself indicated no great understanding of the actual nature of this method or the philosophy from which it stemmed, but the leading philosophers contemporary with Darwin, John Herschel, William Whewell, and John Stuart Mill, were equally adamant in their conviction that the *Origin of Species* was just one mass of conjecture. Darwin had proved nothing! From a philosophical point of view, evolutionary theory was sorely deficient. Even today, both Darwin's original efforts and more recent reformulations are repeatedly found philosophically objectionable.[6] Evolutionary theory seems capable of offending almost everyone.

In the nineteenth century, "to be scientific" meant to be like John Herschel's extension of physical astronomy to the sidereal regions.[7] Thus, Darwin was especially anxious to hear the opinion of Herschel, the "great philosopher" referred to in the opening paragraph of the *Origin*. He sent Herschel a copy of his book and wrote to Lyell to pass on any comments which the great physicist might make since "I should excessively like to hear whether I produce any effect on such a mind."[8] Herschel's opinion was rapidly forthcoming. Darwin wrote to Lyell, "I have heard, by a roundabout channel, that Herschel says my book 'is the law of higgeldy-piggeldy.' What this exactly means I do not know, but it is evidently very contemptuous. If true this is a great blow and discouragement."[9] In the face of such a rejection by the most eminent philosopher-scientist of the century, it is easy to understand Darwin's pleasure when he discovered in an equally roundabout way that another great philosopher, John Stuart Mill, thought that his reasoning in the *Origin* was "in the most exact accordance with the strict principles of logic."[10] On closer examination, however, Mill's

---

6. See, for example, Woodger (1929), Kneale (1949), von Bertalanffy (1952), Grene (1958), Himmelfarb (1959), Manser (1965), and Moorehead and Kaplan (1967).

7. Cannon (1961), p. 238.

8. Darwin, November 23, 1859, *Life and Letters* (1887), 2:26.

9. Darwin, December 12, 1859, ibid., 2:37.

10. Darwin, Henry Fawcett to C. Darwin, January 16, 1861, *More Letters* (1903), 1:189.

endorsement can be seen to be not nearly reassuring. Darwin had properly used the Method of Hypothesis, but this method belonged to the logic
of discovery, not proof. In spite of twenty years' labor, Darwin had failed
to provide proof for his theory of evolution by natural selection.

Darwin's own views on the nature of science exhibited the conflicts and
inconsistencies typical of his day. He evidenced the usual distrust of "hypotheses" while grudgingly admitting their necessity. For example, in a
letter to Hooker, Darwin claimed that he looked upon "a strong tendency
to generalize as an entire evil"[11] and yet admitted in his *Autobiography:*
"I cannot resist forming one on every subject."[12] In the opening paragraph
of the *Origin of Species* Darwin sketched the following history of the development of his theory:

> When on board H.M.S. "Beagle," as naturalist, I was much struck
> with certain facts in the distribution of the inhabitants of South America,
> and in the geological relations of the present to the past inhabitants
> of that continent. These facts seemed to me to throw some light on
> the origin of species—that mystery of mysteries, as it has been called
> by one of our greatest philosophers. On my return home, it occurred
> to me, in 1837, that something might perhaps be made out on this ques-
> tion by patiently accumulating and reflecting on all sorts of facts which
> could possibly have any bearing on it. After five years' work I allowed
> myself to speculate on the subject, and drew up some short notes; these
> I enlarged in 1844 into a sketch of the conclusions, which then seemed
> to me probable: from that period to the present day I have steadily
> pursued the same object (p. 1).

In his *Autobiography* Darwin recalls roughly this same sequence of events
but adds that he "worked on true Baconian principles and without any
theory collected facts on a wholesale scale."[13] But Darwin was well aware
that the possibility of species evolving had occurred to him soon after his
return from his voyage on the *Beagle* (if not before) and his principle
of natural selection not much later, in October 1838. Five years may have
elapsed before he permitted himself to write an essay on the subject, but
he had speculated and collected facts in the light of these speculations
from the very first.

The source of this fabrication is easy to uncover. One of the most prevalent confusions in the work of even the best scientists and philosophers
was between the temporal order of an actual scientific investigation and

11. Darwin, January 11, 1844, *More Letters* (1903), 1:39.
12. Darwin (1958), p. 141.
13. Darwin (1958), p. 119.

the logical order of a reconstruction of scientific method. According to empiricist epistemology, all knowledge has its foundation in experience. This tenet was mistakenly taken to imply that all scientific investigation has to *begin* with observation. The true inductive scientist began collecting data indiscriminately, with no preconceived ideas, and gradually evolved broader and broader generalizations. The process of scientific investigation assured both the truth and the empirical meaningfulness of the resultant propositions. Deductivists approached nature with a hypothesis already in mind, and speculators flew too quickly to generalizations of too great a scope. Thus, Darwin can be found saying of his coral reef paper, "No other work of mine was begun in so deductive a spirit as this; for the whole theory was thought out on the west coast of S. America before I had seen a true coral reef."[14] Similarly with respect to his theory of evolution, he conceded to Asa Gray, "What you hint at generally is very, very true: that my work will be grieviously hypothetical, and large parts by no means worthy of being called induction, my commonest error being probably induction from too few facts."[15]

In a letter to Henry Fawcett, however, Darwin indicated that he realized that data cannot be gathered efficiently without some hypothesis in mind:

> About thirty years ago there was much talk that geologists ought only to observe and not to theorize; and I well remember someone saying that at this rate a man might as well go into a gravel-pit and count the pebbles and describe the colours. How odd it is that anyone should not see that all observation must be for or against some view if it is to be of any service![16]

What mattered was that the hypothesis be an empirical hypothesis, one that could be verified or refuted by observation, and that serious attempts be made to gather the relevant data. Immediately after saying that his coral reef paper was begun in a deductive manner, Darwin adds, "I had therefore only to verify and extend my views by a careful examination of coral reefs."[17] Darwin's own experience as a scientist forced him to recognize that the order in which hypotheses were formed and the relevant data gathered was not rigidly set. It helped to have a hypothesis in mind, but hypotheses had to be changed as the investigation proceeded.[18] His theory of the formation of coral reefs was formulated before he had collected

14. Ibid., p. 98.
15. Darwin, November 29, 1859, *More Letters* (1903), 1:126.
16. Darwin, September 18, 1861, ibid., 1:195.
17. Darwin (1958), p. 98.
18. Ibid., p. 141.

very much of the relevant data and turned out to be largely factual. His theory of evolution by natural selection was formulated after several years of observations as a naturalist, followed by two decades of additional, selective investigation, and it too was basically correct. His theory of the Parallel Roads of Glen Roy had much the same history on a smaller scale as his theory of evolution, but was mistaken. In the case of inheritance, Darwin actually did amass facts wholesale before he succeeded in working out his theory of pangenesis. Its fate was even more unhappy than that of the Parallel Roads theory. Clearly there was no connection between the temporal order of fact-gathering and hypothesizing and the truth of the hypothesis.

However, Darwin did think that the temporal order of the verification and the publishing of a hypothesis was important in the sociology of science, both for the sake of the scientist's own reputation and for the sake of the acceptance of his theory by the scientific community. To a young scientist, Darwin advised, "I would suggest to you the advantage, at present, of being very sparing in introducing theory in your papers (I formerly erred much in Geology in that way); *let theory guide your observations,* but till your reputation is well established, be sparing of publishing theory. It makes persons doubt your observations."[19]

Darwin also had a reason for advising restraint in publication which was less personal and more significant to the progress of science. He was aware of the fate of Lamarck and Chambers. Chambers had received considerable popular acclaim but scathing denunciation from the academic community, including T. H. Huxley (1854). Lamarck received nothing but ridicule from all sides. Darwin criticized Lamarck and Chambers, not for suggesting mechanisms for evolution which he thought were mistaken, but for foisting their views on the scientific community without sufficient effort at careful formulation and verification. Lamarck and Chambers looked upon the process of scientific verification as a very casual affair. Darwin looked upon these matters as of utmost gravity. Lamarck published radically new theories in a variety of fields as diverse as mineralogy and meteorology. He felt that the originality of his impressionistic sketches would be enough to motivate others to undertake the subordinate task of filling out and verifying his theories. As Burkhardt (1970) has aptly observed, public neglect and private ridicule were the fate of his theories. Prior to knowing that Chambers was the author of the *Vestiges of Creation,* Darwin

19. Darwin, 1863, *More Letters* (1903), 2:323.

chided him for objecting to the skepticism of scientific men. "You would not fulminate quite so much if you had had so many wild-goose chases after facts stated by men not trained in scientific accuracy."[20]

Darwin's conservative views on publication were rewarded to some degree. His *Origin of Species* did not suffer the same fate as those works on evolution that had preceded it. It was treated as a serious work of science even by those who denounced it. But a similar reticence on the part of Gregor Mendel resulted in his laws of heredity being overlooked for almost forty years. Mendel published his laws in 1865, soon after the appearance of the *Origin*. He too wanted to avoid being branded a speculator. In his original paper he barely alludes to his unobservable "factors," though the cogency of his entire argument rested upon their existence. In a letter to Carl Nägeli (Mendel, 1867), he claimed that, "as an empirical worker" he had to "define constancy of type as the retention of a character during the period of observation." Perhaps Darwin's advice might protect scientists from being engulfed in half-baked scientific publications, but it also contributed to the obscurity of Mendel's potentially great contribution to science. Because Lamarck neglected the niceties of the sociology of science, his work was ignored with disdainful embarrassment. Because Mendel was too scrupulous in observing them, his work was overlooked. Darwin struck an appropriate compromise.

Darwin's beliefs on the ethics of publication go a great way toward explaining his reticence in publishing his own views on the origin of species until his big book was complete. Darwin emphasized the importance of providing empirical evidence for scientific hypotheses for still another reason. The science of Darwin's day was filled with metaphysical principles enunciated as if they were scientific laws. Darwin found Herbert Spencer to be one of the chief offenders in this regard. Because of the jargon of the period, Darwin's discussions of the differences between scientific laws, the axioms of mathematics, and metaphysical principles are none too clear, but in these matters he was no worse than the leading empiricist philosophers of the day. He tended to distinguish the formal from the empirical sciences by the prevalence of deductive reasoning in the former. Of Spencer, Darwin complained:

> His deductive manner of treating any subject is wholly opposed to my frame of mind. His conclusions never convince me; and over and over again I have said to myself, after reading one of his discussions,—

20. Darwin, April 30, 1861, *More Letters* (1903), 2:443; see also, Darwin (1871), p. 606.

"Here would be a fine subject for half-a-dozen years' work." His fundamental generalizations (which have been compared in importance by
some persons with Newton's Laws!) which I daresay may be very valuable
under a philosophical point of view, are of such a nature that they
do not seem to me to be of any strictly scientific use. They partake
more of the nature of definitions than of laws of nature. They do not
aid me in predicting what will happen in any particular case. Anyhow
they have not been of any use to me.[21]

The distinction which Darwin discerned in the works of Spencer was
between a scientific hypothesis which has not yet been verified and a purely
metaphysical principle which, by the methods of science, cannot be confronted by empirical data at all. He felt "inclined to cavil at speculation
when the direct and immediate effect of a cause in question cannot be
shown."[22] Darwin doubted the merits of Spencer's scientific work, not
merely because the evidential basis was frequently lacking but because
Darwin doubted that there was any.

Although Darwin envied mathematicians their "sixth sense," he also mis-
trusted them. For example, in the sixth edition of the *Origin,* he had
to change the number of descendants of a single pair of elephants which
would still be alive after 750 years from 15 million to 19 million. "I got
some mathematician to make the calculation, and he blundered and caused
me much shame."[23] Again to Hooker, he remarked, "You will be amused
to observe that geologists have all been misled by Playfair, who was misled
by two of the greatest mathematicians! And there are other such cases;
so we could turn round and show your reviewer how cautious geologists
ought to be trusting mathematicians."[24]

Darwin could not help but know the crucial role which mathematics
had played in physics, since Herschel had repeatedly emphasized it in his
Discourse, but it did not seem to be in the least useful in his own work
in biology. On occasion he had to make some simple computations. When
these proved too much for him, he asked for help. In later life his son
Francis took over this duty. What seemed to bother Darwin was that the
kind of mathematical techniques he needed most were not available. For
Darwin, mathematics consisted of deductive reasoning, and he distrusted
greatly "deductive reasoning in the mixed sciences." In his own work,

21. Darwin (1958), p. 109; see also letter to J. Fiske, December 8, 1874, *Life
and Letters* (1887), 2:371.
22. Darwin, March 4, 1850, *More Letters* (1903), 2:133.
23. Darwin, April 15, 1872, ibid., 1:336.
24. Darwin, August 7, 1869, *More Letters* (1903), 1:315.

he seldom was presented with a situation in which he could use such deductive reasoning. He was constantly forced to deal in probabilities, and no one could tell him how to compute and combine such probabilities. He agreed with Lyell about "mathematicians not being better able to judge of probabilities than other men of common-sense."[25] Darwin tended to extend his view of mathematicians to physicists, especially in their attempts to calculate the age of the earth. They had been wrong in their geological calculations so many times in the past, Darwin was not overly terrified by the specter of Lord Kelvin's estimation of the age of the earth.[26]

One reason which Darwin frequently gave for the *Origin*'s being so maligned and misunderstood was that Wallace's joint discovery had forced him into premature publication. No similar pressure forced Darwin to publish his theory of pangenesis, and it met with even more resistance than the *Origin* had. In reality, all of Darwin's theoretical work had much the same structure. As Darwin put it: "The line of argument often pursued throughout my theory is to establish a point as a probability by induction and to apply it as hypotheses to other parts and see whether it will solve them."[27] If it did solve them, then Darwin thought that there was good reason to accept the hypotheses tentatively. As Darwin wrote of F. W. Hutton's review of the *Origin:*

> He is one of the very few who see that the change of species cannot be directly proved, and that the doctrine must sink or swim according as it groups and explains phenomena. It is really curious how few judge it in this way, which is clearly the right way.[28]

And in his autobiography he says of his theory of pangenesis:

> Towards the end of the work I give my well-abused hypothesis of pangenesis. An unverified hypothesis is of little or no value. But if anyone should hereafter be led to make observations by which some such hypothesis could be established, I shall have done good service, as an astonishing number of isolated facts can thus be connected together and rendered intelligible.[29]

Darwin produced several theoretical works in his career: two were in error (pangenesis and his explanation of the Parallel Roads of Glen Roy)

25. Darwin, June 17, 1860, ibid., 1:154. For several amusing anecdotes about Darwin's mathematical ineptness see Himmelfarb (1959).

26. For a further discussion see the comments to the reviews of Haughton and Fawcett.

27. DeBeer (1960), 3:142.

28. Darwin, April 23, 1861, *Life and Letters* (1887), 2:155.

29. Darwin (1958), p. 120.

and two correct (evolutionary theory and his explanation of the formation of coral reefs).[30] His procedures in formulating, articulating, and verifying these theories and explanations were roughly the same, yet no one complained of the methodology of his Glen Roy or coral reefs papers. If Darwin's methodology was faulty in the *Origin,* then it should have been equally faulty in all of his scientific works.

What was wrong with Darwin's theories? Was his methodology faulty? What is the nature of inductive formulation and proof which eluded Darwin? These are the questions which occur to the modern reader when first confronted by the reviews of the *Origin of Species* and Darwin's later work, especially the *Descent of Man.* Some of the reviewers were obviously biased. Some were merely mouthing undigested platitudes. But many of the reviewers were competent scientists honestly trying to evaluate a novel theory against the commonly accepted standards of scientific excellence, and evolutionary theory consistently came up wanting. The solution to these puzzles can be found in the philosophies of science promulgated by such philosophers as Herschel, Whewell, and Mill and their most important predecessors, Aristotle, Bacon, and Newton. Darwin was caught in the middle of a great debate over some of the most fundamental issues in the philosophy of science—the difference between deduction and induction and the role of each in science, the difference between concept formation and the discovery of scientific laws, the relation between the logic of discovery and the logic of justification, the nature of mathematical axioms and their relation to experience, the distinction between occult qualities and theoretical entities, and the role of God's direct intervention in nature. Before philosophers of science had thoroughly sorted out these issues, they were presented with an original and highly problematic scientific theory to evaluate. That they rejected evolutionary theory, a theory which has outlasted many of the theories judged to be exemplars of scientific method, says something about the views of science held by these philosophers and scientists.

In the succeeding chapters, the interconnections between Darwin's theory of evolution and the philosophies of science current in his day will be traced. It is hoped that this discussion will aid the reader in understanding the various criticisms in the ensuing reviews and in placing them in their appropriate conceptual setting. Kuhn (1962) has argued in his influential

30. For a discussion of some of Darwin's other scientific theories see Ghiselin (1969).

book that scientific revolutions are so traumatic that no reasons can be given for preferring one theory to another. Either one accepts the old theory and its standards of meaningfulness and truth or the new theory and its differing standards. The transition is almost mystical. Needless to say, considerable controversy has followed upon Kuhn's reconstruction of scientific revolutions as fundamentally arational processes. However, all of the discussions in this dispute have centered on revolutions in physical theory. The revolutions in biology have been all but ignored.[31] The essays in this anthology should provide ample source material for the disputants on the various sides of this controversy.

In Chapter 2 the logic of justification and its relation to discovery will be examined. What did it take to *prove* a scientific theory? Why had Newton's theory supposedly fulfilled the requirements of proof while Darwin's failed? In Chapter 3 the problems surrounding concept formation and meaning will be discussed. Early scientists were deeply impressed by the failure of Aristotelian science in the hands of the Scholastics. They identified this failure with the "deductive method" and with the central role of "occult qualities" in early scientific theories. The fear of occult qualities was so strong in many scientists that any reference to unobservables was looked upon with extreme suspicion. But coexisting with this hard-nosed empiricism was an equally strong belief in God as an underlying first cause and his occasional direct intervention in natural processes. This schizophrenic attitude of scientists proved to be one of the strongest impediments to the acceptance of evolutionary theory. As Kuhn (1962) has observed: "For many men the abolition of that teleological kind of evolution was the most significant and least palatable of Darwin's suggestions. The Origin of Species recognized no goal set either by God or nature" (p. 171).

Chapter 4 will deal with the effect that essentialism had on the acceptance of evolutionary theory. Although this topic has been discussed extensively elsewhere[32] it is too intimately connected with the problems at issue to allow omission. Seldom in the history of ideas has a scientific theory conflicted so openly with a metaphysical principle as did evolutionary theory with the doctrine of the immutability of species.[33]

31. See Ruse (1970 and 1971), Greene (1971), and Ghiselin (1971).
32. Mayr (1959a) and Hull (1965).
33. The following works are recommended for further reading in the philosophies of Aristotle, Bacon, Herschel, Whewell and Mill: Blunt (1903), Ducasse (1951 and 1960), Anschutz (1953), Heathcote (1953), Hochberg (1953), Strong (1955), Cannon (1961), Walsh (1962), Grene (1963), and Butts (1968).

2 · The Inductive Method

In the light of the great advances in physics in the seventeenth and eighteenth centuries, nineteenth-century philosophers and scientists collaborated to produce a philosophy of science consistent with these achievements. In their informal remarks these authors tended to provide useful insights into the scientific enterprise, but when they tried to be rigorous in their characterization of the scientific method, they tended to make pronouncements which frustrated the rational evaluation of newly emerging scientific theories, specifically evolutionary theory. In this chapter, the philosophies of science of Aristotle, Bacon, Herschel, Whewell, and Mill will be sketched, both as these men set them out and as they were popularly misconceived. It will be argued that evolutionary theory did fail to meet the standards of proof established by these philosophers for the simple reason that no theory could possibly fulfill them, including the physical theories which these authors had chosen as paradigm.

In Darwin's day Aristotle was looked upon by scientists as the author of infinite error and Bacon as the man who fashioned the method which Newton was to use to unlock the mysteries of the universe. Bacon's characterization of ancient philosophers was typical:

> From a few examples and particulars (with the addition of common notions and perhaps of some portion of the received opinions which have been most popular) they flew at once to the most general conclusions, or first principles of science. Taking the truth of these as fixed and immovable, they proceeded by means of intermediate propositions to educe and prove from them the inferior conclusions; and out of these they framed the art. After that, if any new particulars and examples repugnant to their dogmas were mooted and adduced, either they subtly molded them into their system by distinctions and explanations of their rules, or else coarsely got rid of them by exceptions; while to such particulars as were not repugnant they labored to assign causes in conformity

with those of their principles. But this was not the natural history and experience that was wanted; far from it.[1]

As far as Aristotle's professed epistemology is concerned, Bacon's characterization is anything but fair, though it accurately depicts the level to which it frequently sank in actual practice. However, in singling out Aristotle's emphasis on deduction at the expense of induction, Bacon failed to identify the most significant source of error in Aristotle's epistemology. In fact, deduction will continue to play a central role in scientific procedures throughout the history of science. The real culprits are the quest for *absolute certainty* in the acquisition of knowledge and Aristotle's use of *intuition* to accomplish this end. Future philosophers of science like Bacon, Herschel, and Mill will join Aristotle in his quest for absolute certainty but will attempt to obtain it by a strict application of inductive logic.

Confusion over the nature of deduction and induction is one of the most conspicuous features in attempts to set out a philosophy of science from Aristotle to John Stuart Mill. The metaphor that permeates the literature is that of a pyramid. For example, Adam Sedgwick (1860) complains of Darwin's theory:

> But I must in the first place observe that Darwin's theory is not *inductive,*—not based on a series of acknowledged facts pointing to a *general conclusion,*—not a proposition evolved out of facts, logically, and of course including them. To use an old figure, I look on the theory as a vast pyramid resting on its apex, and that apex a mathematical point.

Deduction proceeded from the most general propositions at the apex of the pyramid to the particulars at the bottom, induction from particulars to generalizations of broader and broader scope. A more misleading metaphor is hard to imagine. It is inapplicable to Aristotelian logic. Both Whewell and Mill were well aware that it was inapplicable to logical inferences as they viewed them. Yet the metaphor persisted, and has done so to this day in spite of the efforts of generations of logic professors. The pyramid which people have in mind when they contrast deduction with induction is not a pyramid of inferences but a pyramid of concepts like the Linnaean hierarchy.[2] At the peak of such a pyramid is the class of greatest generality, e.g., animals. Animals are divided into additional classes of reduced scope, e.g., protozoans, round worms, and chordates. Each of these classes is subdivided in turn until the lowest level is reached—species,

1. Bacon (1620), bk. I. Aphorism CXXV.
2. For example, see J. J. Baker and G. Allen (1968).

subspecies, or varieties. At the bottom of the pyramid are arrayed individual organisms.

When we turn to a consideraiton of statements of greater and greater generality and inferences among them, the whole idea of descending and ascending in a pyramid ceases to be appropriate. For example, the following hypothetical syllogism is the classic example of an Aristotelian deductive argument:

> (1)   All hairy quadrupeds are animals.
>   All horses are hairy quadrupeds.
>   _____
>   Hence, all horses are animals.

Although this argument is deductive, no decrease in generality accompanies the inference from the premises to the conclusion. Horses are included in hairy quadrupeds, and hairy quadrupeds are included in animals, but the conclusion is no less general than either of the premises. It retains reference to the class of smallest scope (horses) and largest scope (animals). Only the middle term has dropped out. Even when the deduction is to the individual, no decrease in generality accompanies the inference from the premises to the conclusion:

> (2)   All men are mortals.
>   Socrates is a man.
>   _____
>   Hence, Socrates is a mortal.

Once again, the conclusion contains reference to the class of greatest generality. One has not descended in any pyramid of inferences. The story is somewhat different for induction. The paradigm example of an inductive inference is induction by simple enumeration. In such inferences, one reasons from a series of singular statements about individuals to a universal generalizaton:

> (3)   This crow is black.
>   This crow is black.
>   This crow is black.
>   This crow is black.
>   .
>   .
>   .
>   .
>   .
>   _____
>   Hence, all crows are black.

Here the increase in generality is obvious. The conclusion is more general than the premises taken separately or together. But there has been no ascent in the pyramid of concepts. Crows and blackness are mentioned in all the constituent statements. To use such an inductive argument one must already know how to recognize crows and the color black.

The important difference between deduction and induction which has emerged through the years turns out to concern not generality but *certainty*. In a valid deductive argument, the conclusion follows *necessarily* from the premises. If the premises are true and the argument valid, the conclusion *must* be true. Both arguments (1) and (2) are deductive in this sense. In an inductive argument, the conclusion follows from the premises, if at all, with only a certain degree of probability. If the premises are true and the argument warranted, then the conclusion is *probably* true. In an inductive argument there is always the possibility that true premises will lead to a false conclusion. For example, in argument (3) it turns out that not all crows are black (nor all ravens; nor are all swans white). It is easy to show that the difference between deductive and inductive arguments does not depend on any putative decrease or increase in generality between premises and conclusions. For example, the following argument is analogous to the deductive argument (2), yet it is inductive. It is inductive because the conclusion does not follow necessarily from the premises:

> (4)   Almost all swans are white.
>   The Brookfield Zoo owns three swans.
>   _____
>   Hence, they are probably white.

In the ensuing discussion of the philosophies of Aristotle, Bacon, Herschel, Whewell, and Mill, we will use "deduction" and "induction" in the modern sense of demonstrative and nondemonstrative inference. Our anachronistic usage produces less distortion than one might think, since all the arguments which these philosophers considered deductive are also deductive in this modern sense and all those that were thought inductive remain inductive. The difference between the two usages stems from the rationale used to make the distinction and from the fact that certain types of inferences which counted as neither deductions nor inductions in some of the earlier systems of logic now are included in one group or the other. Even the difference in rationale does not introduce any change in emphasis, since one of the chief concerns of philosophers from Aristotle to Mill was to evade the uncertainty inherent in inductive inferences. Bacon, Herschel, and Mill attempt to eliminate it by trying to make inductive inferences more rigorous, Aristotle by recourse to intuition, and Whewell by reference to self-evident truth.

For Aristotle, knowledge began with the observation of individuals like Callias, Socrates, Alcibiades, and Ajax and proceeded by simple enumeration until one intuited that which is universal in them—their essence, or formal cause. This process of concept formation preceded any inductive inferences in the strict sense, since it preceded the formation of statements. It cannot be represented as in argument (3), since in this argument one already has the concepts, crow and black. One merely observes individuals and intuits their essence. For example, one might intuit that all the individuals listed above were men, or human beings, or animals. Intuition need not begin with the lowest universal and proceed only later to universals of higher generality. It is possible to observe only one member of one species of a universal of high generality and still correctly intuit that universal. Usually more extensive observations are necessary. The number and variety of observations needed for intuition vary from person to person and from subject matter to subject matter. Fewer observations are usually necessary in mathematics than in such empirical sciences as biology. Nor did Aristotle think that the preceding enumeration of individuals could be complete, since he believed that there were indefinitely many individuals in any species. It is not possible to observe all human beings and then merely by mechanically summing their properties to discover their essence. Intuition was necessary.

Little has been said thus far about intuition. There is very little to say. Although intuition is the source of all certain knowledge, Aristotle hardly mentions it. According to Aristotle, the essence of mankind is to *nous,* to know, to intuit. Man has the capacity for the direct apprehension of the essence of things, to see the universal in the individual—and in this capacity intuition is infallible![3] Aristotle gives no justification for the existence of infallible intuition. He merely observes that without it, the acquisition of certain knowledge would be impossible.

For Aristotle, scientific knowledge had to be universal, immutable and absolutely certain. The existence of essences assured that it would be universal and immutable. Intuition guaranteed that it would be certain. Aristotle can be found saying that, whenever observation and theory conflict, credit must go to observation[4] and warning against speculation from too

3. Aristotle, *Posterior Analytics,* pt. 2, ch. 19.
4. Aristotle, *The Generation of Animals,* 760^{b}28.

small induction,[5] but as Aristotle's own efforts and those of his followers give ample proof, these precautions tend to be ignored in an epistemology based on the infallible intuition of immutable essences. The source of error in Aristotle's epistemology is not the deductive organization of knowledge. After all, verification by the deduction of observational consequences is one of the best ways of checking the truth of a scientific law, but deduction to the individual as exemplified in argument (2) has no place in Aristotelian epistemology. Science deals only with that which is universal, with essences. After essences have been genuinely intuited, checking by observation is superfluous.

By the time of Francis Bacon (1561–1626), the scientific revolution was well under way with the work of Copernicus, Galileo, Tycho Brahe, Kepler, Gilbert, and Harvey, but to anyone living at the time the trend would have been difficult to discern. Bacon hoped to replace the Aristotelian deductive method with his method of *true induction,* or *induction by exclusion.* Just as Aristotle had devised a set of rules for deduction, Bacon was going to do the same for inductive logic. He hoped to produce an inductive machine for grinding out true, scientific laws.

Bacon's plan was to begin with extensive gathering of data without any speculation.[6] He felt that the best way to preclude any premature speculation was to have this labor performed by unlettered underlings. Besides, he believed that it was "somewhat beneath the dignity of an undertaking like mine that I should spend my own time in a matter which is open to almost every man's industry."[7] No doubt comments like these prompted Harvey to say that Bacon wrote philosophy "like a Lord Chancellor." Bacon was never to have his army of workers. Instead, in the last year of his life he himself tried to make a beginning in the extensive gathering of data completely on his own. Even if Bacon had been given his army of workers, it is doubtful that his enterprise would have been very successful. It is now commonplace to recognize, as Darwin recognized over a century ago, that "the observer can generalize his own observations incomparably better than anyone else" and that "observation must be for or against a view if it is to be of any service."

Bacon hoped to assure the truth of his generalizations by extensive fact-gathering and induction. As data were gathered, they were to be entered on three tables—a table of essence, a table of deviation, and a table of

5. Ibid., 756[a]1.
6. Bacon (1620), bk. 2, Aphorism XI.
7. Ibid., p. 271.

degree. For example, if the nature of heat is under investigation, those things that seemed hot would be listed on the first table, those that lacked heat on the second, and variations in heat on the third. Only after these tables were filled out did induction actually begin. True induction for Bacon was induction by exclusion or, as it is sometimes called, complete elimination. If the relevant concepts have already been formulated and if natural phenomena can be divided into a finite number of discrete natural kinds, then induction by complete elimination is possible. All the possible causes of a kind of event can be listed and all but one eliminated. But under such circumstances induction by complete enumeration is also possible.

Bacon thought that the only induction recognized by Aristotle and practiced by his followers was induction by simple enumeration as exemplified in argument (3). But Aristotle did not consider the process by which the universals were abstracted from the individual as induction; it was intuition. Induction of any kind could occur only at the level of universals, and at this level both induction by complete enumeration and induction by complete exclusion (or elimination) are possible, since Aristotle believed in the existence of a finite number of discrete natural kinds. Argument (5) below is an example of an induction by complete exclusion.

(5) Scalene triangles have three sides.
Isosceles triangles have three sides.
Equilateral triangles have three sides.
_____
Hence, all triangles have three sides.

(6) Man, the horse, and the mule are long-lived.
Man, the horse, and the mule are all bileless animals.
_____
Hence, all bileless animals are long-lived.

The conclusion to argument (5) seems to follow from the premises necessarily because of the suppressed premise that these are all the species of triangle that there are. All species of triangle have been enumerated. When the suppressed premise is added, the argument becomes deductive. Argument (6) is actually taken from Aristotle. Here, too, the deductive nature of the inference is readily apparent. Only if the three species mentioned are all the bileless animals which exist is the argument valid. Both induction by complete enumeration and induction by complete elimination are possible only in a world which is divisible into a finite number of natural kinds and are deductive in nature. Aristotle was cognisant of this fact.[8] Many later philosophers were not. This conflict between the desire to make induction certain while still remaining inductive is evident in Bacon's writings. Induction by exclusion was not only a method of discovery but it afforded absolutely certain knowledge. "Now what the sciences stand in need of is a form of induction which shall analyse experience and take it to pieces, and by a due process of exclusion and rejection lead to an inevitable conclusion."[9]

Bacon maintained that scientists were to proceed from sense perception of the individual to the lowest universals and from there gradually, without omitting a step, to axioms of greatest generality.[10] Taken literally, this view of induction would completely eliminate deduction as anything but an after-the-fact exercise, but on occasion, Bacon seems to think otherwise. Both deduction and induction have roles to play in science—"from the new light of axioms, which have been educed from those particulars by a certain method and rule, shall in their turn point out the way again to new particulars, greater things may be looked for. For our road does not lie on a level, but ascends and descends; first ascending to axioms, then descending to works."[11]

It is difficult to reconcile these remarks concerning the role of deduction in science with Bacon's earlier pronouncements about gradual ascent. If the formation of concepts or the discovery of laws is to proceed by true induction, step by step, without any jumps, how can a concept or a law be of wider scope than the evidence from which it was derived? To the extent that true induction is possible, the collateral security of deducing new confirming instances is impossible.

Herschel, Mill, and Whewell all concurred at least verbally in Bacon's confused distinction between deduction and induction in terms of increasing and decreasing generality, and much of Herschel and Mill consists merely in tidying up Bacon's methods of induction. Both authors, however, emphasized the role of deduction a good deal more strongly than Bacon had. For example, Herschel says:

> It is to our immortal countryman Bacon that we owe the broad announcement of this grand and fertile principle; and the development of the idea, that the whole of natural philosophy consists entirely of

8. Aristotle, *Posterior Analytics*, 68^{b}15.
9. Bacon (1620), p. 19.
10. Ibid., bk. 1, Aphorism CIV.
11. Bacon (1620), bk. 1, Aphorism CIII.

a series of inductive generalizations, commencing with the most circum-
stantially stated particulars, and carried to universal laws, or axioms,
which comprehend in their statements every subordinate degree of generality, and of a corresponding series of inverted reasoning from generals
to particulars, by which these axioms are traced back into their remotest
consequences, and all particular propositions deduced from them . . .
   But it is very important to observe, that the successful process of scientific enquiry demands continually the alternate use of both the *inductive*
and *deductive* method.[12]

Herschel resolves the conflicting roles of deduction and induction by
observing that "in the study of nature, we must not, therefore, be scrupulous
as to *how* we reach to a knowledge of such general facts: provided only
we verify them carefully when once detected, we must be content to seize
them wherever they are to be found."[13] Careful induction was advisable
but not necessary. Hence, deductive verification would be more than an
empty exercise.
   Mill had much the same story to tell but did so with considerably
more force:

   Bacon's greatest merit cannot therefore consist, as we are so often
   told that it did, in exploding the vicious method pursued by the ancients
   of flying to the highest generalization first, and deducing the middle
   principles from them; since this is neither a vicious nor an exploded,
   but a universally accredited method of modern science, and that to which
   it owes its greatest triumphs.[14]

The distinction between the temporal order of discovery and the logical
order of a reconstructed logic had surfaced at last. Unfortunately, an awareness of this crucial distinction came in the midst of the great controversy
between Whewell and Mill and did little to improve the coherence of
the debate. Each side accused the other of confusing proof with discovery.
In following the course of this controversy, one gets the impression that
neither man appreciated the importance of this distinction until after he
had committed himself on too many issues. Thereafter, each tried to shore
up his position without openly admitting error, while trying to expose the
other's evasive maneuvers. The situation was not helped by an ambiguity
in the key term "induction." Whewell called all processes which contributed
to the construction of a scientific theory "induction," whereas Herschel and
Mill limited the term to inferences from the known to the unknown.

12. Herschel (1830), pp. 104, 174–175
13. Ibid., p. 164.
14. Mill (1874), pp. 68–69.

Herschel and Mill thought that both the logic of discovery and the logic of justification could be analyzed in terms of deduction and induction, whereas Whewell thought that no simple rules could be given for the process of discovery.[15] For Herschel and Mill the key tool in the logic of discovery and justification was induction by elimination.

Darwin repeatedly made use of induction by elimination. For example, in his paper on the Parallel Roads of Glen Roy (1839), he reasoned that these roads or shelves were caused by either lakes or arms of the sea. If they were caused by lakes, then huge barriers had to be erected and removed successively at the mouth of the glen. Since Darwin could not see how such huge quantities of rock could be moved about, he eliminated the lake hypothesis. Hence, the shelves were remnants of former shores of the sea, produced as the land was gradually raised above sea level.

As Darwin's critics have been happy to point out, Darwin was mistaken. Soon after publication of the Glen Roy paper, Agassiz's glacier theory was brought to Darwin's attention, and Darwin immediately realized that huge barriers of ice provided just the sort of dams needed by the lake hypothesis. But not until 1861 was he finally forced to admit defeat, writing to Lyell, "I am smashed to atoms about Glen Roy. My paper was one long gigantic blunder from beginning to end. Eheu! Eheu!"[16] In his autobiography, Darwin added:

> During these two years I took several short excursions as a relaxation, and one longer one to the Parallel Roads of Glen Roy, an account of which was published in the 'Philosophical Transactions.' This paper was a great failure, and I am ashamed of it. Having been deeply impressed with what I had seen of the elevation of the land of South America, I attributed the parallel lines to the action of the sea; but I had to give up this view when Agassiz propounded his glacier-lake theory. Because no other explanation was possible under our then state of knowledge, I argued in favour of sea-action; and my error has been a good lesson to me never to trust in science to the principle of exclusion.[17]

Darwin says that the Glen Roy paper taught him a good lesson—never to trust the principle of exclusion—precisely that principle which empiricist philosophers were touting as the key feature both in the logic of discovery and in the logic of justification. Darwin was correct to doubt this principle but wrong in thinking that he could do without it. He continued to use the principle of exclusion throughout his later works. What he had learned

15. Ibid., pp. 199, 284; Herschel (1841), pp. 196–197.
16. September 6, 1861, *More Letters* (1903), 2:188.
17. Darwin (1958), p. 84.

was not that the principle of exclusion could *never* be trusted but that it could never be trusted *completely*. It could, however, confer some probability on a hypothesis. It could not in actual practice make a hypothesis absolutely certain, since in the natural sciences it is never possible to eliminate *all* alternatives.

In his Glen Roy paper Darwin had obviously overlooked the possibility of ice barriers, but there were other possibilities as well. For example, even if the shelves had been formed by sea action, a gradual decrease in sea level would have explained them as well as a gradual elevation of the land. (In point of fact, the sea level had decreased 400 feet in the preceding 1000 years.) In his coral reefs paper (1842) Darwin explained the growth of coral reefs by reference to the gradual sinking of the ocean floor. The facts, he contended, were inexplicable by any other theory. In this instance Darwin's theory was basically correct, but he had not eliminated all the possibilities—to name one, a gradual increase in sea level. (As mentioned previously, the opposite had actually occurred. Darwin also mentions certain phenomena which would have had the same effect as a drop in sea level.)

Darwin also used this line of reasoning in the *Origin of Species* (p. 33) and *The Descent of Man* (p. 269). Again he had not eliminated all possible alternatives, but he did think that he was justified in discounting the current explanation in terms of special creation. He dismissed it, not because it was an incorrect scientific explanation, but because it was not a proper scientific explanation at all. "I would give absolutely nothing for the theory of Natural Selection, if it requires miraculous additions at any one stage of descent."[18] Darwin's reasoning on this issue will be examined later in Chapter 4.[19] It might also be mentioned that Bacon was no more successful than Darwin in eliminating all alternative explanations for phenomena he was investigating. For example, his explanation of the tides was erroneous because, in opposition to Galileo, he discounted the rotation of the earth. Galileo's explanation, however, was equally erroneous.[20]

The method of exclusion, when incomplete, failed to provide absolute proof for a hypothesis, but could incomplete inductions provide at least some support for a theory? Darwin thought they could. The feature of the philosophies of science professed by Herschel and Mill which would

18. Darwin, *Life and Letters* (1887), 2:7.
19. See also the reviews by Hopkins and Wright in this volume.
20. Bacon (1620), bk. 2, Aphorisms XXXVI and XLVI.

be of greatest use to Darwin's critics was their contention that only *logically necessary modes of inference afforded proof*. Although Mill recognized the importance of approximations in science, they were merely stages on the road to something better.[21] Until this final stage was reached, proof had not been provided. The only reason for resorting to approximations was that the correct definitions had not yet been discovered, the phenomena had not been correctly reduced to natural kinds. When they had been discerned, universal correlation was guaranteed.[22] In abstract sciences like zoology, botany and geology "a probability is of no account; it is but a momentary halt on the road to certainty, and a hint for fresh experiments."[23]

Mill has received considerable credit for not dismissing evolutionary theory out of hand as Herschel and Whewell had done,[24] but the credit is not deserved.[25] Mill discusses evolutionary theory in a footnote to his *System of Logic,* but only after treating Broussais' hypothesis about the localization of disease, Gilbert's contention that the earth is a magnet, and the suggestion that the brain be looked upon as akin to a voltaic pile. Finally, Mill says:

> Mr. Darwin's remarkable speculation on the *Origin of Species* is another unimpeachable example of a legitimate hypothesis. What he terms "natural selection" is not only a *vera causa,* but one proved to be capable of producing effects of the same kind with those which the hypothesis ascribes to it: the question of possibility is entirely one of degree. It is unreasonable to accuse Mr. Darwin (as has been done) of violating the rules of Induction. The rules of Induction are concerned with the conditions of Proof. Mr. Darwin has never pretended that his doctrine was proved. He was not bound by the rules of Induction, but by those of Hypothesis. And these last have seldom been more completely fulfilled. He has opened a path of inquiry full of promise, the results of which none can foresee. And is it not a wonderful feat of scientific knowledge and ingenuity to have rendered so bold a suggestion, which the first impulse of every one was to reject at once, admissible and discussable, even as a conjecture?[26]

Mill's endorsement was a double-edged knife. While admitting the legitimacy of Darwin's procedure as a method in the logic of discovery, he

21. Mill (1874), p. 387.
22. Ibid., p. 388.
23. Mill (1865), 2:308.
24. DeBeer (1964), p. 164, and Ellegård (1958), p. 195.
25. Himmelfarb (1959).
26. Mill (1874), p. 328.

denied that Darwin had proved anything. Not only had Newton used the method of hypothesis in discovering his laws, but he had also gone on to supply the necessary inductive proof.[27] Darwin had not. Thus, both Darwin and Fawcett were mistaken in their interpretation of Mill's remarks. T. H. Huxley, in the following quotation, reveals no deeper understanding:

> There cannot be a doubt that the method of inquiry which Mr. Darwin has adopted is not only rigorously in accord with the canons of scientific logic, but that it is the only adequate method. Critics exclusively trained in classics or in mathematics, who have never determined a scientific fact in their lives by induction from experiment or observation, prate learnedly about Mr. Darwin's method, which is not inductive enough, not Baconian enough, forsooth for them. But even if practicle acquaintance with the process of scientific investigation is denied them, they may learn, by the perusal of Mr. Mill's admirable chapter "On the Deductive Method," that there are multitudes of scientific inquiries in which the method of pure induction helps the investigator but a very little way.[28]

In his *Logic,* Mill distinguished between the Method of Deduction and the Method of Hypothesis. The Deductive Method consisted of three parts: direct induction, deduction, and verification. The ancients had omitted the final step. The Method of Hypothesis omits the first. To provide *proof* for a hypothesis, it had to be tested by the canons of legitimate induction. For Mill, this meant the deduction and checking of additional consequences of the hypothesis and the exclusion of all other possible causes:

> This process may evidently be legitimate on one supposition, namely if the nature of the case be such that the final step, the verification, shall amount to and fulfill the conditions of a complete induction. We want to be assured that the law that we have hypothetically assumed is a true one; and its leading deductively to true results will afford this assurance, provided the case be such that a false law cannot lead to a true result—provided no law except the very one which we have assumed can lead deductively to the same conclusions which that leads to.[29]

Mill contended that Darwin had used the Method of Hypothesis, not the Deductive Method, as Huxley implied. The Method of Hypothesis provided no proof. For proof, a complete induction was required.

Given all that has been said in this chapter concerning the philosophies

27. Ibid., p. 323.
28. Huxley (1896), p. 72.
29. Mill (1874), p. 323; see also ibid., p. 318n.

of science set out by Herschel, Whewell, and Mill, it is no surprise that Darwin's theory was found wanting. No scientific theory could hope to fulfill the unrealistic standards which these men set. The paper by Hopkins, included in this volume, clearly shows how inapplicable early philosophies of science were to evolutionary theory. As Darwin said of Hopkins' views, "On his standard of proof, *natural* science would never progress."[30] There is certainly nothing wrong with philosophers appropriating the term "proof" and giving it a technical meaning. But they also went on to claim that certain scientific theories—most notably Newton's theory of universal gravitation—had actually been proved in this sense of "proof" while other theories—for instance, Darwin's theory of evolution—had not. It is certainly true that Darwin had not provided a complete induction, but neither had Newton or any of his followers. Those sections in the writings of these early philosophers of science in which they claim to show that Newton's logics of discovery and justification accord with their own exposition have almost no relation either to Newton's theory or to historical fact.

The major difference was that Newton's theory was accepted. It was part of the received doctrine. Doubts about action-at-a-distance and aether had largely subsided through familiarity. Any reconstruction of scientific methodology had to be made to fit it or—and this second alternative was the more usual—Newtonian theory had to be reinterpreted to fit the reconstruction. Incredible as it may seem, W. C. Wells' "Essay on Dew" (1818) was repeatedly cited as the paradigm of scientific method. A more tedious, prosaic exercise is hard to imagine, but it was experimental, exhaustive, of limited scope, and above all safe. It concerned only observable features of the empirical world and challenged no accepted views. Darwin's explanations of the Parallel Roads of Glen Roy and the formation of coral reefs were more on this pattern and elicited no irate challenges concerning methodology. His theory of evolution was new on the scene and was measured against the Newtonian myth. It did not fit, and there was no special reason to rework the myth to fit it.

Contemporary philosophers of science commonly concede that early philosophers of science were hopelessly confused on such key issues as the differences between discovery and proof, between deduction and induction, and between inference and concept formation. Herschel, Whewell, and Mill raised important questions and made significant contributions to our understanding of science, but theirs was not the last word in the philosophy of science. If a theory failed to measure up to the standards these early

30. Darwin, June 1, 1860, *Life and Letters* (1887), 2:108.

philosophers set, it might not be the theory that was at fault. What then is the current view on the philosophy of science? How does evolutionary theory, both in its original form and in its modern reformulations, measure up to today's standards?

The most widely accepted model of scientific method among logical empiricists is termed the hypothetico-deductive model (H-D model). On this model, a hypothesis is formulated in some unspecified manner and then checked by deducing various consequences from it. If the deductions turn out to be true, then the hypothesis is gradually confirmed. No scientific hypothesis, however, is ever completely confirmed or verified. Scientific laws are universal in form, and not all consequences can be checked. However, if one of the deductive consequences of the hypothesis turns out to be false, the hypothesis is immediately shown to be false and must be rejected. Hence, falsification is very easy.

The H-D model has the virtue that it does not prescribe any rigid procedures in discovery. There is no cut-and-dried logic of discovery. Certainly there may be methods of scientific investigation which have proved fruitful in the past and are more likely to produce results than others. For example, analogies and models are two very effective means of discovery, while simple trial-and-error is not very efficient. But the chief virtue of the H-D model is that it recognizes the impossibility of absolute proof for any scientific theory. According to the method of "inverse deduction" incorporated in the H-D model, a scientific theory can become increasingly more probable; that is all. In inverse deduction, one reasons from the truth of specific inferences made from a law to the increased likelihood that the law is true. From the point of view of deductive logic such inferences are fallacious—the fallacy of affirming the consequent. Of course, early philosophers of science recognized the method of inverse deduction but thought it was only a temporary heuristic device.

As might be expected from its brevity, the foregoing description of the H-D model is overly simple. Hypotheses are not tested in isolation but embedded in theories. Thus, if a deductive consequence of the theory turns out to be false, a single hypothesis is not falsified. Rather, any one of several hypotheses and observations may be at fault. Instead of abandoning a theory upon a negative deduction, the theory is usually modified, and considerable latitude is permitted in the degree of modification permissible while still terming the theory the "same" theory.

The elegance and simplicity of the H-D model stems from its deductive character. In order for singular observation statements to be deducible from a law, that law must be universal in form. Neither deducibility nor the universality of the relevant laws can be abandoned without losing this elegance and simplicity. However, many putative laws in science, especially in biology and specifically in evolutionary theory, are not universal in form. They are statistical statements, trend and tendency statements, and approximations. From the statement that blue-eyed couples almost always produce blue-eyed children, it is impossible to deduce the eye-color of a particular child. However, it can be inferred that it is highly likely that all of the children of a blue-eyed couple, *ceteris paribus*, will have blue eyes. The inference is inductive.

If all less-than-universal statements are interpreted merely as particular statements which assert only that at least one subject has the predicate indicated, then the relation between verification and falsification is reversed. Verification becomes easy; falsification impossible. But scientists do not treat all less-than-universal statements as particular statements. There is considerable difference between the claim that 99 percent of smokers will die of lung cancer and the claim that 1 percent of them will. This difference is the subject matter of statistics and probability theory. In this type of inference, neither verification nor falsification is easy. In neither case will a single observation suffice—even in principle. Populations have to be defined and samples taken. Currently, philosophers of science are trying to formulate what might be called a hypothetico-inductive model of science, but it is proving to be a very frustrating task. Numerous troublesome paradoxes have arisen, and few solutions have been forthcoming.

How appropriate are current schematizations for evolutionary theory? Darwin maintained that the *Origin* was one long argument and a convincing one at that:

> Some of my critics have said, "Oh, he is a good observer, but has no power of reasoning." I do not think that this can be true, for the *Origin of Species* is one long argument from the beginning to the end, and it has convinced not a few able men. No one could have written it without some power of reasoning.[31]

An argument it may be, but the traditional notions of deduction and induction were completely inadequate for characterizing it. For example, Darwin says in the opening paragraphs of the *Origin:*

> For I am well aware that scarcely a single point is discussed in this volume on which facts cannot be adduced, often apparently leading to

31. Darwin (1958), p. 140.

conclusions opposite to those at which I have arrived. A fair result can be obtained only by fully stating and balancing the facts and arguments on both sides of each question; and this cannot possibly be here done.[32]

In a letter to H. G. Bronn, Darwin admits, "You put very well and very fairly that I can in no one instance explain the course of modification in any particular instance."[33] If evolution by natural selection was to be accepted at all, it would have to be "chiefly from this view connecting under an intelligible point of view a host of facts. When we descend to details, we can prove that no one species has changed; nor can we prove that the supposed changes are beneficial, which is the groundwork of the theory. Nor can we explain why some species have changed and others have not."[34]

As Huxley observed, the *Origin* is "a mass of facts crushed and pounded into shape, rather than held together by the ordinary medium of an obvious logical bond; due attention will, without doubt, discover this bond, but it is often hard to find."[35] The modern reader frequently grows impatient with Darwin's method in the *Origin* of piling example on example, but this was the only method open to him given the structure of evolutionary theory. This format is still characteristic of works in evolutionary theory.[36] As Darwin formulated it, evolutionary theory did not permit verification or falsification by the simple expedient of deducing a single observational consequence and checking it. Given Newton's equations and a few reasonably unproblematic simplifying assumptions and boundary conditions, a physicist could deduce some fairly precise observational consequences which could be used to confirm or disconfirm his theory. He could deduce where Mars should be at a precise time. Darwin's theory was neither mathematical in notation nor deductive in form. Given evolutionary theory and the kind of data available to a biologist, no precise, unequivocal inferences were possible. One could not predict how an extant species would evolve or retrodict how an extinct species did evolve. At best, only a range of some more or less possible outcomes could be generated. In this respect, Darwin's theory was at a decided disadvantage. The two means of verification recognized in Darwin's day were the deductive subsumption of one general law under a more general law and the deduction of specific observational con-

32. Darwin (1859), p. 2.
33. October 5, 1860, *More Letters* (1903), 1:172–173.
34. Darwin, *Life and Letters* (1887), 2:210.
35. Huxley (1896), p. 25.
36. See Dobzhansky (1937), Mayr (1942 and 1963).

sequences of the law. Darwin's basic principles were related inferentially
but not in a strict deductive hierarchy. Evolutionary theory was basically
statistical in nature. The relevant inferences were inductive.

Charles S. Peirce was one of the few philosophers who recognized and
emphasized the statistical nature of evolutionary theory. He saw Darwin's
theory as one more application of the statistical method which Maxwell
(1831–1879) had used so successfully in physics and Quetelet (1796–1874)
in sociology.[37] The work of these men and others had prepared the scientific
community for "the idea that fortuitous events may result in a physical
law."[38] Peirce's accurate understanding of the structure of evolutionary
theory is sufficiently rare to warrant reproducing in its entirety:

> The Darwinian controversy is, in large part, a question of logic. Mr.
> Darwin proposed to apply the statistical method to biology. The same
> thing has been done in a widely different branch of science, the theory
> of gases. Though unable to say what the movements of any particular
> molecule of gas would be on a certain hypothesis regarding the constitu-
> tion of this class of bodies, Clausius and Maxwell were yet able, by
> the application of the doctrine of probabilities, to predict that in the
> long run such and such a proportion of the molecules would, under
> given circumstances, acquire such and such velocities; that there would
> take place, every second, such and such a number of collisions, etc.;
> and from these propositions they were able to deduce certain properties
> of gases, especially in regard to their heat-relations. In like manner,
> Darwin while unable to say what the operation of variation and natural
> selection in every individual case will be, demonstrates that in the long
> run they will adapt animals to their circumstances. Whether or not exist-
> ing animal forms are due to such action, or what position the theory
> ought to take, forms the subject of a discussion in which questions of
> fact and questions of logic are curiously interlaced.[39]

Although Peirce was almost alone in approving of the form of Darwin's
theory, he joined the majority in rejecting selected parts of the content.
He preferred Agassiz's *Essay on Classification* (1859).[40] In 1893 he passed
the following judgment on evolutionary theory:

> What I mean is that his hypothesis, while without dispute one of
> the most ingenious and pretty ever devised, and while argued with a

37. James Clerk Maxwell was a Scottish physicist whose famous equations revolu-
tionized physics. M. A. Quetelet used statistics in a rudimentary way in compiling
figures for crimes and suicides; see Quetelet (1846). Herschel reviewed this work
in the *Edinburgh Review* (1850), 42:1–57.
38. Peirce (1935), 6:297.
39. Peirce (1877).
40. Peirce (1931), 1:203–205.

wealth of knowledge, a strength of logic, a charm of rhetoric, and above
all with a certain magnetic genuineness that was almost irresistible, did
not appear, at first, at all near to being proved; and to a sober mind
its case looks less hopeful now than it did twenty years ago.[41]

At the turn of the century, the fortunes of evolutionary theory had
reached their lowest ebb. Physicists to a man had proved that the earth
was not nearly as old as Darwin's theory required. Geneticists such as
de Vries and Morgan were claiming that mutations were discrete, not
gradual; thus, Darwin had been wrong in thinking that evolution was
gradual. The excesses of evolutionists like Haeckel and the disputes concerning orthogenesis and neo-Lamarckianism had earned for evolution a decidedly bad reputation. It was more metaphysics than science.[42] One would
expect the situation to improve considerably when physicists and geneticists
discovered their errors, and it did, but the problem of the form of evolutionary theory remained unchanged. It was still, at best, probabilistic. In the
1930's Fisher, Haldane, and Wright attempted to use the recent advances
in genetics to produce a theory of evolution which was both mathematical
and genetical. The unit of evolution became the gene, and evolutionary
change was to be measured in terms of changes in gene frequencies. Even
so, the situation with respect to verification did not improve much. Given
modern evolutionary theory, knowledge of mutation rates, rates of differential reproduction, and so forth, the best that a modern evolutionist could
do was to provide a distribution function of possible outcomes.[43]

Two alternatives seem open to us at this juncture: either revert to a
hypothetico-inductive model of science or argue that evolutionary theory
after a century is still inadequately formulated and that in a more finished
form will conform to the H-D model. The problem with the first alternative
is that there is no H-I model. Thus, this alternative reduces to the admission
that there is no reconstruction of science appropriate to evolutionary theory
as it now stands. Biologists are currently working on the second alternative.
Some are attempting to reformulate macro-evolutionary theory more rigorously so that deductive confirmation or disconfirmation is possible.[44]
Others find evolutionary theory in terms of organisms and their interactions

41. Peirce (1935), 6:297.
42. Unfortunately, space prohibits us from pursuing the history of the reception
of evolutionary theory during this period. Peter Vorzimmer (1970) has written
an informed but uneven treatise on these "years of controversy." See also Provine
(1971).
43. See, for example, Lewontin (1969); for further discussion see Hull (1974).
44. MacArthur and Wilson (1967) and Williams (1970).

too crude to permit an adequate formulation of evolutionary theory. Instead, they call for a molecular version of evolutionary theory, hoping in this manner to fulfill the requirements of the H-D model.[45] This reductive maneuver is quite common in science. Whenever laws at one level of analysis appear indeterministic, analysis of the phenomena at a lower level often reveals relations which are closer to universality. The only thing that stands in the way of true, universal laws is accuracy of measurement and the level of analysis. It should be remembered, however, that after years of success in physics, the reductive program was frustrated at the subatomic level.

On the first alternative, philosophers of science have more work to do. They have to work out the details and general outline of a reconstruction of science appropriate to evolutionary theory in its present form. After all, notice how much better evolutionary theory looked when Peirce switched from the Newtonian to the Maxwellian paradigm. On the second alternative, biologists have more work to do. They have to produce a scientific theory consisting of universal laws which permit specific predictions. On either view, evolutionary theory and philosophy of science are still at odds.

Most of this chapter has concerned the detrimental effects of early philosophies of science on the reception and acceptance of evolutionary theory. A few words must be said in defense of these systems. When early philosophers of science were not concerned with rigid proof and certainty, much that they said was applicable to science in general and to evolutionary theory in particular. (Compare Hopkins' strict interpretation with Fawcett's more informal treatment in this volume.) For example, Bacon provides a list of auxiliary devices to help in scientific investigation. These *praerogative instantiarum* are much more original and applicable to actual scientific investigation than his methods of true induction. Almost everything that Herschel had to say about the discovery and verification of scientific theories—when he was not being rigorous—fits evolutionary theory perfectly. Mill is similarly relevant when he is discussing such things as multiple causation, approximate generalizations, and probable inference.

Even the extensive fact-gathering encouraged by these empiricist philosophies of science was not entirely detrimental to evolutionary theory, given its statistical nature. Darwin did not waste the twenty years preceding the publication of the *Origin*. Only such a wide sampling of data had the slightest chance of supporting it. The relation between Whewell's phi-

45. Schaffner (1969).

losophy of science and evolutionary theory is even more equivocal. Nothing
in Whewell's second-level philosophy of science was incompatible with evolutionary theory. He did not concur in the popular conviction that a strict
logic of discovery would assure truth. In the colligation of ideas why not
superinduce the idea of evolution by natural selection on the facts? Certain
ideas are self-evident. Why not evolution? The answer can be found at
a deeper level in a belief which Whewell shared with Aristotle, Bacon,
Herschel, and Mill, a belief in the existence of natural kinds definable
by a single set of necessary and sufficient conditions, a belief in essences.
Empiricists wanted to eliminate so-called "occult qualities" from science
and yet they retained that element in earlier philosophies which contributed
the most to the prevalence of metaphysical entities—essentialism.

3 · Occult Qualities

Nineteenth-century scientists were in agreement that all references to occult qualities should be excluded from science, but they were not clear as to what it was about a quality that made it occult. Bacon thought that such Aristotelian notions as the hot and the moist were occult qualities but that his own explanation in terms of the movement of unobservable particles was perfectly acceptable. Newton thought of light as being corpuscular but excluded all reference to such theoretical entities from his published writings. Such hypotheses as light corpuscles and the aether had no place in science. Darwin found Forbes's explanation of the distribution of living creatures in terms of polar principles absolute rubbish, on a par with magnetism moving tables. Owen's Law of Irrelative Repetition in terms of a polarizing force opposing the Platonic *ratio* was scarcely better. Yet his own theories necessitated unobservable gemmules and the spontaneous generation of the original life forms in the distant past. Mendel claimed that he defined constancy of type merely as the retention of a character during the period of observation, yet could not avoid postulating unobservable "factors" as the cause of his observed ratios. In the preceding chapter, the Victorian aversion to speculation in the sense of large inductive leaps was examined. In this chapter we will deal with speculation in the sense of unobservable entities, substances, and properties. The crucial distinction is between theoretical entities for which indirect evidence can be obtained and metaphysical entities which are unobservable in principle.

Bacon intended the notion of gradual induction and complete elimination to insure the *truth* of the generalizations derived, but induction from experience also served a second function in empiricist epistemology—to insure the *empirical content* of scientific terms. If scientific concepts are abstracted from experience, regardless of how large the inductive leap, they will still be grounded in experience and not empty verbiage. From Bacon and Newton to Herschel and Mill, scientists and empiricist philosophers had a deep-seated aversion to what they viewed as Scholastic occult qualities. One of the primary goals of scientists during and after the scientific revolution was to make certain that scientific concepts and generalizations were not empirically meaningless. The issue was not whether a scientific generalization was true but the more basic question of whether or not it was empirically meaningful.

Aristotelian science, in the hands of the Scholastics, had degenerated to such an extent that many putatively empirical claims were actually immune to experimental disproof. Such notions as the hot and the moist had begun in Aristotle as observational terms. If something felt hot, then it was hot. But gradually these terms took on additional dimensions. Maybe one body felt hotter than another, but that did not guarantee that it contained more of the hot. The gradual evolution of observational terms to something else is common throughout the history of ideas, but the nature of this "something else" varies. One possibility is that the term becomes a theoretical term in a well-articulated scientific theory. For example, "force" may well have begun its career as an observational term referring to the effort a human being expends in holding a heavy object but eventually such reference became all but irrelevant. Theoretical terms frequently refer to unobservable entities, but because these terms occur embedded in a well-articulated scientific theory, indirect evidence can be obtained for the existence of the theoretical entities to which they refer. Another fate for an observational term is to become purely metaphysical. There is nothing wrong with metaphysical terms, just so long as they are openly metaphysical and do not masquerade as something else. For example, when a metaphysician refers to such things as bare particulars, he is not claiming that they exist in the same sense that trees and planets do. Too often, however, philosophers and scientists have confused metaphysical terms with those that have empirical referents. The consequences have been less than salutary.

The terms usually singled out by Aristotle's critics as being "occult" were those designed to characterize Aristotelian essences. Aristotle would have been dismayed by the claim that his essences were unknowable, since for him essences were the only things that were knowable in the true sense of the word. Perhaps essences were furthest from the level of gross observation, but they were the most knowable in the sense of intuitive knowledge. Aristotle's critics, of course, would reply that intuition itself was a metaphysical notion. Aristotle claimed that, at bottom, all certain knowledge was known by intuition, and intuition was infallible. Observation might well excite intuition to act, but only intuition was capable of apprehending the true objects of scientific knowledge—eternal, immutable essences. The most obvious objection to Aristotle's postulation of infallible intuition is that people do make mistakes. In fact, almost everything which Aristotle himself claimed to have intuited turned out to be false. Several evasive maneuvers are possible for saving the notion of infallible intuition. The extent to which one uses one of these maneuvers determines exactly how metaphysical the notion of intuition itself becomes.

One way to salvage the notion of infallible intuition is to deny the status of intuition to any process that eventuates in falsehood. "You say that you intuited that all crows are black? Well, here is a white crow." "Oh, well, that must not have been intuition." Intuition was intended to assure the truth of the premises in deductive arguments so that the truth of the conclusions would be assured, but all that this maneuver does is to shift the problem back one step. Now the problem is deciding which claims of intuition are factual and which mistaken.

A second way to save intuition is to discount all exceptions as monsters. Aristotle says that science deals with what happens always or for the most part. This latter qualification did not mean that scientific knowledge could be less than universal in form. Rather, it was included in recognition of the existence of accidents and monsters, and science does not concern itself with accidents or monsters. "That bird is white? Well, then it isn't a crow." Among normal individuals, Aristotle distinguished between essential attributes and accidental properties. Essential attributes were those characteristics which made a thing what it was—its defining characteristics. Accidental properties were those characteristics which varied in a haphazard way from individual to individual—the individual differences which were to be the raw material of evolution.[1] In addition to these accidental characteristics which occurred quite as a matter of course, Aristotle also recognized rare individuals completely outside the natural order—monsters.

According to Aristotle, neither accidental properties nor monsters allowed of explanation and hence were not the proper subject matter of science. All natural laws were true universal generalizations. What of exceptions? They were merely accidents or monsters and could easily be dismissed. It was this Scholastic habit of monsterbaiting which proved the greatest obstacle in the development of Scholastic science. On the modern view of science, it is precisely the "accidents" and "monsters" which call for

1. See comments on Jenkin's paper.

explanation.[2] It was Darwin's attention to so-called accidental variations which led to his theory of evolution, and it in turn demolished the Aristotelian distinction between essential and accidental characteristics. No characteristic is so "accidental" that it cannot become in time typical of a group. And what would have happened in genetics if Morgan had dismissed his one-in-a-million white-eyed male fruit-fly as an inexplicable accident?

Darwin was aware of the importance of exceptions in science and also of the human propensity to ignore them:

> I had, also, during many years, followed a golden rule, namely, that whenever a published fact, a new observation or thought came across me, which was opposed to my general results, to make a memorandum of it without fail and at once; for I had found by experience that such facts and thoughts were far more apt to escape from the memory than favourable ones.[3]

He had gathered this admonition from Herschel, who had said in his *Discourse* (1830:165):

> . . . in any case any exception occurs, it must be carefully noted and set aside for re-examination at a more advanced period, when, possibly, the cause of the exception may appear, and the exception itself, by allowing for the effect of the cause, be brought over to the side of our induction; but should exceptions prove numerous and various in their features, our faith in the conclusion will be proportionally shaken, and at all events its importance lessened by the destruction of its universality.

Philosophers and scientists through the ages have been nearly unanimous in their opinion that scientific laws had to be universal in form. One way to save the universality of scientific laws is to discount all exceptions as monsters, as not properly fitting the definition of the names of the classes of entities related in the law. Once the phenomena have been properly reduced to natural kinds, universal laws must be forthcoming.[4]

In their investigations, scientists tend to concentrate on exceptions to their laws with the hope that such an emphasis will lead to their improvement. Like essentialists, they hope to discover true, universal generalizations; but, unlike essentialists, they tend to change the law to accommodate the exception rather than to dismiss the exception as inexplicable. The end result of the essentialist distinction between essential and accidental at-

2. This view in biology can be traced back at least to William Harvey, who in 1657 observed that the laws of nature can often be discerned by careful investigations of rareties.
3. Darwin (1958), p. 123.
4. Mill (1874), p. 388.

tributes is that essential attributes tend to become occult, unobservable, unknowable. Because this distinction concerns the development of concepts within a system, not some intrinsic quality of a concept from the beginning, and because there is a strong tendency to dismiss erroneous explanatory schemata as empty metaphysics, two examples will be discussed at some length in preparation for the discussion of evolutionary theory and occult qualities. These examples are Aristotle's polar principles, elements, and humors—and Newton's light corpuscles, aether, and action-at-a-distance.

In the preceding chapter, Aristotle was quoted as saying that man, the horse, and the mule were all bileless animals and assuming that these are all the bileless animals that there are. But he can also be found saying that the ass, deer, roe, camel, seal and dolphin lack gall bladders (676^b26, 677^a30). It would seem reasonable to infer that in a teleological system like Aristotle's there should be some correlation between lacking a gall bladder and lacking gall. These references might reasonably be dismissed as a later development of his ideas, or as oversights and errors in transcription or translation, but even if they accurately reflect Aristotle's view, he need not stand convicted of contradicting himself. For Aristotle, bile was not merely the fluid secreted by the liver and stored in the gall bladder. It was one of the four humors and, as such, part of an interconnected system of concepts.

The basic elements in Aristotle's science were two sets of polar principles—the hot and the cold, the dry and the moist. All phenomena, whether in physics, physiology, or meteorology, were explained in the last analysis by reference to these four qualities. The four basic elements of physics were fire, air, earth, and water. Whatever properties fire had were due to its being hot and dry, air hot and moist, and so on. The four humors formed the foundation of Aristotelian physiology. Yellow bile (or choler), like fire, was hot and dry, black bile (or melanocholer), like air, was hot and moist, and so on. In a healthy person these four humors were balanced, but when one of them gained ascendancy, illness resulted. Therapy consisted in attempts to reinstate this balance.

This system cannot be elaborated any more fully here. The purpose of introducing it at all is to show that as a system of interconnected concepts it is not different in kind from modern scientific theories. Changes in one element in the system have ramifications throughout the system. For example, in Aristotle's physics of natural place, the natural place for earth was at the center. Copernicus, by placing the sun (fire) at the center, considerably simplified astronomical computation, but he overturned all of Aristotelian physics, and indirectly physiology, medicine, meteorology, and so on. Aristotelians had good reason for rejecting Copernicus' suggestion.

Because bile was part of this loosely connected system of concepts, its presence or absence was not purely a matter of observation. Some animal might possess bile, though none was discernible, and others lack it, though a bitter fluid could be extracted from the liver. Similarly, some animal might feel no colder or moister than another animal, but because it was lower on the scale of being, it had to be colder and moister. "Cold" and "moist" became theoretical terms connected only indirectly to our feelings of cold and moisture. Because of the subsequent history of Aristotelian science and its distance from us in time, modern readers might be tempted to dismiss the whole system as metaphysical nonsense, but Aristotle's procedure of inventing theoretical entities which cannot be observed directly is not significantly different from modern scientific method.

Today we explain the recurrent fever and chills of malaria by reference to the life cycle of a species of *Plasmodium* rather than by the cyclical overpredominance of blood and phlegm. We believe the action of this protozoan on the blood cells can be explained in terms of the molecular structure of DNA and RNA, and this structure in terms of molecules, atoms, and various subatomic particles. Aristotle stopped at the level of the four qualities. Today we stop at the level of quantum mechanics. There is no doubt that our explanation is far superior to that of the ancient Greeks. Care must be taken, however, in identifying the features of the modern explanation which make it preferable to earlier explanations. It does not lie, as scientists in Darwin's day contended, in the paucity of unobservable entities.

The elements in Aristotelian natural science became "occult," not because of the happenstance that the human organism cannot sense them directly (after all we cannot perceive ultraviolet light) but because the system in which the concepts were embedded was so loosely organized that little in the way of indirect evidence was possible either. For example, even though we cannot see ultraviolet light, this light has certain effects on our skin which we can detect. Numerous predictions can be made about what ultraviolet light should do, given current theory. The four qualities and other elements in Aristotelian physics started out as observables and progressed gradually to the level of unobservables, but this development was not accompanied by a corresponding increase in precision in the systemic interrelations of the more theoretic concepts and between them and possible observations. For example, a newly discovered disease could be attributed to predominance of phlegm by one physician and of bile by another, and there would be no experiential way of resolving the dispute. Extreme reliance on authority became more and more the order of the day.

Bacon and Newton attributed the degraded state of Scholastic philosophy both to a penchant for rash, inductive leaps and to a weakness for occult qualities. In order to avoid reference to occult qualities, they proposed that science should make reference only to those entities, substances, or qualities whose character or effects could be immediately shown. For Bacon, this proposal took the shape of fulminations against forms. Newton expressed the same message in his cryptic *hypothesis non fingo.*[5] The manifesto was easier to proclaim than to practice. Bacon realized that his tables could be used to abstract forms as easily as causes, and he went to great lengths to assure his readers that this was not what he was doing. Although he referred to "forms," he did not mean anything like Aristotelian essences:

> I cannot too often warn and admonish men against applying what I say to those forms to which their thoughts and contemplations have hitherto been accustomed . . . For when I speak of forms, I mean nothing more than those laws and determinations of absolute actuality which govern and constitute any simple nature, as heat, light, weight, in every kind of matter and subject that is susceptible of them.[6]

The distinction toward which Bacon was groping was between concepts and laws, between defining a concept in terms of a set of covarying characters and combining concepts thus defined into empirical laws. In actual practice, Bacon's efforts at discovering causes by his inductive method differed little from those of the Scholastics. To overlook the "grand distinction" between the coexistence of characters and the succession of phenomena in time was, as it seemed to Mill, "the capital error in Bacon's view of inductive philosophy."[7] The conceptual shift which made the recognition of this distinction more obvious was the change from an ontology of substances and attributes to an ontology of events. The paradigm of causation had become one billiard ball hitting another.

In Bacon's work, as well as the Scholastics', the search for laws of nature too often resulted in the enunciation of definitions,[8] but of equal importance

5. See discussion in the review by Wright included in this volume.
6. Bacon (1620), bk. 2, Aphorism XVII.
7. Mill (1874), p. 381.
8. The confusion of definitions and laws is still quite common. For example, D. R. Cressey, using von Hartmann's method of analytic induction, ends up defining "criminal violation of financial trust" instead of discovering its causes. See Cressey (1950) and Robinson (1951).

was the nature of the attributes referred to in these definitions. Although
Bacon tried to restrict himself to observable facts and nothing but observable
facts, he was no more able to do so than scientists of other ages. In fact,
many of Bacon's most significant errors stemmed from his over-reliance
on superficial appearances. Like Aristotle, he argued that the earth did
not seem to move; hence, it did not move. Like Aristotle, he believed
in the spontaneous generation of macroscopic organisms because careful
observation showed it to occur "in animalculae generated from putrification
as in ants' eggs, worms, flies, frogs after rain, etc." Bacon did concede
that more careful observation was still called for.[9] In Darwin's day, Sedgwick argued that we should accept the fossil record as we find it,
not interpolate unknown fossils and missing strata. His catastrophic description of fossil forms and geological strata was supposedly the history
of Creation. "It is not the dream of a disordered fancy," he asserted, "but
an honest record of successive facts that were stamped by Nature's hand
on the chronicle of the material world."[10]

Could unobservables play an important role in science and, if so, in
what sense unobservable? One of the commonest objections to evolutionary
theory was that it rested on hypotheses, on unobservable and unknowable
phenomena. In reply to these criticisms, Darwin wanted to distinguish between theories and theoretical entities and once again took refuge in the
unimpeachable authority of Newton. In a letter to Gray he objects:

> Your distinction between an hypothesis and theory seems to me very
> ingenious; but I don't think it is ever followed. Every one now speaks
> of the undulatory *theory* of light; yet the ether is itself hypothetical,
> and the undulations are inferred only from explaining the phenomena
> of light. Even in the *theory* of gravitation is the attractive power in any
> way known, except by explaining the fall of the apple, and the movements
> of the Planets? It seems to me that an hypothesis is *developed* into
> a theory solely by explaining an ample lot of facts.[11]

He continues along much the same vein in a letter to F. W. Hutton:

> I am actually weary of telling people that I do not pretend to adduce
> direct evidence of one species changing into another, but that I believe
> that this view in the main is correct, because so many phenomena can
> be thus grouped together and explained. But it is generally of no use,
> I cannot make persons see this. I generally throw in their teeth the

9. Bacon (1620), bk. 2, Aphorism XL and XLI.
10. Sedgwick, *Life and Letters* (1890), 2:190.
11. Darwin, February 18, 1860, *Life and Letters* (1887), 2:80.

universally admitted theory of the undulation of light,—neither the undulation, nor the very existence of ether being proved—yet admitted because the views explain so much.[12]

Finally, he wrote to Lyell:

> With respect to Bronn's objection that it cannot be shown how life ← arises, and likewise to a certain extent Asa Gray's remark that natural selection is not a *vera causa*, I was much interested by finding accidentally in Brewster's *Life of Newton*, that Leibnitz objected to the law of gravity because Newton could not show what gravity itself is. As it has chanced, I have used in letters this very same argument, little knowing that any one had really thus objected to the law of gravity. Newton answers by saying that it is philosophy to make out the movements of the clock, though you do not know why the weight descends to the ground. Leibnitz further objected that the law of gravity was opposed to Natural Religion! Is this not curious? I really think I shall use the facts for some introductory remarks for my bigger book.[13]

Without realizing it, Darwin had chosen an especially inappropriate champion, both because gravity, interpreted as an innate attraction between every pair of particles of matter, *was* an occult quality on a very subtle level and because Newton himself was one of the most influential advocates of the view of science which Darwin was being forced to combat. Newton stated his ideas on science very grudgingly and with more than a little petulance. They were elicited by attacks first on his theory of light and then on his law of universal gravitation.

Early in his life, Newton set out a theory of light in which he claimed to prove that white light resulted from the commingling of light rays of various colors. As his notebooks show, Newton viewed these rays as being composed of "globules," or corpuscles, but when he published the results of his investigations, he was even more careful than Mendel had been in avoiding reference to the theoretical entities which he was postulating. He did, however, claim to have performed a crucial experiment which had refuted the currently popular optical theories, including Robert Hooke's undulation theory. The numerous letters published in reply to Newton's optical theory had one thing in common—they called it a "hypothesis." The polemics that ensued clearly reveal a systematic ambiguity in the term which was still prevalent in Darwin's day and which continues down to the present. In the case of Newton, Ignatius Gaston Pardies, a professor

12. April 20, 1860, *More Letters* (1903), 1:184.
13. February 23, 1860, *Life and Letters* (1887), 2:83–84.

of mathematics, began his letter by referring to Newton's ingenious hypothesis concerning light. Newton replied in a model of overkill:

> I do not take it amiss that the Rev. Father calls my theory an hypothesis, inasmuch as he was not acquainted with it. But my design was quite different, for it seems to contain only certain properties of light, which, now discovered, I think easy to be proved, and which if I had not considered them as true, I would rather have them rejected as vain and empty speculation, than acknowledged even as an hypothesis.[14]

This terminological confusion was quickly cleared up. In his next letter Pardies apologized for his use of "hypothesis." It had been done "without any design, having only used that word as first occurring to me; and therefore request it may not be thought as done out of any disrespect."[15] "Hypothesis" had been used to mean "whatever is explained in philosophy." Newton accepted Pardies' apology but went on to enunciate an additional sense of the term "hypothesis." Even if hypotheses were only statements concerning substances which one might postulate as the causes of observable phenomena, Newton was opposed to them. In his *Optics* he had provided the laws which governed the behavior of light, regardless of whether it was made up of particles, waves, or something else. Scientists did not speculate about things they could not prove. Similarly, in his *Principia,* Newton believed that he had provided the laws which governed the movements of material bodies. His reply to those scientists who wanted to know the nature of gravity was the same as he had made to those who had wanted him to speculate about the nature of light:

> So far I have explained the phenomena of the heavens and of the sea by the force of gravity, but I have not yet assigned the cause of gravity . . . I have yet been able to deduce from the phenomena the reasons for these properties of gravity and I invent no hypotheses. For everything which is not deduced from the phenomena should be called an hypothesis, and hypotheses, whether metaphysical or physical, whether of occult qualities or mechanical, have no place in experimental philosophy. In this philosophy propositions are deduced from the phenomena and rendered general by induction. It is thus we have come to know the impenetrability, the mobility, the *impetus* of bodies and the laws of motion and of gravity. It is enough that gravity really exists, that it acts according to the laws we have set out and that it suffices for all the movements of the heavenly bodies and of the sea.[16]

14. Newton (1958), p. 92.
15. Ibid., p. 105.
16. Newton (1960), p. 547.

It is certainly true that no theory can explain everything. Certain features will always, for a time, have to be assumed. But Newton's contemporaries wanted to know what kind of considerations led Newton and might lead them to accept gravity. Gravity really exists, but what is it that exists? What characteristics might it have in addition to those necessitated by its role in the theory? Although these were precisely the questions which Newton asked himself, he objected to them when asked by others. He wanted to limit public debate just to those propositions "deduced from the phenomena." A more problematic phrase is hard to imagine, though it was to become commonplace in the literature.[17] Deduction in the strict sense is a necessary inference, to be contrasted with induction. To deduce something from the phenomena sometimes meant that it was observed (no formal inference being involved) or that it was inferred from observations (induction in the broad sense). In what sense had Newton deduced his optical and gravitational laws from the phenomena? He had hardly observed either the various colored light rays in white light (though he had seen the diffracted spectrum when white light was passed through a prism) or gravity (though he had seen bodies move in certain paths and at certain velocities). From these observations in the context of his general beliefs about the makeup of the universe, he was able to infer his laws. This inference, however, fitted none of the models of induction set out in his day. In short, Newton would have been hard put to distinguish between his laws, which had been deduced from the phenomena, and physical hypotheses about mechanical qualities, since at best these were precisely what his laws were.

Newton hated to admit that certain of his principles were physical hypotheses about mechanical qualities, but his critics, especially in France, went so far as to accuse him of introducing into science metaphysical hypotheses about occult qualities. Cartesians such as Leibnitz tirelessly reminded him that gravity looked very much like such occult qualities as sympathy, antipathy, attraction, and hostility, the very kinds of qualities which Newton wished to exclude from science. In his eulogy to Newton, Fontenelle assessed the situation as follows:

17. For example, Herschel (1841:179) claims that perception is philosophical deduction from experiment; Charles Babbage (1837:57), speaks of an observor deducing a law from an induction of a hundred million instances; James Clerk Maxwell (1890) claimed that his law of the distribution of electricity on the surface of conductors was analytically deduced from experiment; and see the title of Lyell (1851).

He declares very freely that he lays down this attraction, only as a cause which he knows not, and whose effects he only considers, compares and calculates; and in order to avoid the reproach of reviving the *Occult qualities* of the Schoolmen, he says, that he establishes none but such Qualities as are *manifest* and very visible by their phenomena, but that the causes of these Qualities are indeed occult, and that he leaves it to the other Philosophers to search into them.[18]

But the Schoolmen also claimed that their qualities were manifest. After all, what is more manifest than the hot and the cold, the dry and the moist? In fact, of course, the technical terms "hot," "moist," and so on, had lost their original experiential connotations, but so had the notion of force in Newtonian theory. It had little to do with the feeling one gets when lifting a heavy object.

The feature of Newton's gravitational theory which his contemporaries found most objectionable was its apparent reliance on action at a distance. Gravity acted instantaneously in the absence of contact throughout all space. One way to eliminate such a counter-intuitive notion was to postulate a universal ether pervading all space. Prior to Newton, scientists had frequently postulated subtle fluids to explain natural phenomena, especially in biology, but such fluids were looked upon with some degree of suspicion. With the tremendous success of Newton's theory, the postulation of subtle fluids as explanatory principles became respectable. For instance, in his review of the *Origin* included in this volume, Hopkins justifies Lamarck's vital fluids by comparing them with the luminiferous ether. Thus, Newton contributed unwittingly to two opposing trends in science—the rigid, overly restrictive, ultra-empirical inductive method and the almost unlicensed introduction of subtle fluids.[19]

There was a difference between the occult qualities of the Schoolmen and Newton's light corpuscles and gravity, and it was the difference which Newton himself pointed out, but for different reasons. When attacked, Newton retreated to mathematics. He invented no hypotheses. He offered no mechanisms. He simply supplied the equations which accounted for various phenomena mathematically. (A similar retreat has occurred in this century in quantum mechanics.) Newton need not have retreated. Because he could and did provide laws which accurately accounted for a wide range of phenomena with a greater degree of accuracy then any previous theory, there was good reason to believe in the existence of the postulated

18. Newton (1958), p. 463.
19. See Thomas Hall (1968).

entities and qualities and to attribute to them the characteristics the theory required.

Justified or not, Newton did not like to have to postulate anything which even slightly resembled an occult quality. On this score, Darwin was at a considerable advantage. Darwin had to postulate *no new forces or substances*. He merely showed the possible, almost the inevitable, consequences of fairly common observations about the organic world. Offspring vary slightly from each other and from their parents. More offspring are produced than can survive. It is reasonable to expect differential mortality as a function of these variations. And so on. Darwin introduced nothing that faintly approached an occult quality, metaphysical entity, or vital force. Although an occasional author with extreme effort was able to read his references to natural selection as a reification of Nature selecting, Darwin's theory was purely naturalistic.[20]

When Darwin's critics objected to his evolutionary theory as hypothetical and speculative, they had in mind three features of his formulation: (1) evolution occurring so slowly that no one had observed or was likely to observe one species evolving into another; (2) the initial spontaneous origin of life being in the distant past and unknowable; and (3) Darwin's uncautious assertion that *all* species had evolved and most likely from a *single* form.

Darwin grew weary of telling people that he did not pretend to address direct evidence of one species changing into another. Among those who contributed most to Darwin's weariness was Huxley. Throughout their collaboration, Huxley steadfastly maintained that intersterility infallibly distinguished species and "until selective breeding is definitely proved to give rise to varieties intersterile with one another, the logical foundation of the theory of natural selection is incomplete."[21] Darwin believed, on the other hand, that it was difficult "to make a marked line of separation between fertile and infertile crosses."[22] Moreover, those groups distinguished by constant characters did not always coincide with those distinguished by sterility. As the years went by Darwin became resigned to the disagreement:

> I did not understand what you required about sterility: assuredly the facts given do not go nearly so far. We differ so much that it is no use arguing. To get the degree of sterility you expect in recently formed

20. See, for example, the review by Wollaston included in this volume.
21. Huxley (1896), p. vi.
22. August 1, 1860, *More Letters* (1903), 1:166.

varieties seems to me simply hopeless. It seems to be almost like those naturalists who declare they will never believe that one species turns into another till they see every stage in the process.[23]

You will say Go to the Devil and hold your tongue. No, I will not hold my tongue; for I must add that after going, for my present book, all through domestic animals, I have come to the conclusion that there are almost certainly several cases of two or three or more species blended together and now perfectly fertile together.[24]

You are so terribly sharp-sighted and so confoundly honest! But to the day of my death I will always maintain that you have been too sharp-sighted on hybridism.[25]

Two points are at issue in the preceding quotations: the scientific issue of the sufficiency of sterility as a criterion for species status and the philosophical question concerning the need for direct confirmation of theories. We cannot treat the scientific question here, since the literature is voluminous. All that can be mentioned is that one of the cornerstones of the synthetic theory of evolution is the biological definition of "species" in terms of interbreeding, potential interbreeding, and reproductive isolation.[26] Darwin argued against intersterility as a criterion for species status and for the constancy of characters. Modern evolutionists argue against the criterion of constancy of characters and for reproductive isolation (which in several important respects is different from intersterility).[27]

The question that interests us here, however, is whether or not Huxley was justified in requiring direct evidence of evolution by natural selection before accepting the theory without reservation. Either Huxley is wrong in his views on the relation between direct evidence and proof or else the majority of scientists have been mistaken throughout the history of science. Scientific theories are accepted long before anything like direct proof is provided. For example, when Copernicus enunciated his heliocentric system, scientists immediately selected the observation of stellar parallax as the most significant test of the theory. None was observed until the 1830's—long after all reasonable men had accepted the heliocentric system.

23. December 28, 1862, ibid., 1:225.
24. January 10, 1863, *More Letters* (1903), 1:232.
25. January 30, 1868, ibid., 1:287.
26. Mayr (1963).
27. For further discussion, see the comments on the papers by Wollaston and Fawcett in this volume. Darwin mentioned isolation in the *Origin* (1859:105) but later largely ignored it. Not so Mayr (1963). Peter Vorzimmer (1970) discusses these issues in his book on Darwin.

Similarly, not until recently has anything like the observation of the evolution of new species of multicellular organisms been observed. (The development of species by polyploidy has been observed on occasion, but this is not a Darwinian method of evolution, and the evolution of new species of bacteria is commonplace.) Yet, few reasonable men withheld consent to the theory that species have evolved and that natural selection is the chief (if not the only) mechanism involved.

For example, even in Darwin's own day, F. J. Pictet contended in his review of the *Origin of Species* (1860) that he would not accept Darwin's deductions until he saw for himself the evolution of a new organ. By 1864 Pictet had been converted and in 1866 published a paper in support of evolutionary theory (Pictet and Humbert, 1866). Needless to say, the direct proof he required had not been supplied. Instead, he had been convinced by the numerous indirect proofs of evolutionary theory. On this score special creation and evolution by natural selection were on different footings. No one had seen a new species evolve anymore than they had seen one specially created, but unlike the special creationists, Darwin had presented a mechanism for evolution. Certain implications of his theory could be checked and his theory as a whole gradually confirmed or disconfirmed without the direct observation of species evolving. Special creation was little more than a bald assertion. Whether as a natural event lacking any scientific explanation or as a supernatural event, special creation could be checked only by direct observation. As Darwin was to discover in his attempts to argue against the "theory" of special creation, it was even less well-organized than Aristotle's doctrine of polar principles and humors.

Confirmation of a scientific theory by reasonably direct means is always welcome, but such confirmation comes long after the theory has been established beyond all reasonable doubt. Empiricists had performed a worthwhile service in emphasizing the continuing role of experience in science, but, as Herschel observed: "Some things we come to believe not through observation but by demonstration—speed of light, action at a distance in gravity, the size of the sun, etc."[28] As this list indicates, the things we come to believe through demonstration, or indirect proof, are a mixed bag. The difficulty lies in distinguishing between theoretical entities and those that lack any empirical referents. If the issues had been more clearly drawn at the time, it would have been difficult for Darwin's critics to charge that natural selection was other than observable—as observable as a phenomenon can get. The situation was somewhat more equivocal with respect

28. Herschel (1830), pp. 23–24.

to the initial spontaneous generation of living creatures from nonliving substances—but not very.

Before Darwin, the common ploy in dealing with the origin of species by natural or supernatural means was to dismiss the question as dealing with events that occurred too long in the past ever to be known. Hence, scientists were free to believe as they saw fit, and they generally saw fit to believe in accordance with revealed religion. "The mystery of creation is not within the legitimate territory of science." Theology must be assiduously excluded from the dynamical sciences but was a necessary adjunct to the historical sciences.[29]

Darwin was presented with a dilemma. The currently popular doctrine of the miraculous creation of species, a flashing together of elemental atoms into fully formed plants and animals, ran counter to the naturalistic tenor of evolutionary theory and its author.[30] He maintained that if he were forced to introduce "new powers" or "principle of improvement," he would reject his theory of natural selection as "rubbish."[31] Darwin felt himself committed to a naturalistic explanation of the initial origin of species. He had none. To make matters worse, the naturalistic explanation for the origin of life, the spontaneous generation of life, had a decidedly bad reputation in scientific circles.[32] The spontaneous generation of multicellular organisms had been disproved after years of laborious effort, just in time for its advocates to switch their allegiance to microbes. It was Darwin's misfortune to propose his theory of evolution, which seemed to necessitate the spontaneous generation of the original forms of life, at the same time that Pasteur was beginning to route spontaneous generation from its last refuge. "All life comes from pre-existing life" was the popular dictum.

In the *Origin* Darwin bowed to protocol and referred to the first creation of life by means of a supernatural metaphor. "There is a grandeur in this view of life, with its several powers, having been originally breathed into a few forms or into one."[33] Later he added, "by the Creator." Such a reference to the diety, however, did not necessarily commit Darwin to a miraculous creation of life. "To my mind it accords better with what we know of the laws impressed on matter by the Creator, that the produc-

29. Whewell (1840), 3:639; see also Wollaston's review in this volume.
30. For evidence that this was the tacitly accepted view, see Carpenter, Hooker, Wollaston, and Fawcett in this volume.
31. Darwin, *Life and Letters* (1887), 2:6; see also Darwin (1859), p. 409.
32. See, for example, the papers by Wollaston, Pictet, Sedgwick, Owen, and Agassiz in this volume.
33. Darwin (1859), p. 490.

tion and extinction of the past and present inhabitants of the world should have been due to secondary causes, like those determining the birth and death of the individual."[34] In this opinion Darwin agreed with the leading authorities of the day. With one or two exceptions, the author of every review in this collection concurred with this view of God as acting through exceptionless laws, not by miraculous special creations. Even so, Darwin was unhappy that he had gone so far and wrote to Hooker, "But I have long regretted that I truckled to public opinion, and used the Pentateuchal term of creation, by which I really meant 'appeared' by some wholly un-known process."[35]

Darwin was never to resolve this dilemma, but there is little doubt as to which alternative he preferred. As late as 1870, he can be found saying:

> Spontaneous generation seems almost as great a puzzle as preordination, I cannot persuade myself that such a multiplicity of organisms can have been produced, like crystals, in Bastian's solutions of the same kind. I am astonished that, as yet, I have met with no allusion to Wyman's positive statement that if the solutions are boiled for five hours no or-ganisms appear; yet, if my memory serves me, the solution when opened to air immediately becomes stocked. Against all evidence, I cannot avoid suspecting that organic particles (my gemmules from the separate cells of the lower organisms!) will keep alive and afterwards multiply under proper conditions.
> What an interesting problem it is.[36]

It is frequently argued that evolutionary theory was wasted motion unless Darwin could also show how the first forms of life originated. In his defense, Darwin once again sought refuge in the authority of Newton:

> [Bronn] seems to think that till it can be shown how life arises it is no good showing how the forms of life arise. This seems to me about as logical (comparing very great things with little) as to say it was no use in Newton showing the laws of attraction of gravity and the consequent movement of the planets, because he could not show what the attraction of gravity is.[37]

In this instance Darwin's defense is apt. Newton, like Darwin, felt that he had accomplished something by his theory even though he had left some questions unanswered; e.g., questions concerning the nature of light and the cause of gravity. It is interesting to note that Mill, at least, agreed with

34. Ibid., p. 488.
35. March 29, 1863, *Life and Letters* (1887), 2:202–203.
36. July 12, 1870, *More Letters* (1903), 1:321.
37. February 18, 1860, ibid., 1:140–141.

Darwin, saying, "I do not think it an objection that [evolutionary theory] does not, even hypothetically, resolve the question of the first origin of life, any more than it is an objection to chemistry that it cannot analyse beyond a certain number of simple or elementary substances."[38]

Newton was condemned for not stating explicitly certain assumptions of his theory. Darwin was condemned for doing just the opposite (by Hopkins, for example). Darwin answered that "though perhaps it would have been more prudent not to have put it in, I would not strike it out, as it seemed to me probable, and I give it on no other grounds."[39] Here was Darwin's sin: he had not exercised enough caution. Science must be safe. The inductive method was calculated to make it safe. Many scientists were willing to admit that varieties may have arisen through natural selection, but one must not carry things too far. These varieties never strayed from the confines of their species (Pictet). Or perhaps, as Forbes and others suggested, there were several centers of creation, and then dispersal, but no evolution beyond certain bounds—say, the generic level. Like Tycho Brahe's compromise between the geocentric and heliocentric systems, in which the sun revolved around the earth while all the other known planets revolved around the sun, hybrid systems like Forbes's failed to gain widespread acceptance. There was considerable wishful oratory about moderation in all things—a little bit of evolution tempered with a little bit of special creation, a little bit of natural selection and a little bit of divine guidance—but at bottom the two positions were incompatible. As soon as a natural explanation was acknowledged for the origin and evolution of some species, nothing but time stood in the way of extending it to all species. Conversely, as soon as a supernatural explanation was allowed for the origin of some species, there was nothing to prevent a supernatural explanation being given for all species. Unlike Wallace, Darwin insisted on telling a totally consistent naturalistic story or none at all.

38. Mill (1910), 2:181; see also Fawcett in this volume.
39. December 25, 1859, Life and Letters (1887), 2:46.

4 · Teleology

Teleology had been part of the conceptual framework of Western science from ancient Greece until the time of Darwin. Natural phenomena might be governed by scientific laws, but it was the Creator who had instituted them. Galileo and Newton replaced one physical theory with another, but they left the teleological world-picture intact. The findings of paleontologists and geologists had necessitated a reinterpretation of Genesis. But it was Darwin who finally forced scientists to realize just how trivial teleology had become in their hands. The change in scientific thought marked by the appearance of the *Origin of Species* was so fundamental that it certainly deserves the title of a conceptual revolution. The problems which arose from species evolving will be discussed in the next chapter. In this chapter we will be concerned with the difficulty which scientists had in reconciling their belief that the order in the world must be in some sense teleological with their growing awareness that the only sense in which it could be termed teleological was a trivial one.

Teleology, the belief that things in the empirical world "strive" to attain ends, is one of the most influential and misunderstood doctrines in the history of philosophy. In one interpretation, teleology entails a universal conscious being ordering everything for the best. As Plato put it: "I heard some one reading, as he said, from a book of Anaxagoras, that mind was the disposer and cause of all, and I was delighted at this notion, which appeared quite admirable, and I said to myself: if mind is the disposer, mind will dispose all for the best, and put each particular in the best place."[1]

For the Platonic doctrine of external teleology, Aristotle substituted the subtler view of immanent teleology. For Aristotle the ends were internal

1. Plato, *Phaedo*, St. 96–99.

to the subject. Individuals did what was best for themselves by themselves. Every natural kind (or species) had a static, immutable essence as its formal cause. These essences made a thing what it was. These essences in turn also served as final and efficient causes. The end toward which all things strove was to realize their own essence as fully as possible. "[For] any living thing that has reached its normal development and which is unmutilated, and whose mode of generation is not spontaneous, the most natural act is the production of another like itself, an animal producing an animal, a plant a plant, in order that, as far as its nature allows, it may partake in the eternal and divine. That is the goal towards which all things strive, that for the sake of which they do whatsoever their nature renders possible."[2]

Both of these versions of teleology can be interpreted to be exciting, substantive claims about the empirical world. In these interpretations they have been shown to be untenable. For example, according to external teleology, given a knowledge of the ordering mind, one should be able to infer what the empirical world must be like, and conversely, given knowledge of the empirical world, one should be able to infer the character of the ordering mind. If the universe is a perfectly running mechanism with a place for everything and everything in its place, then one type of mind is implied. If it is a shoddily constructed mechanism which needs constant tinkering, another type of mind is implied. What external teleologists did not like about evolutionary theory was the type of mind it implied. God could have constructed the world so that species evolved by natural selection in the struggle for existence, but that kind of God did not seem especially worthy of love and veneration. Evolutionary theory had even more devastating consequences for immanent teleology. If essences are static and if in general the goal toward which all things strive is to realize their essence, then the wholesale progressive change entailed by evolutionary theory is impossible. If evolution has occurred, then either essences are not static or else things must strive, not to fulfill their own essence, but the essence of some other species.

Both versions of teleology also have more sophisticated interpretations, but with these interpretations they become vacuous. An advocate of external teleology might be led to say that whatever the laws of nature turn out to be, God did it. He never suspends these laws in divine intervention, and these laws reveal nothing of his character. An advocate of immanent teleology might assert that all it takes for the world to be teleological

2. Aristotle, *De Anima,* 415[a] 26.

is for regularities to exist. The emptiness of these positions is all too apparent, but they were the positions to which the proponents of teleology were at last driven. Darwin's theory was one of the chief instruments in the final trivialization of teleology.

Thanks largely to the efforts of Thomas Aquinas, Aristotelian metaphysics, physics, and natural history were incorporated into Church dogma. Through the years, however, Aristotle's metaphysics took on a decidedly Platonic caste. God was not merely a detached, impersonal first and final cause, but an overseer who actively participated in the affairs of the world. He both instituted the laws of nature and periodically suspended them in miracles. Copernicus and Galileo had to combat the Aristotelian physics of natural place, but they did not attack either the metaphysics or theology of Church doctrine. God still instituted the laws of physics and still suspended them on occasion, but these laws were the laws of Newton, not Aristotle. But of even greater significance, Newton's laws required the continued and continuing efforts of God in their operation. Newton was anxious to reconcile physics with theology, and his instrument was Richard Bentley (1662-1742).[3]

Bentley did for Newtonian physics what Aquinas had done for the physics of Aristotle. In "A Confutation of Atheism from the Origin and Frame of the World" (1693), he argued that Newtonian physics was not only compatible with the existence of God but also necessitated it. Bentley's arguments all follow the same format. Certain phenomena were derivable from Newton's laws, e.g., the earth traveling around the sun in an ellipse. Other phenomena were not derivable from Newton's laws, e.g., the sun being luminiforous and the other bodies in the solar system opaque, the distances of the various planets from the sun, and the inclination of the earth's axis of rotation to its plane of revolution. Now that Aristotle's physics of natural place had been abandoned, these phenomena had no explanation in terms of natural law. They had no secondary causes. They appeared to be "accidental" features of the universe. Yet they had to be precisely as they were in order for the earth to support life. If the earth were farther away from the sun, then it would be too cold to support life. If nearer, it would be too hot. If the earth's path had been more elliptical, the extremes in temperature on earth as it revolved around the sun would be too great. If the earth's axis had not been inclined to the plane of its orbit, parts of the earth would be too hot to support life and others too cold. If the earth revolved around the sun more slowly, the seasons

3. Like Whewell, Bentley was Master of Trinity College, Cambridge (1700-1742).

would be too long. If more quickly, too short. And so on. All these phenomena had to be as they were if life—especially man—was to exist. Their
disposition did not follow from Newton's laws. Hence, either they occurred
by accident or else God did it. The chance coincidence of so many phenomena was extremely implausible. Hence, God was responsible. The existence of life was good. Hence, God was good.

This Neoplatonic world-view was to find favor among such biologists
as Cuvier, Owen, Agassiz, and von Baer. Space was God's sensorium, gravity his conscious will, and each species a divine thought. Even an empiricist
like Herschel concurred in this ontology, hoping to combat the more radical
empiricism of Locke and Hume and the materialism of Laplace. According
to Herschel, falling bodies are urged to the earth's surface "by a force or
effort, the direct or indirect result of a *consciousness* and a *will* existing
somewhere, though beyond our power to trace, which force we term
gravity."[4]

In his Bridgewater Treatise, Whewell updated Bentley's arguments for
God's existence. In reading Herbert Mayo's *The Philosophy of the Living*
(1838), Darwin was incensed over Mayo's contention that Whewell was
profound "because he says length of days adapted to duration of sleep
of man!!! whole universe so adapted!!! and not men to Planets.–instance
of arrogance!!"[5] The Bridgewater Treatises were published because scientists had been gaining the reputation of being materialists, determinists,
and atheists, especially in France with the work of D'Alembert, Diderot,
de la Mettrie, Laplace, and Voltaire. Whewell was pleased to do what
he could to combat the growing slander by showing the mutual dependence
of science and religion. "My prescribed object is to lead the friends of
religion to look with confidence and pleasure on the progress of the physical
sciences, by showing how admirably every advance in our knowledge of
the universe harmonizes with the belief of a most wise and good God."[6]

Periodically, proponents of teleology, whether of the immanent or the
external variety, entertained a possible alternative explanation for the adaptation of living creatures to their environments—the survival of the fit.

4. Herschel (1833), pp. 221–222.
5. DeBeer (1960), p. 134.
6. Whewell (1833), p. i; though Whewell's Bridgewater Treatise was Volume
III of this series, it was the first to appear in print. The complete title of this
series was *The Bridgewater Treatises on the Power, Wisdom and Goodness of God,
as Manifested in the Creation.* Each contribution will be cited separately by the
title of that volume. We have only a fragment of Charles Babbage's (1837) efforts
which does not form a part of the Bridgewater Treatises, even though it bears
the title *The Ninth Bridgewater Treatise.*

As Aristotle observed, instead of organisms' being so well-organized and well-adapted because it was of their essence to be so, "such things survived, being organized spontaneously in a fitting way; whereas those which grew otherwise perished and continue to perish." Aristotle immediately dismisses such a ridiculous contention. "Yet it is impossible that this should be the true view. For teeth and other material things either universally or normally come about in a given way; but of not one of the results of chance or spontaneity is this true."[7] Whenever anything occurs regularly, it must be because the formal and final causes coincide. Because an organism is what it is, it produces teeth as they should be. The haphazard production of numerous alternatives and the elimination of those that are not fit occur too irregularly to be part of the natural order.

An identical story can be told for William Whewell, with one minor complication. He somehow had to reconcile his own advocacy of final causes with Bacon's "barren virgin" quip. Whewell's solution was to argue that final causes should be excluded from physical inquiry but were perfectly legitimate as the results of such inquiry. Final causes were barren because they themselves were the fruits of the labor.[8] God ordered the universe according to his divine plan, and it was the goal of science to discover this plan. Any explanation in terms of the perishing of the unfit and the survival of the fit was inconceivable:

> If the objector were to suppose that plants were orginally fitted to years of various lengths, and that such only have survived to the present time, as had a cycle of a length equal to our present year, or one which could be accommodated to it; we should reply, that the assumption is too gratuitous and extravagant to require much consideration.[9]

Like a true philosopher, Whewell went on to hedge his bet. Even if adaptations did come about by such a gratuitous and extravagant mechanism, "it does not remove the difficulty. How came the functions of plants to be *periodical* at all?"

Prior to the publication of the *Origin,* philosophers and scientists had some justification for rejecting nonteleological explanations of adaptation, since there was no alternative explanation. Lamarck's theory came closest to supplying a mechanism for adaptation, but few were willing to accept it, both because of its own weaknesses and because of the inordinate influence of Cuvier. After the *Origin,* however, the reactions of philosophers

7. Aristotle, *Physics,* bk. II, ch. 8.
8. Whewell (1833), p. 180.
9. Ibid., p. 27.

and scientists were not much different. Mill, in his *Three Essays on Religion* written during the decade following the publication of the *Origin* and published posthumously in 1874, still found the argument from design preferable to evolution by natural selection:

> I regret to say, however, that this latter half of the argument is not so inexpugnable as the former half. Creative forethought is not absolutely the only link by which the origin of the wonderful mechanism of the eye may be connected with the fact of sight. There is another connecting link on which attention has been greatly fixed by recent speculations, and the reality of which cannot be called into question, though its adequacy to account for such truly admirable combinations as some of those in Nature, is still and will probably long remain problematical. This is the principle of "the survival of the fittest."

Mill continues:

> Of this theory when pushed to this extreme point, all that can now be said is that it is not so absurd as it looks, and that the analogies which have been discovered in experience, favourable to its possibility, far exceed what any one could have supposed beforehand. Whether it will ever be possible to say more than this, is at present uncertain. The theory if admitted would be in no way whatever inconsistent with Creation. But it must be acknowledged that it would greatly attenuate the evidence for it.[10]

Darwin's arguments in the *Origin* are mere analogies, whereas the argument from design was a genuine inductive argument. "In the present state of our knowledge, the adaptions in Nature afford a large balance of probability of creation by intelligence."

Many philosophers and scientists were willing to accept evolutionary theory if only Darwin would admit divine providence. Without it they felt trapped between inexorable law and blind chance. The terminology of this position varied, but the message was always the same. Some maintained that some phenomena were law-governed, other phenomena not. Others maintained that all phenomena were law-governed. In either case, God instituted these laws and guided their action. A universe governed by divinely instituted law did not seem so cold and barren. Add periodic miracles and the world-picture became even more intimate. Exclude God, and both accidental and law-governed phenomena became equally "accidental," given the peculiar terminology of the time.[11]

10. Mill (1969), pp. 172, 174.
11. For example, see the review by von Baer and the exchange between Mivart and Wright in this volume.

Herschel's position on these matters was typical. In his 1861 edition of *Physical Geography of the Globe,* he added the following comment on Darwin's theory:

> This was written prior to the publication of Mr. Darwin's work on the *Origin of Species,* a work which, whatever its merit or ingenuity, we cannot, however, consider as having *disproved* the view taken in the text. We can no more accept the principle of arbitrary and casual variation of natural selection as a sufficient condition, *per se,* of the past and present organic world, than we can receive the Laputan method of composing books (pushed *à outrance*) as a sufficient account of Shakespeare and the Principia.[12]

Herschel is repeating his charge quoted earlier that Darwin's theory was the law of higgledy-piggledy. Evolution by random variation and natural selection did not exhibit the mathematical regularity and simplicity characteristic of physical law. The laws of physics were worthy of God's authorship, just as *Hamlet* and *Macbeth* were worthy of Shakespeare and the *Principia* worthy of Newton. Darwin's theory implied a God who would compose a book by randomly striking the keys of a typewriter until something turned out. Herschel did not want to deny that evolution might occur by law, but it had to be a law worthy of God.

> Equally in either case, an intelligence, guided by a purpose, must be continually in action to bias the directions of the steps of change—to regulate their amount–to limit their divergence–and to continue them in a definite course. We do not believe that Mr. Darwin means to deny the necessity of such intelligent direction. But it does not, so far as we can see, enter into the formula of this law; and without it we are unable to conceive how far the law can have led to the results. On the other hand, we do not mean to deny that such intelligence may act according to a law (that is to say, on a preconceived and definite plan). Such law, stated in words, would be no other than the actual observed law of organic succession; or one more general, taking the form when applied to our own planet, and including all the links of the chain which have disappeared. But the one law is a necessary supplement to the other, and ought, in all logical propriety, to form a part of its enunciation. Granting this, and with some demur as to the genesis of man, we are far from disposed to repudiate the view taken of this mysterious subject in Mr. Darwin's book.[13]

12. Herschel (1861), p. 12; see von Baer's review in this volume for an expansion on this reference to *Gulliver's Travels.*
13. Herschel (1861), p. 12.

A quick reading of the preceding paragraph might lead the reader to think that Herschel is saying much more than he is. Herschel pleads for a supplementation of Darwin's laws by some reference to intelligent direction, but what does he suggest? A description of phylogeny! Darwin was at a loss as to the manner in which he could reply to criticisms such as these. The request for a law of divine providence could be taken literally, in which case Darwin felt obligated to deny it. If God were all-good, all-knowing, and all-powerful, then one would expect that organisms would tend to vary in directions advantageous to their survival. Darwin found no evidence of such directed variation. As far as Darwin could tell, variations were "chance," they occurred "in all directions." By these claims Darwin did not mean that variations were uncaused. Rather he meant that whatever laws governed variation, they were currently unknown and to the best of Darwin's knowledge they were not teleological. That is, there was no correlation between those variations which an organism might need and those it would get. The God implied both by evolution and the current state of the organic world was capricious, cruel, arbitrary, wasteful, slipshod, and totally unconcerned with the welfare of his creations, a god reminiscent of the Old Testament, not nineteenth century natural theology.

> There seems to me too much misery in the world. I cannot persuade myself that a beneficent and omnipotent God would have designedly created the Ichneumonidae with the express intention of their feeding within the living bodies of Caterpillars, or that a cat should play with a mouse. Not believing this, I see no necessity in the belief that the eye was expressly designed.[14]

One might argue that God moves in strange ways and that he is guiding things, though we cannot discern his plan. However species evolve, that is the way he planned it. Darwin found this position empty verbiage. He saw no reason to utter such pious inanities. No mention is made of intelligent direction in the formula of evolution by natural selection, but neither did it enter into the formula of the law of universal gravitation. In a series of letters, Darwin tried, with little success, to make his position clear to Lyell:

> No astronomer, in showing how the movements of planets are due to gravity, thinks it necessary to say that the law of gravity was designed that the planets should pursue the courses which they pursue. I cannot

14. Darwin, May 22, 1860, *Life and Letters* (1887) 2:105.

believe that there is a bit more interference by the Creator in the construction of each species than in the course of the planets. It is only
owing to Paley and Co., I believe, that this more special interference
is thought necessary with living bodies. But we shall never agree, so
do not trouble yourself to answer.[15]

When you come to "Deification," ask yourself honestly whether what
you are thinking applies to the endless variations of domestic productions,
which man accumulates for his mere fancy or use. No doubt these are
all caused by some unknown law, but I cannot believe they were ordained
for any purpose, and if not so ordained under domesticity, I can see
no reason to believe that they were ordained in a state of nature. Of
course it may be said, when you kick a stone, or a leaf falls from a
tree, that it was ordained, before the foundations of the world were
laid, exactly where that stone or leaf should lie. In this sense the subject
has no interest for me.[16]

Will you honestly tell me (and I should be really much obliged)
whether you believe that the shape of my nose (eheu!) was ordained
and "guided by an intelligent cause"? By the selection of analogous and
less differences fanciers make almost generic differences in their pigeons;
and can you see any good reason why the natural selection of analogous
individual differences should not make a new species? If you say that
God ordained that at some time and place a dozen slight variations
should arise, and that one of them alone should be preserved in the
struggle for life and the other eleven should perish in the first or few
generations, then the saying seems to me mere verbiage. It comes to
merely saying that everything that is, is ordained.[17]

At one time God had played an important role in physics, but gradually
his function had been eroded, until reference to him was little more than
pious honorifics. It is often said that evolutionary theory brought an end
to the practice of including God as a causal factor in scientific explanations.
A more accurate characterization is that it demonstrated forcefully that
this day had already passed. The architects of the demise of teleology were
not atheistic materialists but pious men like Herschel, Whewell, and Mill,
who thought they were doing religion good service by limiting the domain
of the accidental and of the miraculous. To them the more the empirical
world was shown to be governed by secondary causes acting according
to God-given laws, the more powerful and ominiscient God was shown
to be. With rare exception, every author included in this volume, whether

15. June 17, 1860, *More Letters* (1903), 1:154.
16. August 13, 1861, ibid., 1:192.
17. August 21, 1861, *More Letters* (1903), 1:193.

in favor of evolutionary theory or against it, strongly supported this view
of God's relation to natural phenomena.

Accidental occurrences could be shown not to be accidental, either because of the direct intervention of God or by subsuming them under law.
Physics since Newton had made constant inroads on the domain of the
accidental, thereby limiting the need for God's direct intervention. As Mill
put it, there were two conceptions of theism, one consistent with science,
one inconsistent. "The one which is inconsistent is the conception of a
God governing the world by acts of variable will. The one which is consistent, is the conception of a God governing the world by invariable laws."[18]

Darwin used this position on the relation between science and religion
in an attempt to gain a fair hearing for his theory. He prefaced the first
edition of the *Origin* with quotations from Whewell and Bacon to the
effect that no impediment should be placed in the path of scientific inquiry
because no law could be discovered which was contrary to God's will. In
the sixth edition he added a third quotation to this same effect by Bishop
Butler. All Darwin wanted to do was to extend the domain of secondary
causes to include the creation of species. Although in the past the creation
of species had been considered miraculous and outside the realm of law,
so had many other phenomena which had been shown to be law-governed.
Darwin was acting within the currently accepted tradition of expanding
the realm of law.

Whewell wanted to retain the creation of species as miraculous but he
recognized that such an explanation was not a tenet of physical science
but of natural theology.[19] Whewell's bare recognition that the origin of
species was even susceptible to a naturalistic explanation gave heart to
Lyell, who had proved so uncharacteristically timid on the evolution of
species. In a letter to Herschel, he wrote, "Whewell, in his excellent treatise
on the Inductive Sciences, appears to me to go nearly so far as to contemplate the possibility at least of the introduction of fresh species being governed by general laws."[20]

There were those like Sedgwick and Agassiz who wished to retain special
creation to lend significance to their teleological claims. A majority of philosophers and scientists, however, gradually abandoned the notion of special
creation but still wished to retain teleology as a significant doctrine. White-
head, Dewey, and Peirce all espoused creative evolution. Wallace, Asa Gray,

18. Mill (1969), p. 135.
19. Whewell (1837), 3:640.
20. Lyell, May 24, 1837, *Life of Lyell* (1881), 2:12.

and Lyell belabored Darwin with arguments for admitting divine providence, especially with respect to man's mental and moral faculties. For
example, Lyell hailed Wallace's suggestion that "there be a Supreme Will
and Power which may not abdicate its function of interference but may
guide the forces and laws of Nature." What these men did not realize
was that by pushing God further and further in the background as the
unknowable author of natural law, regardless of the nature of these laws,
they had prepared the way for his total expulsion. Like Kant's *Ding an
sich,* he was remote, obscure, unknowable, somehow underlying everything
and very important but of no conceivable consequence to any endeavor.
Teleological supplements were as otiose to evolutionary theory as they were
to contemporary versions of Newtonian theory. The superficiality of teleological claims in a deterministic universe was not totally lost on the educated
public. For example, one writer observed with respect to the habit of naturalists to interpolate their writings with references to God: "It appears
to them that, unless they drag the Creator into every second paragraph,
their essay will not possess the necessary religious veneering for the public
taste."[21]

Darwin was not impervious to the pathos in the dilemma which confronted his colleagues. Neither a world ruled by deterministic laws nor
one in which chance played an important role seemed bearable. Even
though Darwin had not intended to write atheistically, his theory robbed
the traditional solution to this dilemma of all its plausibility. The closest
Darwin came to a resolution of the conflict between teleology and the
facts of natural history was his suggestion that God instituted the general
laws, whereas the details were left to chance:

> My theology is a simple muddle; I cannot look at the universe as
> the result of blind chance, yet I can see no evidence of beneficent design,
> or indeed of design of any kind, in the details. As for each variation
> that has ever occurred having been preordained for a special end, I
> can no more believe in it than that the spot on which each drop of
> rain falls has been specially ordained.[22]

> On the other hand, I cannot anyhow be contented to view this wonder-
> ful universe, and especially the nature of man, and to conclude that
> everything is the result of brute force. I am inclined to look at everything
> as resulting from designed laws, with the details, whether good or bad,

21. Anonymous, *Popular Science Review* (1866), 5:215.
22. July 12, 1870, *More Letters* (1903), 1:321.

left to the working out of what we may call chance. Not that this notion
at all satisfies me.[23]

Bentley and Whewell had argued that God had instituted natural laws
and taken care of the details. Darwin believed that perhaps God instituted
natural laws, but he could not be responsible for specific applications both
because of their triviality in some cases and because of their cruelty in
others. But it should be recalled that by "chance" Darwin meant governed
by laws not as yet known. Darwin's solution was a circuitous statement
of determinism. Interpreted realistically, teleology was incredible. Given a
sufficiently sophisticated interpretation, it was of no relevance to science.[24]

23. Darwin, May 22, 1860, *Life and Letters* (1887), 2:105.
24. The issue of teleology has had a recent resurgence but in entirely different
terms. Several philosophers and biologists have attempted to analyze teleological
systems naturalistically in terms of genetic programs, negative feedback, and so
on. See Mayr (1961) and Canfield (1966).

5 · Essences

If teleology permeated Western thought, essentialism was even more pervasive. The number and variety of metaphysical systems generated on the continent and in England was staggering. Realists maintained that the universal existed in nature, perhaps in individuals themselves (Aristotelians), or both in nature and in the mind of the creator (Neoplatonists). Neoplatonists such as Cuvier, Owen, and Agassiz are often referred to as "idealists" in this respect, but the term is not very apt. Idealism is usually contrasted with materialism, idealists maintaining that only ideas exist (material objects are just clusters of ideas), and materialists maintaining that only matter exists (minds are merely brain processes). Although Huxley and Darwin might legitimately be called materialists, Cuvier, Owen, and Agassiz were hardly idealists in this sense. Rather they were idealists in the sense that they tended to explain the order in nature by reference to ideal types. In short, they were essentialists. As metaphysically aloof as nominalists might be, they too were strongly predisposed to essentialism. Individuals were the only independent existences. A scientist could divide individuals into classes as he pleased—just so long as these classes were discrete. Nominalists were as much essentialists as the Aristotelians and Neoplatonists, but without their metaphysical justification.

The empiricist-rationalist dichotomy cuts across the preceding distinctions. Empiricists maintained that all knowledge had its source in experience and could be analyzed completely in terms of it. Rationalists maintained either that sense experience played no role in the acquisition of knowledge or else that truth could not be decided completely by reference to sense experience. The mind played an active role. As one might expect, empiricists tended to be nominalists and materialists, whereas rationalists were frequently disposed to idealism, though numerous exceptions can be found to both generalizations. In this chapter, the effect of these metaphysical notions on the acceptance of evolutionary theory will be traced.

Alvar Ellegård (1957 and 1958) sees the disagreement over the immutability of species as a conflict between empiricist and "idealist" philosophies of science. Although our story would be much neater if this were true, it is not. Empiricists and "idealists," while not equally opposed to the concept of species evolving, were still both opposed. Evolution by random variation and natural selection conflicted with teleology. Any evolution at all, regardless of the mechanism, as long as it is gradual, conflicted with the essentialist notion of natural kinds—and on the necessity of such natural kinds, empiricists and "idealists" were in agreement. The doctrine of the existence of static, immutable natural kinds underlay all of the previously discussed topics—inductive proof, occult qualities, and teleology—and evolutionary theory struck a strong blow against this doctrine.

One might expect Whewell to argue for the existence of natural kinds, since his metaphysics depended on it. In writing his *History* Whewell was happy to discover that the most eminent physiologists of his day believed that indefinite divergence from original type was impossible. *"Species have a real existence in nature, and a transition from one to another does not exist."*[1] The impossibility of evolution—or transmutation, as it was usually called—was not a generalization from experience for Whewell but a necessary prerequisite for knowledge. "Our persuasion that there must needs be characteristic marks by which things can be defined in words, is founded on the assumption of the *necessary possibility of reasoning."*[2] In Whewell's view, if evolutionary theory were true, knowledge was impossible—not a small drawback to the theory. Peirce (1877) recognized that the "Darwinian controversy is, in large part, a question of logic" and Dewey (1910) that "the real significance of Darwinian evolution was to introduce a new 'mode of thinking,' and thus to transform the 'logic of knowledge'."

Contrary to Ellegård's belief, the empiricist philosophies of Herschel and Mill both depended upon the existence of discrete natural kinds. Mill distinguished between two types of uniformities in nature: those of succession in time and those of coexistence. The former were expressed in causal laws, the latter in definitions. Mill's four (sometimes five) methods of induction were designed to discover succession of kinds of events in time, on the assumption that these kinds had already been discovered. Mill concurred in the essentialist position that natural kinds were those classes which are "distinguished by unknown multitudes of properties and not solely by a few determinate ones—which are partitioned off from one another by

1. Whewell (1837), 3:626.
2. Whewell (1840), 1:476.

an unfathomable chasm . . . The universe, so far as known to us, is so
constituted, that whatever is true in any one case, is true in all cases of
a certain description; the only difficulty is, to find the description. Kinds
are Classes between which there is an impassible barrier . . ."[3]

Though Mill's empiricist ontology did not directly require the existence
of discrete natural kinds, his logic of justification did. At bottom, proof
was supplied by eliminative induction. "Either A, B, or C can cause E.
A and B are absent. Hence, C must cause E." For this inference to be
valid, all of the alternative causes of a certain kind of event must be specified and all but one eliminated. As Darwin discovered in his use of this
type of argument, two difficulties present themselves—the actual elimination
of the alternatives and their complete specification. The consequence that
evolutionary theory had for Mill's notion of proof was that if the kinds
in question were species of plants and animals, then there were not a
finite number of sharply distinguishable kinds. Induction by complete elimination was impossible. Not only did Darwin not "prove" his theory, but
also, given evolution by gradual variation, such a proof would be forever
impossible.

Mill justified his belief in the existence of causal laws by reference to
the law of universal causation. It alone was justified by enumerative induction. He could find no justification for a law of universal coexistence, but
he accepted it anyway on the authority of the most eminent biologists
of his day. Mill's belief in a finite number of discrete kinds permitted
him to retain his notion of induction by complete elimination. It also permitted him to maintain that all causal laws were universal in form. The
only reason for a causal law to be less than universal in form was a
failure to distinguish natural kinds accurately. Once natural kinds had
been accurately distinguished, universality was guaranteed. Thus, immutable
essences were as central to Mill's philosophy as they were to Whewell's.

Whewell was quick to point out that Mill's four methods of induction
"take for granted the very thing which is most difficult to discover, the
reduction of phenomena to formulae." Induction, in Whewell's interpretation, provided these formulae. Whewell believed that observations were
necessary for scientific knowledge. Though Whewell did not think that
the fundamental ideas of science were innate, he did think that they were
necessary. They might start out as observed facts, but in the progress of
science they would come to be known as necessary truths. As necessary
truths they could not be derivable just from experience, nor could they

3. Mill (1874), pp. 80, 201, 471.

be verified solely by reference to it. Rather they could be known to be necessary truths only by intuition. The mind superinduces concepts on the facts. These concepts are appropriate if they are clear and distinct (the traditional rationalists' criterion) and if they fit the facts closely (but considerable latitude is permitted for reinterpreting the facts to fit the concepts). One of the notable features of Whewell's philosophy of science is that he includes no mechanism for periodic conceptual revolutions. He seems to assume that in each area of science there will be one and only one scientific revolution and that in physics that revolution had already occurred. In physics, at least, no future revolutions could occur. Although Whewell's views on science do not preclude a series of conceptual revolutions in a particular branch of science, they are certainly biased toward a linear interpretation of scientific progress. In point of fact, Whewell himself had an extremely reactionary attitude toward any advances in science after his formative years.

A belief in a finite number of discrete kinds of entities underlay both Mill's and Whewell's notions of verification. It also was one of the major factors in the prevalence of occult qualities in science. Science deals with classes of entities. On the essentialist view, these classes must be distinguishable by characters which universally covary. Time and again no such set of universally covarying observable characters could be found. Hence, these classes had to be distinguishable by unobservable characters. Sometimes this maneuver met with considerable success (e.g., in distinguishing elements by their atomic weight and number). Sometimes it did not (e.g., distinguishing life from non-life by the presence of a vital force).

In reading the scientific and philosophic works of the period, the modern reader often finds himself wondering how essentialism could have had the slightest appeal or could have seemed in the least plausible. The answer can be found in its apparent applicability to physical geometry. Here, there was no doubt as to which characters were essential and which were accidental. Maybe all reptiles did not have three-chambered hearts, but all triangles had three sides. Geometric figures were the chief examples of natural kinds in essentialist philosophies from Plato and Aristotle to Whewell and Mill. Species of plants and animals were the second most popular examples. When the diversity of the organic world was only very poorly known, organic species seemed to be as distinct as those of geometry. By the time of Darwin, however, biologists had been forced to recognize that they were not quite so distinct. In fact, the scandalous state of taxonomy was one of the chief arguments used by Darwin against the doctrine of the permanence of species. (See the comments by Carpenter in his review.) For essentialists, such difficulties and disagreements were only technical problems to be overcome by more careful examination of the data. There *had* to be diagnostic characters. For Darwin no such characters existed for species extended in time.

Inorganic elements also, on occasion, were cited as examples of natural kinds, though they ran a very poor third to those of geometry and natural history. (See Jenkin's review.) What must be kept in mind is that for essentialists *all* natural kinds were *equally* immutable. Some appreciation of how incredible Darwin's theory must have sounded to his contemporaries can be obtained by realizing that for them one species of living creature evolving into another was as impossible as a pentagon evolving into a hexagon or lead into gold. Not only did evolutionary theory seem to presuppose spontaneous generation, but also it smacked of alchemy. One can understand Darwin's reticence about making public his views on the evolution of species when one discovers that half a century later, after scientists had become used to the notion of evolution, Frederick Soddy (1877–1956) feared to make public his conviction that transmutation of chemical elements was possible lest the chemists "have his head off." In 1917, when one chemical element was actually transmuted into another, a Darwin had his hand in this evolutionary adventure—Darwin's grandson, C. G. Darwin.[4]

Although philosophers generally agree that science can have no implications for true metaphysics, theoretical innovations in science from Darwin to Einstein have necessitated a complete re-evaluation of the notion of natural kinds. The result of this re-evaluation has been the recognition that separate stories must be told for geometric, biological, and chemical species. Philosophers of science contemporary with Darwin tried to tell one story which would be adequate both for the "deductive sciences" like geometry and arithmetic and the "inductive sciences" like physics and biology.

Whewell in the Kantian tradition assimilated the axioms of physics to the axioms of geometry. All were a priori. Experience might serve as an impetus to the formation of such ideas, but once conceived, their truth could be known a priori. This position seemed plausible enough for general spatial ideas, such as triangles having three sides and parallel lines never meeting, but it began to become somewhat strained when the ideas of causation and design were included on the list of a priori truths about the empirical world and broke down entirely under such empirical laws as those of Kepler

4. Schonland (1968), p. 159.

and Newton. Mill, on the other hand, assimilated the axioms of geometry to those of physics, and in this view, he was joined by Herschel. All were inductions from experience. Certain ideas spring from experience, almost spontaneously, like very basic notions of space, time, and causation, but they nevertheless were empirical claims, whose truth could be decided by scientific investigation. Kepler's laws certainly seemed straightforwardly empirical. A body could fulfill all the requirements for being a planet and still fail to move in accordance with Kepler's laws, but the basic axioms of geometry seemed immune to verification or falsification by experiment or observation. The parallel postulate for sure was thought to be a posteriori, yet there seemed to be no way to confirm or refute it.

Although the current received doctrine in history is that no historical dispute should be elucidated or resolved by contrasting it with the present views on the subject, it seems grossly unfair to leave the reader in the same state of confusion in these matters as that which prevailed in the nineteenth century. The question is whether the axioms of the formal and empirical sciences are self-evident truths superinduced on the facts by the mind, or truths inductively collected from the facts. For both, the answer is neither. Both sides were mistaken, but the mistake was conceptual, not factual. According to the current view, a sharp distinction must be drawn between pure, uninterpreted systems, such as pure geometry, and the theories of empirical science, such as physical geometry. All the statements in pure geometry, whether axioms, definitions, or theorems, are tautological and in no sense about physical space. Nor are they truths of reason or of the mind. In Euclidean geometry, parallel lines never meet, and no physical or psychological observation is relevant to this claim. However, axiomatic systems can be given physical interpretations. For example, the phrase "straight line" in Euclidean geometry can be taken to denote light rays, transforming it into a physical geometry. In such an interpretation, the parallel line postulate is not a priori true. It is false. As an axiom in pure geometry, the parallel line postulate is a priori true but in no sense about physical space. As an axiom in physical geometry, it is about physical space and false a posteriori. Until this radical distinction was made, no philosopher could hope to understand the relation between the formal and empirical sciences.[5]

Within empirical science a further distinction must also be made. Species

5. The problem of the relation between mathematics and natural science has had a long history. Periodically, early thinkers set forth ideas very close to those now held on the subject; e.g., Buffon. However, Buffon viewed mathematics as being truths of the mind.

of inorganic objects differ in several important respects from species of plants and animals. Earlier the transmutation of chemical elements was compared to the evolution of species, but the two processes are very different. Physicists change a small quantity of one element into a like quantity of another element. Lead and gold, defined in terms of their atomic number, still remain lead and gold. Even if all the lead in the universe were transmuted into other elements, a spot would still be reserved for lead in the classification of elements. If eventually lead were to form once more, it would be lead, pure and simple. The process by which a physical element is generated and its past history are irrelevant to its identity.

The situation is quite different with respect to species of living creatures. Biologists commonly maintain that species must be monophyletic. Various definitions of "monophyly" have been suggested, but the two most prevalent definitions are (a) a taxon is monophyletic if it is derived by descent from a single immediately ancestral species; and (b) a taxon is monophyletic if it is derived by descent from a single immediately ancestral taxon of its own or lower rank. On interpretation (a) a family is monophyletic only if every species included in the family is derived from another species in the family or at most one immediately ancestral species not included in the taxon. In interpretation (b) a family is monophyletic only if every species included in the family is derived from another species in the family or from one or more species not included in the family, as long as all of these species belong to the same family.

In either interpretation species become historical entities. Having the appropriate past history is necessary for a speciess to be the species that it is. For example, if a species of dinosaur were to evolve from present-day reptiles which was in every respect, save its phylogenetic history, identical to an extinct species of reptile, it would still be a new, separate species. This view of species as being monophyletic evolutionary units is the currently dominant view.[6] There is also a minority view that animal species should be treated like chemical elements defined symptomatologically,[7] and at least one philosopher agrees with this position.[8] Thus, Darwin's theory had two seemingly contrary implications for the relation of biology to physics. Its naturalism brought biology and physics into closer accord, but its apparent reliance on historical concepts tended to drive the two apart again.

6. See Simpson (1961) and Mayr (1969).
7. Sokal and Sneath (1963).
8. Smart (1963).

Darwin dismissed theological explanations of the origin of species because they were not properly scientific. He also dismissed explanations in terms of "plans," divine or otherwise. They, too, were empty verbiage. Explanations in terms of coherence of plan had been common in biology for well over a century before Darwin, especially in the writings of ideal morphologists like Goethe, Cuvier, Owen, and Agassiz. As long as one believed in God, and these plans could be interpreted literally as thoughts in the mind of the creator, then such explanations had some explanatory force, but if reference to God is left out of the explanatory picture, then all that is left are the plans. Rather than being explanations, the existence of such "plans" calls for explanation. In England, at least, Darwin's aversion to scientific explanations in terms of explicit reference to God and conformity to plan was shared by a growing number of scientists. For example, Edward Forbes,[9] in 1854, published an explanation of the distribution of living creatures in geological time in terms of polarity, "a manifestation of force of development at opposite poles of an ideal sphere." Forbes divided geological time into two periods—the Neozoic and Paleozoic. The distribution of generic types then formed a figure eight with its constriction during the Triassic and Permian epochs. Although Forbes argued that this distribution occurred in time, the explanatory ideal plan was atemporal. To put Darwin's reaction to Forbes's explanation simply, it made him sick. Either it was nothing more than a description of the distribution of generic types in time or else it was like magnetism turning a table.[10] Darwin could not see how references to ideal plans could explain anything. To say that the hand of a man or a monkey, the foot of a horse, the flipper of a seal, the wing of a bat, and so forth, have all been formed on the same ideal plan "is no scientific explanation."[11] One of the purposes of evolutionary theory was to provide a naturalistic explanation for such homologies.[12]

Modern readers of the *Origin of Species* tend to grow impatient with Darwin for going to such lengths to argue against special creation and divine plans. They are no longer considered even viable candidates for scientific explanations. The decision was not so clearcut in Darwin's day. Many philosophers and some scientists still looked upon miracles and ideas in the mind of the creator as acceptable elements in a scientific explanation, though most scientists were loath to discuss the issue. These "explanations"

9. Forbes (1854), p. 430.
10. Darwin, July 2, 1854, *More Letters* (1903), 1:77.
11. Darwin (1871), pp. 31–32.
12. See Hopkins, Mivart, and Wright reviews in this volume for additional opinions of Forbes's principle of polarity.

were not so much refuted by Darwin as discounted. Similarly, the theological and metaphysical objections to evolutionary theory gradually subsided, not because their authors had been converted, but because no one whose opinion mattered was listening any more. The change can be seen in the difference between the aggressive confidence of Agassiz's first review of the *Origin of Species* in 1860 and the petulant timidity of his final comments in 1874. Even in his original review of the *Origin,* Agassiz acknowledged that some naturalists looked upon his idea of creation by God's mental exertion as a kind of bigotry and his holding the view a sign of *non compos mentis.* Fourteen years later he hardly dared mention his pet theory of creation by repeated acts of divine cognition. Owen went so far as to assert that the leading naturalists of the day, referred to by Darwin as special creationists, had *never* maintained the view that species were specially created by divine action. Rather, he along with the other so-called special creationists had always maintained that some unknown law governed the introduction and extinction of species. Lyell, to the contrary, had the candor to admit that he "formerly advocated the doctrine that species were primordial creations, and not derivation."[13] He differed from the catastrophists only in his belief that species were created and extinguished serially, rather than in wholesale lots.

Naturalistic explanations had become the order of the day. The authors included in this anthology by and large agreed that scientific theories were to be judged only by scientific standards. Even the philosopher primed by Agassiz for his review of the *Origin,* Francis Bowen, admitted "all that has been claimed for the proper independence of true physical science,–that its conclusions are to be tested by their own evidence, and not by their agreement or want of agreement with the teachings of Scripture, with received doctrine in theology or philosophy, or with any foreign standard whatsoever."[14] Like Newton's *hypothesis non fingo,* this proclamation was easier to declare than to implement because the boundaries between science, philosphy, and theology began to blur when one approached the problem of man's mental powers and moral nature. These faculties had always been the provinces of philosophy and theology. While perusing the following reviews, the modern reader may be puzzled by the disproportionate attention given to Darwin's explanations of the hive-making habits of bees and the slave-making habits of ants. The emphasis was not misplaced. Darwin was

13. Lyell (1873), p. 469; see also the reviews of Hooker, Carpenter, Wollaston, and Fawcett in this volume.
14. Bowen (1860), p. 476.

taking his first steps toward the explanation of man's higher faculties. His intentions were not lost on his reviewers. The conclusions of *true* physical science were to be judged only by scientific standards, but Darwin was trespassing on the property of philosophy and theology and deserved to be criticized by their standards, and this authors like Sedgwick and Bowen proceeded to do. In doing so, they were not intruding on the proper domain of science. Rather, in Bowen's words, the "intrusion, if any, comes from the other side. It is now the naturalist, the pure physicist, who, quitting his own territory, but, as he professes, still relying exclusively on physical evidence, seeks to build up metaphysical conclusions."[15]

Even after the boundary between philosophy and science was redrawn to distinguish between psychology and epistemology, the adjustments in philosophy necessitated by evolutionary theory were of the most fundamental kind. Either one of the chief examples of natural kinds had to be dismissed as not actually being natural kinds, or else the notion of natural kind had to be reworked. Some philosophers chose the former alternative and retreated to geometry. *Bos bos* was not properly a natural kind. Equilateral triangles were. Relativity theory drove the adherents of natural kinds out of their last sanctuary. Some philosophers took the second alternative and tried to produce a metaphysics which did not depend on the essentialist notion of natural kinds. The extreme difficulty of this undertaking can be seen in the development of philosophy during the last fifty years.[16]

Later criticisms of evolutionary theory tended to be directed at Darwin's mechanisms—chance variation and natural selection. Perhaps species evolved and a naturalistic mechanism is called for, but Darwin's explanation was inadequate. The two modifications of evolutionary theory most frequently suggested were the substitution of evolution by saltation for Darwinian gradual evolution and the addition of a force or principle internal to organisms to direct them in their evolutionary development. The former served to salvage the discreteness of natural kinds, the latter to save teleology.

Even Darwin's strongest supporters took great pride in keeping one toe dry. Huxley, for example, thought Darwin had loaded himself "with an unnecessary difficulty in adopting *Natura non facit saltum* so unreservedly."[17] By and large, Darwin was right. There are genetic mechanisms which result in new species' evolving in a single generation (polyploidy),

15. Ibid., p. 447.
16. D. L. Hull (1974).
17. L. Huxley (1902), p. 189.

but they occur rarely. Given what we know of genetics and population biology, gradual evolution is the rule. Huxley's suggestion has been taken up periodically by scientists (e.g., Goldschmidt and Schindewolf). From a purely scientific point of view there is no a priori reason for preferring one type of evolution over the other. From a metaphysical point of view, there is. Speciation by saltation would permit the retention of the discreteness of natural kinds, a tenet central to essentialism.

Gray, Wallace, and Lyell all argued for the inclusion of some vestige of teleology in evolutionary theory. On this score, they were joined by a host of other scientists, including Owen and Mivart. Mivart, for instance, reasoned that just as there was a principle internal to an individual organism which determined its regular embryological development, there must also be a similar principle to determine the evolutionary development of species. By reasoning in this manner, these scientists were returning to something like the Aristotelian notion of immanent teleology. In this case, however, the end was not a cyclical returning to the same type but a progression from one type to another. Darwin could find no evidence of such a directing internal influence and steadfastly refused to admit it. Once again, Darwin seems to have been right. We now know why organisms go through their regular cycles of development—the information coded in their DNA. This information has been built up through eons of natural selection. No hint of a mechanism by which the future development of a species may also be incorporated into this code has been discovered. The truly amazing feature of Darwin's intellect was the frequency with which he was able to "guess" correctly, even though he lacked the requisite data and anything like an adequate theory governing the phenomena. Modern evolutionary theory is closer to the original Darwinian formulation today than it has ever been.[18]

18. Peter Vorzimmer (1970) comes to conclusions which are almost diametrically opposed to those expressed in this Introduction with regard to Darwin's adherence to gradual evolution and innate directive forces. Vorzimmer thinks that Mivart's criticisms of Darwin's theory "had been the prime instrument in badgering the elderly Darwin into the state of frustrating confusion which marked him on the eve of his retirement" (p. 250). See the discussion of Mivart's review of *The Descent of Man* (1871) in this volume for further examination of Vorzimmer's analysis of the situation.

II
Reviews

Joseph Dalton Hooker (1817–1911)

That was a nice notice in the *Gardeners' Chronicle*. I hope and imagine that Lindley* is almost a convert.—C. Darwin to J. D. Hooker, Down, January 3, 1860 (*Life and Letters,* 2:54)

I heard from Lyell this morning, and he tells me a piece of news. You are a good-for-nothing man; here you are slaving yourself to death with hardly a minute to spare, and you must write a review on my book! I thought it a very good one, and was so much struck with it, that I sent it to Lyell. But I assumed, as a matter of course, that it was Lindley's. Now that I know that it is yours, I have re-read it, and my kind and good friend, it has warmed my heart with all the honourable and noble things you say of me and it. I was a good deal surprised at Lindley hitting on some of the remarks, but I never dreamed of you. I admired it chiefly as so well adapted to tell on the readers of the *Gardeners' Chronicle;* but now I admire it in another spirit."—C. Darwin to J. D. Hooker, Down, [January] 14, [1800] *Autobiography*, p. 237

Review of the *Origin of Species*†

[JOSEPH DALTON HOOKER]

To how many of our gardening readers has it ever occurred to investigate the origin of any of the "favoured races" of plants with which they are familiar in the garden, the orchard, or the forest? Many know or take for granted that the most dissimilar kinds of Strawberries, Apples, and Potatoes have all sprung from one stock, and that most of them have originated within a very recent period; as also that the history of many

* [Joseph Lindley (1799–1865) was the editor of the *Gardeners' Chronicle*. Hooker was fond of imitating Lindley's style when writing anonymous reviews. See, for example, his review of Darwin's *Fertilization of Orchids* in the *Gardeners' Chronicle*, November 1862.]

† From the *Gardeners' Chronicle*. December 31, 1859, pp. 1051–1052; Hooker's review, like most of the reviews in this anthology, was published anonymously.

is attainable, did they think it worth the trouble of searching for. Even this limited knowledge is seldom acquired; and it is rarer still to find a gardener who has ever reflected upon the possible origin of the wild plants that surround him; though he not only knows that these are composed of the very same elements as are contained in the soil and air, but that they are themselves the immediate ancestors of the cultivated plants whose nurture provides him with remunerative labour, and whose increase at the same time supplies him abundantly with food. Little, however, as our gardening friends may have thought of these matters, still the terms "favoured race" and "origin of species," are more or less intelligible to most: but it is very different with "struggle for life." In its application to man civilized and savage, at peace and at war, and in a certain degree to the lower animals, the significance of the phrase is obvious enough; but how it can apply to the vegetable kingdom is not so evident at first sight, and still less evident is its connection with the "orgin of species" and with "favoured races."

Perplexed with these thoughts on reading for the first time the title of Mr. DARWIN'S new volume ON THE ORIGIN OF SPECIES,[1] and knowing that the author is of first-rate standing in science, of great popularity, and a frequent contributor to our columns, we awaited the publication of his book with great curiosity and interest. It is true that we had been referred to a notice in the *Gardeners' Chronicle* (1858, p. 735) of a paper communicated by Mr. DARWIN to the Linnean Society, which gives a clue to the general nature of his work, but rather whetted the curiosity than allayed it; being little more than an indication of one method of research now fully developed, and giving little idea of the extent, variety, and suggestive value of that research, and still less of the comprehensiveness and talent of the present work.

From the above expressions it will be inferred that we have risen from the perusal of Mr. DARWIN'S book much impressed with its importance, and have moreover found it to be so dependent on the phenomena of horticultural operations, for its facts and results, and so full of experiments that may be repeated and discussed by intelligent gardeners, and of ideas that may sooner fructify in their minds than in those of any other class of naturalists, that we shall be doing them (and we hope also science) a service by dwelling in some detail upon its contents. Thus much we may premise, that it is a book teeming with deep thoughts on numberless simple

1. On the Origin of Species by means of Natural Selection; or the Preservation of favoured races in the Struggle for Life. By CHARLES DARWIN, M.A., F.R.S., &c. 8vo, pp. 502. MURRAY.

and complex phenomena of life; that its premises in almost all cases appear
to be correct; that its reasoning is apparently close and sound, its style
clear, and we need hardly add its subject and manner equally attractive
and agreeable; it is also a perfectly ingenuous book, bold in expressions
as in thought where the author adduces what he considers clear evidence
in his favour, frank in the statement of objections to the hypotheses or
conclusions founded on its facts and reasonings; and uniformly courteous
to antagonistic doctrines. In fine, whatever may be thought of Mr. DAR-
WIN'S ultimate conclusions, it cannot be denied that it would be difficult
in the whole range of the literature of science to find a book so exclusively
devoted to the development of theoretical inquiries, which at the same
time is throughout so full of conscientious care, so fair in argument, and
so considerate in tone.

Before entering upon an analysis of those parts of Mr. DARWIN'S work
which more immediately concern our readers, we may explain that it is
devoted to the inquiry of how it is that the world has come to be peopled
by so many and such various kinds of plants and animals as now inhabit
it; and why it is that these do not all present a continuous, instead of
an interrupted series, capable of being divided into varieties, species, genera,
orders, &c. To all careless and many careful naturalists the assertion is
sufficient, that they were so created, and have been endowed with power
to remain unchanged; but there are considerable difficulties in the way
of the adoption of this theory, which all reflecting naturalists allow; whilst
more to our present purpose is the fact, that in Mr. DARWIN'S opinion
the objections to it are more numerous and some of them more conclusive
than the arguments in its favour. In the first there is the fact, that gardeners
and cattle and bird breeders have made races which are not only more
dissimilar than many species in a state of nature are, but if found in a
state of nature would unquestionably be ranked as species and even genera;
while the possibility of their reversion to their parent forms is doubted
by many and denied by some. Then, there is the fact, that in a great
many genera many so-called species present a graduated series of varieties
which have puzzled the best naturalists; and that in such genera at any
rate it is easier to suppose that the species are not separate creations than
that they are so. Again, the whole system of animal and plant classification
into individuals, species, genera, &c., presents a series strictly analogous
to that of the members of the human or any other family; and we further
express our notions of the mutual relations of animals and plants in the
very terms that would represent their affinities as being due to a blood
relationship; so that either Nature has mocked us by imitating hereditary

descent in her creations, or we have misinterrupted her in assuming that she has followed one method for families and another for species. There is then the geological evidence of species not having all been introduced at one period on the globe, but at various successive periods, and in like manner there is abundant evidence that they are being extinguished one by one; and as geologists are now pretty unanimous in believing that the past changes of the globe were the same in kind and degree as those now going on, we can no longer cling to the old vague belief that species were created at a time when all other conditions were different from those now existing; and thus hide one mystery under another. Again, if we hold to original creations we must also admit that it is more likely that a new kind of Strawberry, as different from all the wild as the British Queen is from the Alpine, should naturally originate in England or elsewhere out of air and earth and water, than that it should have originated in accordance with the laws that favoured the gardener's skill in producing the above-mentioned form. Finally, we know that on the one hand all changes of the surface of the globe (of climate, geographical limits, elevation and depression) are extremely slow; and that on the other relations of co-adaptation subsist between the climate, &c., of every country and the animals and plants which it contains; but such co-adaptations could not harmoniously exist if the physical changes were slow, and the creations of species sudden, for in such cases the species must be created either before or after they were wanted, a state of things which no orthodox naturalist would admit; but if, on the contrary, the organisms are supposed to change with the changed conditions of the country they inhabit, perfect adaptation and continuous fitness would be the necessary result.

Such ideas as these, either in whole or in part, have occurred to many naturalists, when in endeavouring to classify their animals and plants, or to account for their distribution, they have found themselves forced to face the difficulty of accounting for their origin. Hitherto most have, as we have already observed, contented themselves with the hypothesis that they were independently created, and the minority have clung to the only other explanation hitherto conceived, that they have all been created by variation from a single first-created living organism, or a few such. Now it must be borne in mind that both these views are mere hypotheses in a scientific point of view; neither has any inherent or prescriptive right to a preference in the mind of the impartial inquirer; but the most superficial observation will show that the hypothesis of original creations is incapable of absolute proof, except the operation be witnessed by credible naturalists, and can only be supported by facts that are either not conclusive against the opposite doctrine, or may be regarded as equally favourable to it; and this hypothesis is hence placed at a disadvantage in comparison with that of the creation of species by variation, which takes its stand on the familiar fact that species do depart from the likeness of their progenitors, both in a wild and domestic state.

On the other hand the hypothesis of creation by variation labours under the disadvantage of being founded upon a series of phenomena, the action of each of which individually, as explained by Mr. DARWIN, would appear to prevent the possibility of species retaining their characters for any length of time, and it therefore appears at first sight opposed to the undisputed fact, that many hundreds of animals and plants have transmitted their characters unchanged (or nearly so) through countless generations, and covered, even within the historic period, many square miles of country each with millions of its exact counterpart. To explain this anomaly and to raise this hypothesis of creations by variation to the rank of a scientific theory, is the object of Mr. DARWIN'S book, and to do this he has endeavoured to invent and prove such an intelligible rationale of the operation of variation, as will account for many species having been developed from a few in strict adaptation to existing conditions, and to show good cause how these apparently fleeting changelings may by the operation of natural laws be so far fixed as to reconcile both the naturalist and the common observer to the idea, that what in all his experience are immutable forms of life may have once worn another guise. The hypothesis itself is a very old one; it was, as all the world knows, strongly advocated by LAMARCK, and later still by the author of the "Vestiges of the Natural History of the Creation," but neither of these authors were able to suggest even a plausible method according to which Nature might have proceeded in producing suitable varieties, getting rid of intermediate forms, and giving a temporary stability to such as are recognised as species by all observers. Mr. DARWIN has been more successful, though whether completely so or not the future alone can show. We shall proceed to examine his method in another article.*

Comments on Hooker

Joseph Dalton Hooker, who succeeded his father, William Jackson Hooker (1785–1865), as Director of Kew Gardens, was a close friend of

---

* The second installment of Hooker's review in the *Gardeners' Chronicle* (January 7, 1860, pp. 3–4) is omitted for lack of interest. It mainly concerns reports of one species of grain being obtained from another by horticultural means.

Darwin and married the daughter of another close friend, John Stevens Henslow (1796–1861). The younger Hooker was also one of Darwin's earliest converts and continued to champion evolutionary theory throughout his life. Although his *Flora of Australia* (1859) could easily have been published without any reference to its theoretical background, he explicitly acknowledged the use which he had made of evolutionary theory. He and Lyell were instrumental in the presentation and publication of the joint papers of Darwin and Wallace. Later, at the Oxford meetings of the British Association, Hooker joined Huxley in opposing the forces of the Bishop of Oxford. Even though Huxley's showmanship gained for him most of the publicity, Hooker's less sensational comments seem to have been more persuasive. (See remarks by Fawcett in the concluding paragraph of his review of the *Origin* below.)

Hooker's publication of a review of the *Origin of Species* in the *Gardeners' Chronicle* was appropriate, since Darwin had made so much use of the data gathered by amateur gardeners and professional horticulturists. The analogy between man producing new species by conscious selection and natural selection in the feral state was one of the main arguments in the *Origin*. Although Hooker's discussion is relatively noncontroversial, he clearly looks upon special creation and evolution by natural selection as equally naturalistic hypotheses to be tested by observation. Hooker's assumption that special creation was to stand or fall purely on observational grounds, rather than being an isolated instance of naturalism, was typical of the scientific tenor of the time. As Hooker pointed out, special creation as a scientific hypothesis was at a disadvantage, since no mechanism had been suggested for the special creation of entire, multicellular organisms (or their eggs). Hence, the only proof of this hypothesis would have to be the direct observation of such an occurrence by a credible naturalist. No one had ever observed anything like the special creation of a new species out of inorganic material. If the catastrophists were right, no one ever could, since the observers would be created simultaneously with other species at the beginning of each epoch. Darwin, on the other hand, had provided a mechanism which could account for the evolution of one species from another. Hence, indirect evidence was available and pertinent. However, because of the slow rate of evolution, proof by direct observation was as much out of the question for evolution by natural selection as it was for special creation.*

---

* See also L. Huxley (1918), Turrill (1953, 1963), and Allan (1967).

William Benjamin Carpenter (1813–1885)

I have had a letter from Carpenter this morning. He reviews me in the 'National.' He is a convert, but does not go quite so far as I, but quite far enough, for he admits that all birds are from one progenitor, and probably all fishes and retiles from another parent. But the last mouthful chokes him. He can hardly admit all vertebrates from one parent. He will surely come to this from Homology and Embryology. I look at it as grand having brought round a great physiologist, for great I think he certainly is in that line.—C. Darwin to C. Lyell, Down, December 5, 1859 (*Life and Letters*, 2:35)

I have just read your excellent article in the 'National.' It will do great good; especially if it becomes known as your production. It seems to me to give an excellently clear account of Mr. Wallace's and my views. How capitally you turn the flanks of the theological opposers by opposing to them such men as Bentham and the more philosophical of the systematists! I thank you sincerely for the *extremely* honourable manner in which you mention me. I should have like to have seen some criticisms or remarks on embryology, on which subject you are so well instructed. I do not think any candid person can read your article without being much impressed with it. The old doctrine of immutability of specific forms will surely but slowly die away. It is a shame to give you trouble, but I should be very much obliged if you could tell me where differently coloured eggs in individuals of the cuckoo have been described, and their laying in the twenty-seven kinds of nests. Also do you know from your own observation that the limbs of sheep imported into the West Indies change colour? I have had detailed information about the loss of wool; but my accounts made the change slower than you describe—C. Darwin to W. B. Carpenter, Down, January 5, 1860 (*Life and Letters*, 2:57)

I have this minute finished your review in the 'Med. Chirurg. Review.'* You must let me express my admiration at this most able essay, and I

* [*British and Foreign Medico–Chirurgical Review.*]

hope to God it will be largely read, for it must produce a great effect. I ought not, however, to express such warm admiration, for you give my book, I fear, far too much praise. But you have gratified me extremely; and though I hope I do not care very much for approbation of the non-scientific readers, I cannot say that this is at all so with respect to such few men as yourself. I have not a criticism to make, for I object to not a word; and I admire all, so that I cannot pick out one part as better than the rest. It is all so well balanced. But it is impossible not to be struck with your extent of knowledge in geology, botany, and zoology. The extracts which you give from Hooker seem to me *excellently* chosen, and most forcible. I am so much pleased in what you say also about Lyell. In fact I am in a fit of enthusiasm, and had better write no more—C. Darwin to W. B. Carpenter, Down, April 6, 1860 (*Life and Letters,* 2:93.)

There has been a plethora of reviews, and I am really quite sick of myself. There is a very long review by Carpenter in the 'Medical and Chirurg. Review,' very good and well balanced, but not brilliant. He discusses Hooker's books at as great length as mine, and makes excellent extracts; but I could not get Hooker to feel the least interest in being praised.

Carpenter speaks of you in thoroughly proper terms.—C. Darwin to C. Lyell, Down, April 10, 1860 (*Life and Letters,* 2:93–94)

Darwin on the Origin of Species*
[WILLIAM BENJAMIN CARPENTER]

It has been calculated by an able naturalist,[1] on data which may be accepted as tolerably satisfactory, that the number of *distinct species* of Animals at present existing on the globe considerably exceed *half-a-million;* about nineteen-twentieths of the whole being insects. Of Flowering Plants the number of species actually known to the botanist is commonly estimated at a hundred thousand; and if we include the Cryptogamia, and make allowance for the imperfect degree in which the botanical treasures of large portions of the surface of the globe have yet been searched out, the number of distinct species furnished by the vegetable kingdom would seem to be fairly set at a hundred and fifty thousand.

But even this aggregation, enormous as it is, would sink into insignificance if all the forms of organic life which have peopled our globe during the long succession of ages chronicled by the geologist, could be brought together within our mental view. For, looking to the large collection of types now

* From the *National Review, January* 1860, 10:188–214.
1. Swainson's *Natural History and Classification of Quadrupeds,* p. 28.

extinct, which have been disinterred from a few scratches made here
and there in the crust of the earth, it cannot be reasonably doubted that
the whole number of fossil species of such animals as leave recognisable
remains behind them must be many times as great as that of the forms
which represent them at the present epoch. And if we further make due
allowance for the fact, that the portion of the palaeontological catalogue
at present known to us really consists of but fragments of a few of the
leaves of the great Stone Book, and that on the pages of that stone book
a vast proportion of the past life of our planet can never by any possibility
have recorded itself, we cannot fairly refuse to admit it as a probability
(to which every new discovery gives additional weight), that the animal
and vegetable life existing at any of that long succession of periods, each
of which is marked out in geological time by a characteristic fauna and
flora of its own, was at least as rich as it is at present, in regard alike
to the number and to the variety of its distinct forms.

Now it seems to be a received article of faith, both amongst scientific
naturalists and with the general public, that all these reputed species have
(or have had) a real existence in nature; that each originated in a distinct
act of creation; and that, once established, each type has continued to
transmit its distinctive characters, without any essential change, from one
generation to another, so long as the race has been permitted to exist.
This idea of the *permanence of species,* embracing those of the common
origin of all the individuals linked together by similarity of characters,
and of the diverse origin of races distinguished by any marked and constant
dissimilarity, is, in fact, embodied, in one shape or another, in every definition of the term which has been framed; and though some bold speculator
like Lamarck, or some ingenious theorist like the author of the *Vestiges,*
has ventured from time to time to question the soundness of its basis,
yet it has given no outward sign of instability, and is commonly regarded
at the present time as one of those doctrines which no man altogether
in his right senses will set himself up seriously to oppose.

Yet there have not been wanting indications, especially during the last
few years, that a re-consideration of the whole subject is felt by several
of the leading minds of our day to be called for by the progress of science;
the difficulty of determining *what are* the characters as to which agreement
shall be held to constitute specific identity, whilst disagreement shall be
accepted as establishing specific diversity, having been found to increase
instead of diminishing with the progress of knowledge. Differences of sufficient constancy and importance for the separation not merely of species,

but of genera, in one group, may be found in another to be so inconstant that they cannot be admitted to rank higher than as individual varieties; and features of diversity which seem so well marked as to leave no room for hesitation when the comparison is limited to two or three individuals which exhibit them under their most pronounced aspect, are often found to shade off so gradationally when a large number of individuals are compared, that no lines of specific demarcation can be drawn among them. It has accordingly come to be recognised by many of our best zoologists and botanists, that no species can be fairly admitted as having a real existence in nature, until its *range of variation* has been determined both *over space* and *through time;* and that the species of the mere collector, who describes every form as new which does not precisely correspond with existing definitions, can only be accepted provisionally, to be verified or set aside by more extended research.[2]

A remarkable example of the results of an inquiry conducted in this spirit has lately made a considerable impression, alike on account of the nature of the subject and the deservedly high reputation of the naturalist by whom it has been conducted. No group of species has been more carefully or completely studied, after the ordinary fashion, than that of the British flowering plants and ferns. In Hooker and Arnott's *British Flora,* 1571 species of these were enumerated and described; whilst by Mr. Babington the number of species was raised to 1708. Within the last eighteen months, however, a new *British Flora* has been published by Mr. Bentham, one of those quiet painstaking workers who, not making fame but truth their goal, are content to spend as many years in the thorough investigation of a subject as other men bestow months. Mr. Bentham has devoted a large part of his time for many years past to the study of the British flowering plants, not as dried in herbaria, but as living and growing in their native habitats; and instead of confining himself to the area of our own islands, he has followed them to every part of Europe through which he has been able to trace them, carefully comparing the forms which they present under different conditions of soil, climate, exposure, &c., and diligently scrutinising with the educated eye of the really scientific botanist into the value of the distinctions, not merely among the species reputed doubtful, but among those commonly considered to be well established. The result has been, that not only has Mr. Bentham been led to add

2. See on this subject Dr. Joseph D. Hooker's *Introduction to the New Zealand Flora,* and Dr. Carpenter's *Memoir on Orbitolites* in the *Philosophical Transactions* for 1855, pp. 277 et seq.

the weight of his authority to the side of those who pleaded for the wide range of variation in such genera as *Salix* and *Rubus,* regarding which there had been the greatest question; but he has shown that a considerable extent of variation is so far from being confined to willows and brambles, that the total number of well-marked species cannot fairly be reckoned at more than 1285; so that about a quarter of the reputed species of the British phanerogamic flora, on which so much pains have been bestowed and so many books written by botanists of the highest reputation, have been thus abolished "at one fell swoop."

Now this result, valuable as it is in itself, has a bearing of far deeper import upon the whole existing method of botanical and zoological systematisation; for it shows how far Nature is from tying herself down by the canons of species-mongers, and how mistaken has been the course of those who, instead of humbly searching for a knowledge of Nature's laws, have arrogated to themselves the right of making laws for Nature. The species of plants and animals which such men have added to our already overloaded catalogues, are of human, not of divine creation; and it is the business of the philosophic naturalist to get rid of all such as soon as possible. He cannot, however, proceed far in his inquiries, without having the question forced upon him as to the extent to which natural species,–that is to say, races which seem to be distinguished by certain constant characters that are transmitted by decent so far as our experience extends,–can be reasonably supposed to have varied *in time,* so as to have undergone in the lapse of ages, under the influence of natural causes, modifications at all corresponding with those which are presented by the races of plants and animals that have been subjected with a comparatively recent period to the influence of man.

This is a problem which Mr. Darwin has been for some years essaying to resolve. His attention was first directed to the inquiry by some facts which struck him in the distribution of the inhabitants of South America, and in the geological relations of the present to the past inhabitants of that continent, during that voyage on board H.M.S. *Beagle* of which he has given us so admirable a Journal. These facts seemed to him to throw some light on the origin of species—that mystery of mysteries, as one of our greatest philosophers has called it; and on his return home it occurred to him, in 1837, that something might perhaps be made out on this question by patiently accumulating and reflecting on all sorts of facts which could possibly have any bearing upon it. After five years' work, he allowed himself to speculate on the subject, and drew up some short notes; these he enlarged in 1844 into a sketch of the conclusions which then seemed to him probable; and from that time to the present he has steadily pursued the same object, with the intention of setting forth, not only his conclusions, but the mass of facts on which they are based, as soon as his imperfect health should permit him to complete a work so extensive. In the mean time it has happened that Mr. Wallace, an intelligent naturalist, who is at present engaged in studying the animal and vegetable productions of the Malay Archipelago, has arrived, without any knowledge of Mr. Darwin's inquiries, at a doctrine essentially the same as his own; namely, that a process of *natural selection* is constantly in operation, on a far grander scale, and with far more perfect results, than man can imitate; and that to this process, operating cumulatively through countless ages, we are justified in attributing an unlimited amount of divergence, not merely between species, but between genera, and, by parity of reasoning, even among the higher groups. A memoir on this subject having been sent to Mr. Darwin by Mr. Wallace, with a request that it should be forwarded to Sir C. Lyell, it was by the latter communicated to the Linnaean Society, and has been printed in its journal, together with extracts from Mr. Darwin's larger work; and as two or three years are likely still to elapse before the latter will be ready for publication, Mr. Darwin has complied with the urgent recommendations of his friends that he should at once put forth his views in a more concise form, so as to benefit the scientific world by such a knowledge of them as should enable them to take root in the minds of those who are not too much hardened by prejudice against their reception, and to bear good fruit by stimulating inquiry in the new direction he has opened up.

As the work before us is to be regarded but as the abstract of the larger treatise which Mr. Darwin has in preparation, we feel it right to limit our discussion of it to an examination of the soundness of the main principle on which it is based. Minute criticism of details would, as it seems to us, be at present altogether misplaced. Indeed, we almost regret that the author has as yet gone into detail at all, since he has laid himself open to a great deal of objection on the score of minor difficulties, which will tend to prevent his fundamental doctrine from finding candid appreciation. And we cannot help thinking that it might have been better if, in this early stage of the inquiry, Mr. Darwin, like Mr. Wallace, had abstained from that explicit avowal of the ultimate conclusions to which it seems to him to lead, which will be pretty sure at once to frighten away many whom he might have otherwise obtained as adherents. Of course, if his principle be firmly based on truth, every thing that is legitimately deducible from it must also be true. But as it is in the nature of things impossible to obtain any thing like positive evidence on the remoter issues of the inquiry, we shall discard for the present all reference to the question whether (as Mr. Darwin thinks probable) men and todpoles, birds and fishes, spiders and snails, insects and oysters, encrinites and sponges, had a common origin in the womb of time, and shall address ourselves only to the arguments urged by Mr. Darwin and Mr. Wallace in support of their doctrine of the modification of specific types by natural selection.

To such as look upon this question from the purely scientific point of view, any theological objection, even to Mr. Darwin's rather startling conclusion, much more to his very modest premises, seems simply absurd. We never heard of any body who thought that a religious question was involved in the inquiry whether our breeds of dog are derived from one or from several ancestral stocks; nor should we suppose that the stoutest believers in the Mosaic cosmogony would be much dismayed if it could be shown that the dog is really a derivation from the wolf. Orthodoxy (on this side of the Atlantic at least) is decidely in favour of the abolition of the two-and-twenty species into which man has been divided by some zoologists, and of the reference of all the strongly-diversified races of man to the Adamic stock. We do not expect to see, even in our "most straitest" sectarian organs, any accusations brought against Mr. Bentham for impiety, because he affirms that three or four hundred of the reputed species of British plants are really descendants of others from which they have gradually diverged; and if he were led by the results of further inquiry to knock off as many more, we believe that he would be left to the criticism of his brother botanists, and that his *British Flora* would not run any risk of being put into the Index Expurgatorius, alongside of Lamarck's *Philosophie Zoologigue* and the *Vestiges of the Natural History of Creation*.

Why, then, should Mr. Darwin be attacked (as he most assuredly will be) for venturing to carry the same method of inquiry a step further; and be accused (in terms which it needs no spirit of prophecy to anticipate) of superseding the functions of the Creator, of blotting out his Attributes from the page of Nature, and of reducing Him to the level of a mere Physical Agency? To our apprehension, the Creator did not finish his labours with the creation of the protoplasts of each species; his work is always in progress; the origin and development of each new being that comes into life, is a new manifestation of his creative power; and the question is simply as to the mode in which it has pleased Him to exercise that power; whether, according to common ideas, He has every now and then swept off a greater or smaller proportion of the inhabitants of the globe, and has replaced them by new forms, brought into existence in some mode altogether unknown to us; or whether, as Mr. Darwin maintains, the apparent introduction of new forms has really been brought about by a gradual and successive modification of the old. For ourselves, we do not hesitate to say that the orderly and continuous working out of any plan which could evolve such harmony and completeness of results as the world of Nature (present and past) spreads out before us, is far more consistent with our idea of that Being who "knows no variableness, neither shadow of turning," than the intermittent action of a power that requires a succession of interferences to carry out its original design in conformity with successive changes in the physical conditions of the globe. And we have no sympathy with those who, to use the admirable language of Professor Powell (whose *Essay on the Philosophy of Creation* contains a masterly refutation of the current theological arguments bearing on this question), maintain that we "behold the Deity *more* clearly in the dark than in the light,–in confusion, interruption, and catastrophe, *more* than in order, continuity, and progress."

Our knowledge as to the Variability of Species in Time is of course mainly derived from observation of the changes induced by the agency of Man, in those species of plants and animals which have been longest subjected to the influence of cultivation and domestication; and although it may be questioned how far the modifications thus induced would tend to perpetuate themselves in a state of nature, yet we cannot shut our eyes to the fact that *the capacity for undergoing such modifications* under conditions how artificial soever, exhibits an elasticity of constitution which would equally tend to adapt these animals to varieties of natural conditions, and thus to originate diversified races which would perpetuate themselves without man's interference. It may be well to adduce a few examples, in which peculiarities of organisation or constitution have sprung up and perpetuated themselves under the notice of competent witnesses. Such peculiarities have been ranked in two categories; those, namely, which are obviously "acquired" under the direct influence of external agencies; and those which, not seeming likely to have been thus occasioned, are spoken of as "spontaneous," or "accidental."

Of the acquirement of peculiarities, we have frequent experience in the changes which animals and plants undergo when they are transported to regions differing greatly in climatic conditions from those they previously inhabited. Thus the longest-woolled sheep that can be brought from Leicestershire or Sussex to the West Indies have their thick matted fleeces replaced in a year or two by short crisp hair; and in the lambs bred in their new country this hair is so brown as to render it somewhat difficult to distinguish them from the kids of the goats with which they are often seen associated. Sometimes the acclimatising process does not modify the character of the parent, but takes effect only on the young, born and bred under its influence. Thus, in a well-known case related by Sir C. Lyell, the English greyhounds that were taken out to hunt the hares which abound on a table-land in Mexico at an elevation of 9,000 feet, were distanced in the chase by want of wind; yet the offspring of these same animals are not in the least incommoded by the rarity of the atmosphere in which they have passed their whole lives, but run down the hares with as much ease as the fleetest of their race in this country.

But peculiarities every now and then appear in the offspring at birth, which, not being traceable to any corresponding change of circumstances, are commonly regarded as "freaks of Nature;" such, for example, as the presence of a sixth finger, or of an additional joint in the thumb, on the human hand; or of that peculiar conformation of the limbs that distinguishes the "ancon" breed of sheep in New England; or the so-called "sporting" varieties of plants. No one, however, who believes in the universality of causation, can fail to perceive on reflection, that any such congenital peculiarities, like the differences among individuals of the same parentage, *must* have had their origin in the condition of the one or both parents at the time of procreation; and this inference is fully borne out by the special tendency of such peculiarities to become hereditary, though they frequently pass over a generation or two, to reappear in a subsequent one. The *latency* of such influences is often extremely remarkable; but sometimes we seem able to trace out their nature, though we cannot comprehend their mode of operation. Thus there is valid scientific evidence that the colour of the offspring of animals whose hue is disposed to vary, is influenced by strong mental impressions on the parents; and that it is in this way that variety of hue was first engendered in races previously of uniform colour, would seem to be indicated by the fact related by Mr. T. Bell, that a litter of puppies born in the Zoological Gardens from a male and female Australian dingo of pure breed, both of which were of the uniform reddish-brown hue that belongs to the race (the mother never having bred before), were all more or less spotted.

Now the art of the breeder consists first in carefully watching for this

spontaneous appearance of any such peculiarities as he may deem it profit-
able to introduce, which then, by taking advantage of their tendency to
become hereditary, especially when they are possessed by both parents,
he establishes as the distinctive feature of a new race; and in this race
he preserves them in full force by a rigorous weeding out of all the individuals which do not possess them. We cannot have a better example of
this process than the recent creation of the Mauchamp breed of sheep,
which produces a fine silky wool, distinguished by the strength as well
as by the length and fineness of its fibre, and specially valuable for the
manufacture of Cashmere shawls. In the year 1828, one of the ewes of
the flock of merino sheep belonging to M. Graux, a farmer of Mauchamp,
produced a male lamb, which, as it grew up, became remarkable for the
silky character of its wool, and for the shortness of its horns; it was of
small size, and of inferior general conformation. Desiring, however, to obtain
other sheep having the same quality of wool, M. Graux determined to
breed from this ram: at first he only obtained it in a single ram and
a single ewe; in subsequent years he got it in a larger proportion of each
progeny; and as his silky-woolled sheep multiplied, he was able to secure
a constant succession by matching them with each other. Amongst the
breed thus engendered, some resembled the ancestral ram in its physical
defects as well as in its wool; but others, while possessing the same character
of wool, reverted to the more symmetrical form of the breed from which
this was an offset; and M. Graux, by a judicious system of crossing and
intercrossing, at last established a race which not only possesses the silky
wool of the first ram without the least deterioration, but is entirely free
from its defects of general conformation.

The agency of Man in this procedure is that of *accumulative selection*.
He can do nothing except on the basis of variations which are first given
to him in some slight degree by nature (the origin of such variations, however, being in the modifications induced by external conditions in the procreative action of the parents); these, which would soon disappear if left
to themselves, by merging in the general aggregate, he not only perpetuates
by selection, but augments by accumulation, adding them up (as Mr.
Darwin felicitously phrases it) in certain directions useful to him. Now
there are instances in which *varieties* have been thus engendered, differing
so much from their original stock and from each other, that, if they were
to be placed before a zoologist ignorant of their genetic relationship, he
would unquestionably rank them not only as *distinct species,* but even as
belonging to *distinct genera.* This is the case, for example, with the various

heads of pigeons, wihch have been closely studied by Mr. Darwin. The English carrier, the short-faced tumbler, the runt, the barb, the pouter, and the fantail, differ from each other not merely in their plumage, but in those points of conformation of beak and skull on which generic distinctions among birds are chiefly founded; and subordinate to each of these breeds are several regularly propagating sub-breeds, which differ as much from each other in characters of minor importance as species do elsewhere. Yet there is no ground for questioning the general belief of scientific ornithologists, that all these breeds have had their origin in the rock pigeon; especially as the comparison of a large number of individuals of these breeds and sub-breeds, including those brought from distant countries, would enable an almost perfect gradational series to be formed between the types that differ most widely in structure.

It is obvious, then, that by such a process what we may designate as *Naturalists' species* might be artificially created to any extent. The question now arises, whether truly *Natural Species* can have been engendered in a similar manner; that is to say, whether any thing like accumulative selection has gone on amongst plants and animals in their feral state, by which the vast multitude of diverse forms now existing may have been evolved from a comparatively small number of original types, and that wonderful series of extinct forms may have been produced, which palaenotology reveals to us as having peopled and repeopled our globe many times through the immeasurable succession of geological ages.

The answer to this question, according both to Mr. Darwin and Mr. Wallace, is to be found in a careful examination of the facts relating to the Struggle for Existence which all wild animals have to maintain. "The full exertion of all their faculties and all their energies," it is observed by Mr. Wallace, "is required to preserve their own existence, and provide for that of their infant offspring. The possibility of procuring food during the least favourable seasons, and of escaping the attacks of their most dangerous enemies, are the primary conditions which determine the existence both of individuals and of entire species." It is remarkable how little the relative fecundity of different species influences their relative abundance or scarcity. We do not find those populations (if we may so use the term) increasing at the greatest rate, whose reproduction is the most rapid. There are many species of fish, for example, in which the eggs annually produced by each female are to be reckoned by the hundred thousand; and if but one in a thousand of her progeny were to come to maturity, the ocean-waters of the whole globe, from their surface to their lowest depths, would in a few years be turned into a mass of finny life. The multiplication of the common flesh-fly, if not kept down by natural checks, takes place at such a rate, that, according to the estimate of Linnaeus, three of these insects and their progeny would devour the carcass of a dead horse more quickly than a lion would do. The Aphides, or plant-lice, have been proved to propagate so rapidly, that in the course of a few months, if all interference were excluded and adequate supplies of food were obtainable, no fewer than 1,000,000,000,000,000,000 would be evolved from a single individual; an amount which only becomes conceivable, when we learn that this mass of life would weigh somewhere about as much as *five hundred millions* of stout men.

Yet we find no reason to believe that there is a permanent increase in the number of any one of these or other productions of the inexhaustible fertility of nature. Our seas are so far from being glutted with fish, that the complaint is rather of their diminished abundance; and though this might be attributed with a show of reason to the hostile influence of man, yet when we bear in mind that even the vast shoals of herrings, pilchards, and mackerel, which he captures, *might* be the offspring of a few score of individuals, it is obvious that these would be replaced at least as rapidly as they are withdrawn, if there were no more powerful check to the increase of their respective tribes. So again, although we are sometimes inconvenienced by the swarms of flesh-flies that have been bred in some neighbouring mass of putrescence which the neglect of man has left to be removed by these scavengers of nature, we are not eaten out of house and home by them, as we very soon should be, if their reproduction were not kept most efficiently in check by the voracity of other animals to which they serve as a prey. Our roses and our hops are most seriously damaged every now and then by the excessive multiplication of the species of plant-lice which respectively attack them; but all the roses and all the hops in the world would soon be destroyed, if the occasional increase of their insect blights were not restrained within bounds by the destruction, through natural agencies, of a vast proportion of every brood engendered. The population of each tribe is, in fact, kept down to a certain general average, subject to occasional excess or reduction, within no very wide limits.

There is no need, however, to go beyond the limits of our own familiar experience in search of evidence of the same general fact; an apposite illustration of which is found, both by Mr. Wallace and Mr. Darwin, in the class of birds. If we take *four* as the average annual production of a single pair, it is easily shown that, if there were no check to the multiplication of its progeny, this would increase within fifteen years to nearly *ten millions*. Yet we have no reason to believe that the bird-population of any country is undergoing an increase. And hence it becomes evident that in each year *as many birds perish as are born;* that is, the progeny being twice as numerous as the parents, whatever be the average number of individuals existing in any given country, *twice that number must perish annually*, either serving as food to hawks and kites, wild cats and weasels, or dying of cold and hunger as winter comes on.

How much more the multiplication of individuals of any type depends upon the general capabilities of the species than upon the number of its progeny, is remarkably illustrated by the case of the American passenger-pigeon, cited by Mr. Wallace. Though this bird lays but one or two eggs, and is said generally to rear but a single young one, yet it is far more abundant than other species which produce two or three times as many young; the vast flocks of these pigeons, whose migrations in search of food are so graphically described by Audubon, being the most extraordinary aggregations of animal life of which we have any knowledge. How are we to account for the maintenance of this wonderful bird-population, which probably exceeds that of any other half-dozen species in the American continent? Obviously by the fact that the food most suitable to this species is abundantly distributed over a very extensive region, offering such differences of soil and climate, that in one part or other of the area the supply never fails; whilst, on the other side, the organization of the bird enables it to take advantage of this wide distribution of its means of sustenance, its powers of flight being remarkable both as to swiftness and endurance, so that it can pass without fatigue over the whole of the district it inhabits, and thus, when the supply of food begins to fail in one place, it is able to migrate in search of a fresh feeding-ground even at a remote distance. In no other birds are these conditions so strikingly combined; for either their food is more liable to failure, or they have not sufficient power of wing to search for it over an extensive area, or during some season of the year it becomes very scarce, and less wholesome substitutes have to be found; and thus, though more fertile in offspring, they can never increase beyond the supply of food in the least favourable seasons.

Nothing is easier, as Mr. Darwin justly remarks, than to admit in words the truth of this universal struggle for life; nothing more difficult than to trace out this general fact in all its bearings. The system of checks and counter-checks provided in the economy of nature is a most complicated one; and we are for the most part only let a little way into the secrets of her arrangement, by some local disturbance in its harmonious working. Thus in Paraguay neither cattle nor horses nor dogs have ever run wild, although they swarm both southward and northward, in a feral state; and this limitation is due to the greater abundance in that area of a certain fly, which lays its eggs in the navels of these animals when first born. The increase of these flies, numerous as they are, must be habitually kept in check by some means, probably by birds. Hence, if certain insectivorous birds (whose numbers are probably regulated by hawks or beasts of prey) were to increase in Paraguay, the flies would decrease; then cattle and horses would become feral, and this would greatly alter the vegetation (as Mr. Darwin has himself observed in parts of South America); this, again, would largely modify the insects; this would affect the insectivorous birds, and so on in ever-extending circles of complexity. "Battle within battle must ever be recurring, with varying success; and yet in the long-run the forces are so nicely balanced, that the face of nature remains uniform for long periods of time, though assuredly the merest trifle would often give the victory to one organic being over another."

A very close connection often exists between agencies that would not at first be supposed to have any mutual relation. How, for example, could the number of domestic cats in a village be imagined to influence the Flora of the neighbourhood? Mr. Darwin shall make answer in his own words:

> I have reason to believe that humble-bees are indispensable to the fertilisation of the heartsease; for other bees do not visit this flower. From experiments which I have tried, I have found that the visits of bees, if not indispensable, are at least highly beneficial to the fertilisation of our clovers; but humble-bees alone visit the common red clover, as other bees cannot reach the nectar. Hence I have very little doubt that if the whole genus of humble-bees became extinct or very rare in England, the heartsease and red clover would become very rare, or wholly disappear. The number of humble-bees in any district depends in a great degree on the number of field-mice, which destroy their combs and nests; and Mr. H. Newman, who has long attended to the habits of humble-bees, believes that 'more than two-thirds of them are thus destroyed all over England.' Now the number of mice is largely dependent, as every one knows, on the number of cats; and Mr. Newman says, 'Near villages and small towns I have found the nests of humble-bees more numerous than elsewhere, which I attribute to the number of cats that destroy the mice.' Hence it is quite credible that the presence of a feline animal in large numbers in a district might determine, through the intervention first of mice and then of bees, the frequency of certain flowers in that district.

One would almost think, in reading Mr. Darwin's string of sequences, that the ingenious author of *The House that Jack Built* had intended to imbue the infant mind with a knowledge of high scientific truths; for the whole of the foregoing extract may be concisely expressed, after the fashion of that immortal legend, in the following brief sentence: "This is the old woman, that kept the cat, that ate the mouse, that killed the humble-bee, that fertilised the clover." And it is obvious that if our country were in the condition of Paraguay, an abundance of wild clover would favour the increase of feral herds of cattle, which, in its turn, would operate extensively in modifying the general flora and fauna of the country.

It is only in a somewhat metaphorical sense that the term "struggle for existence" can be used with respect to plants; but when so understood, it is as applicable to their life as to that of animals. A plant which annually produces a hundred seeds, of which, on an average, only one comes to maturity, may be said to struggle against the other plants which tend to crowd out both itself and its progeny, against the animals which feed upon it in every stage of its development, and against all the physical conditions which are unfavourable to it, whether as to soil or exposure, heat or cold, drought or excessive moisture, or the like. We have many instances of the rapid multiplication and diffusion of particular species, which seem extraordinary only because they are exceptional; the peculiarity depending, not on any unusually rapid production, but simply on the absence of the ordinary repressing agencies. Cases could be given of introduced plants, which have become common throughout whole islands in a period of less than ten years; and in some of the most remarkable of these the extension has not been so much effected by seeds as by the ordinarily slower process of budding. Thus the *Anacharis alsinastrum,* a water-weed, which was imported into this country a few years ago from Canada, has made its way into almost every one of our rivers and canals, and into many of our isolated lakes and ponds, which can only be kept from being choked with it by the employment of artificial means for its destruction; yet this plant has never flowered in Britain, and the thousands of tons of its stalks and leaves which might be annually collected from various parts of our island, are really all extensions of the single individual originally imported. We believe that the like may be said of the Indian couch-grass, which is rapidly extending itself through the cultivated pastures of New South Wales, and which will probably ere long displace other grasses in the remoter parts of that colony, whence it will extend over the rest of Australia. Several of the plants now most numerous over the wide plains of La Plata, clothing square leagues of surface, almost to the exclusion of other plants, have been introduced from Europe; and there are plants now ranging in India from Cape Comorin to the Himalaya, which have been imported from America since its discovery. There is no reason whatever to suppose that in such cases (of which many more might be cited) the fertility of the plants has been temporarily increased in any sensible degree. The obvious explanation is, that, on the one hand, the conditions of life have been sufficiently favourable to promote their vigorous growth and reproduction; whilst, on the other, there has been an absence in their new habitats of those checks which previously restrained their exuberance.

Of the nature of these checks we shall cite an illustration or two from Mr. Darwin. Every one knows that an immense destruction of seeds in process of ripening is occasioned by the voracity of granivorous birds; but even when seeds have fallen into a favourable soil, and have begun to germinate, the young plants have from the first to struggle for their lives, partly against the other plants which already thickly stock the ground, partly against the animals to which they serve as food at that stage of their existence. On a piece of ground three feet long and two wide, dug and cleared, in which there could have been no choking from other plants, Mr. Darwin marked all the seedlings of our native weeds as they came up; and out of 357, no fewer than 295 were destroyed, chiefly by slugs and insects. If turf which has long been mown, or turf which has been closely browsed by quadrupeds, be let to grow, the more vigorous plants gradually kill the less vigorous though fully grown plants; thus out of twenty species growing on a little plot of turf (three feet by four), nine species perished from the other species being allowed to grow up freely.

The competition which it has to sustain with other tribes of plants, is considered by Mr. Darwin to be an agency at least as important as climate in determining the geographical range of any particular species. If we look at a plant in the midst of its range, we see that it is not climate that restrains it from doubling or quadrupling its numbers; for we know that it can perfectly withstand a little more heat or cold, a little more dampness or dryness, since elsewhere it ranges into slightly hotter or colder, damper or drier districts. But towards the confines of its climatic range it will be subject to the competition of other plants, whether of its own kind or of some other, for the spots most favourable to it in regard to warmth or coolness, exposure or protection, dryness or moisture. And thus it seems probable that few plants (and the same will be true also of animals) range so far that they are destroyed by the rigour of the climate alone. It is from this circumstances that so large a proportion of the plants which have been introduced into this country from foreign sources, and which have become perfectly acclimatised, and still restricted to our gardens; being incapable of perfect naturalisation, since they cannot compete with our native plants, nor resist destruction by our native animals.

It is impossible now to pursue the inquiry as to this system of checks and counter-checks in any further detail, so countless are its ramifications, and so complex are its interlacements. And it is far better for our present purpose to fix our minds upon the general fact, about which there can be no kind of doubt or dispute, that each organic being is striving to increase in a geometrical ratio; but that, either through the whole of its life or at particular periods of it, either during each generation or at frequently-recurring intervals, it is subject to heavy destruction, by which its multiplication is so restrained that its numbers are merely kept up to a certain average, without more than temporary excess or diminution, so long as the conditions remain the same. Lighten any check, mitigate the destruction ever so little, and the number of the species will almost instantaneously increase to any amount; as we see in the increase of cockchafer-grubs and wire-worms that ensues on the foolish destruction of a rookery. Where, on the other hand, the excessive multiplication of a newly-introduced species becomes (as in the case of the *Anacharis*) a serious inconvenience, it is obvious that the only effectual check is to be found in the encouragement of its natural enemies, the species of animals (whatever they may be) to which it serves as food.

But how will this constant struggle for existence tend to the modification of specific types? We shall let Mr. Darwin answer this question in his own words, since these words convey, not only with scientific conciseness, but with philosophical caution, the fundamental idea of his whole treatise:

Let it be borne in mind in what an endless number of strange peculiarities our domestic productions, and in a lesser degree those under nature, vary, and how strong the hereditary tendency is. Under domestication it may be truly said that the whole organisation becomes in some degree plastic. Let it be borne in mind how infinitely complex and close-fitting are the mutual relations of all organic beings to each other, and to their physical conditions of life. Can it, then, be thought improbable, seeing that variations useful to man have undoubtedly occurred, that other variations, useful in some way to each being in the great and complex battle of life, should sometimes occur in the course of thousands of generations? If such do occur, can we doubt (remembering that many more individuals are born than can possibly survive) that individuals having any advantage, however slight, over others, would have the best chance of surviving, and of propagating their kind? On the other hand, we may feel sure that any variation in the least degree injurious would be rigidly destroyed. This preservation of favourable variations, and the rejection of injurious variations, I call Natural Selection. Variations neither useful nor injurious would not be affected by natural selection, and would be left a fluctuating element, as perhaps we see in the species called polymorphic.

Now by so much the more diversified the descendants from any one species become in structure, constitution, and habits, by so much will they be better enabled to seize on many and widely diversified places in the polity of nature, and so be enabled to increase in numbers. It is easily shown that the advantage of diversification among the inhabitants of the same region, is really the same as that of the physiological division of labour in the animal body; a greater number of plants and animals of different kinds being able to subsist within a limited area, than could find support if they were all of one type. It is obvious, then, that natural selection will favour divergence of character, and that the tendency of each specific type will be not only to diffuse itself over as wide an area as it is capable of occupying, but to undergo as many diversified modifications as its constitution may permit, in conformity with the diversities of climate, locality, food, and the countless other conditions with which it comes into relation; those diversified forms establishing themselves permanently as *new races,* which are best fitted to fight their way in the struggle for existence. Even a very limited experience affords many instances of natural specialisation of habit, which often bear an obvious relation to specialisation of structure. It is well known that one Cat is prone to catch rats rather than mice, and that this tendency is often inherited; other cats exhibit sporting propensities of different kinds, one bringing home winged game, another hares or rabbits, while another hunts on marshy ground, and almost nightly catches woodcocks or snipes. Recent observations on the Cuckoo have shown that this bird deposits its eggs in the nests of no fewer than twenty-eight different species, and that the cuckoo's egg almost invariably agrees so closely in colour with the eggs among which it is laid, as only to be distinguished from them by a practised eye. This curious fact has been accounted for on the hypothesis that the sight of the eggs amongst which she is about to lay her own, operates through the consciousness of the parent in determining the colour of its shell. But as observation shows that the same individual cuckoo always lays eggs of the same colour; that the cuckoo's eggs are laid upon the ground in the first instance, and are afterwards conveyed in its mouth to the nest of the foster parent upon whose charge the young cuckoo is forced; and that when a nest with eggs of the corresponding colour is not accessible, the egg is deposited in that of some other species,–it seems clear that the colour of the egg is predetermined in the organisation of the parent, and that it is connected with an instinct which leads different individuals constantly to resort to different nests.

Again, the Catskill mountains of the United States are inhabited by two varieties of the Wolf: one with a light greyhoundlike form, which pursues deer; and the other more bulky, with shorter legs, which more frequently attacks shepherds' flocks. Here we can readily see how very divergent varieties might originate from a common stock, and be perpetuated by the operation of natural selection. Wolves inhabiting a mountainous district, and those frequenting the lowlands, would naturally be forced to hunt different prey; any slight change of organisation that might specially adapt an individual wolf to either mode of life would give it the advantage in the struggle for existence, and would favour not only the prolongation of its own life, but its chance of leaving offspring; some of its young would probably inherit the same peculiarities, and would in like manner transmit them to their descendants; and thus, from the preferential preservation of the individuals best fitted to the requirements of each locality, two types distinct in structure and habits, not merely from each other, but also from their common progenitor, would gradually be evolved. Supposing, again, that the conditions of the country changed in such a manner as to affect the supply of food,–the deer, for example, increasing in numbers, and the other prey decreasing in abundance,–during the season of the year when the wolf is most hardly pressed for subsistence; it is quite obvious that a large proportion of the lowland variety must soon perish from starvation, through the failure of their ordinary means of support, the greater abundance of deer being of no use to animals which had lost the capability of profiting by it; whilst, under the same circumstances, the mountain variety would both increase in numbers and would improve in swiftness. And conversely, if the supply of deer were to diminish, and the wolves were obliged to depend upon prey which it requires strength rather than swiftness to master, the mountain variety would be reduced in numbers; whilst the more bulky lowlanders would both increase in population, and would improve in special adaptation to the requirements of their mode of life.

A number of very marked examples of the influence of natural selection

in the establishment and perpetuation of races having special adaptations
to particular climatic conditions would be found, we have reason to believe,
among the feral descendants of the domesticated quadrupeds first introduced
into South America by the Spaniards; and we anticipate that much novel
information on the subject will be made public in Mr. Darwin's more
detailed treatise. But the following instance, recorded by M. Roulin, seems
to us to be so peculiarly apposite to the present inquiry as to deserve
special mention. In some of the hottest provinces of South America, a
race of oxen has spontaneously sprung up, distinguished by the peculiar
clothing of its hide, which consists of a fine but extremely scanty fur. This
race is incapable of maintaining itself elsewhere, the "pelones" (as these
oxen are termed) being too delicate in constitution to bear the cold of
the Cordilleras, to which the cattle are driven for the provision of the
towns situated upon them; and the breed is therefore not encouraged by
human agency. On the other hand, when oxen of other breeds are driven
into the provinces inhabited by these "pelones," they either speedily die
out, or they become gradually and with difficulty acclimatised. In the same
hot provinces another curious variety of oxen presents itself, characterised
by the entire absence of hair; these naked-skinned oxen, termed "calougos,"
are even more delicate in constitution than the "pelones," being utterly
unable to bear a climate colder than their own. Here, then, we have the
spontaneous establishment, within a limited area, of breeds of oxen distinguished by peculiarities quite striking enough to be elsewhere accounted
characteristic of different species. These peculiarities, whilst such as adapt
them to a set of circumstances which are highly unfavourable to the continued existence of the type from which they sprang, on the other hand
render them incapable of maintaining their ground under conditions which
are eminently favourable to their ancestral race. And it seems impossible
to account for the phenomenon upon any other principle than that of
natural selection, as advocated by Mr. Darwin and Mr. Wallace.

This case further illustrates the marked difference in the conditions and
results of *natural* and *artificial* selection; a difference which prevents the
deduction as to the permanence of species, which has been drawn from
the instability of the peculiarities impressed on domesticated races by the
agency of man, from being in the least applicable to divergent forms that
have developed themselves in conformity with the peculiarities of their natural conditions. As Mr. Darwin has justly remarked, "man selects only
for his own food, nature for that of the being which she tends." Many
of the qualities which are most valued in the state of domesticity, are

such as would render the animal utterly unfit to maintain its ground in the struggle for life when left to its own resources. The sleek-coated London carriage-horse, and the stall-fed dairy-cow that gives her twenty or even thirty quarts of milk per day, can only be kept up to such an artificial standard of perfection by a treatment that seriously impairs their constitutional stamina; so that they succumb to depressing influences which would have no effect upon hardy Dartmoor ponies or vigorous Ayrshire cattle: just as the massive bulk and vast strength of the brewer's drayman are often suddenly laid low by the results of an injury which in a truly healthy countryman would be too slight to require attention. Our breeds of quickly-fattening pigs, short-legged sheep, poodle-dogs, pouter-pigeons, and the like, never could have established themselves in a state of nature, because the very first step toward such inferior forms would have led to the rapid extinction of the race; still less could they now maintain a competition with their wild allies. The great speed but slight endurance of the race-horse, the unwieldy strength of the ploughman's team, would both be useless in a state of nature; so that, if turned wild upon the pampas, such animals would either soon become extinct, or, if placed in circumstances sufficiently favourable for their maintenance, would lose in successive generations those extreme qualities that are rather detrimental than beneficial to them, and would revert to that common type in which various powers and faculties are so proportioned to each other as to be best adapted to procure food and secure safety. Thus, then, as Mr. Wallace justly remarks, "domestic varieties, when turned wild, *must* return to something near the type of the original wild stock, or *become altogether extinct.*"

It will not be only on the more important features of the organisation, that Natural Selection will exert its modifying influence; it will often affect characters which we are accustomed to regard as trivial—for instance, *colour*. Among insects and birds we may trace a marked relation between the hue of different tribes and the tints of the spots they respectively frequent, which is obviously a means of passive defense to them against their enemies. Thus leaf-eating insects are green, and bark-feeder mottled gray; the larvae of the Phasmidae (or spectre-insects) can scarcely be distinguished in appearance from the dead stick on which they are commonly found; and the Mantis derives its common name of "walking-leaf" from a resemblance to that object, so close as to deceive any but a near inspection. So, again, the red grouse has very much the colour of heather; the black grouse, that of peaty earth; and the ptarmigan in winter that of snow. These birds would multiply very rapidly if they were not kept under by

birds of prey, which are known to be guided chiefly by eyesight; and any
deviation from the hue most favourable to escape from their notice would
almost certainly ensure the early destruction of the individual that presented
it: so that as complete and constant a uniformity would be maintained
by natural agencies, as the breeder of white sheep maintains among his
flock by the jealous destruction of every lamb that exhibits the faintest
trace of black. So among plants, although the down on the fruit, and
the colour of its flesh, are considered by botanists as characters of the
most trifling importance, yet there is good evidence that in the United
States smooth-skinned fruits suffer far more than those with down from
the attacks of the curculio-beetle; that purple plums are more liable than
yellow plums to a particular disease; whilst another disease attacks yellow-
fleshed peaches far more than those with other-coloured flesh. If, with
all the aids of art, such slight diversities make a great difference in the
success with which the several varieties can be cultivated, assuredly in a
state of nature, where the trees would have to struggle with other trees
and with a host of enemies, such differences would effectually settle which
variety, whether a smooth or a downy, a yellow or a purple-fleshed fruit,
should continue to florish.

The modifying tendency of Natural Selection, then, will be twofold:
on the one hand, to the origination and maintenance of races diverging
in various directions and degrees from the original type; on the other,
to the extinction of all such as are overmastered in the struggle for existence
by the greater energy or more perfect adaptation of other races. And we
think Mr. Darwin quite justified in the conclusion, that "whatever the
cause may be of each slight difference in the offspring from their parents,–and a cause for each must exist,–it is the steady accumulation, through
Natural Selection, of such differences, when beneficial to the individual,
that gives rise to all the more important modifications of structure, by
which the innumerable beings on the face of this earth are enabled to
struggle with each other, and the best adapted to survive." And we fully
agree with him, that *individual* differences, though hitherto accounted as
of small interest to the systematist, are of high importance in any philosophical inquiry into the origin of species, as being the first step towards those
slighter varieties which are barely thought worth recording in works on
natural history. So varieties which are in any degree more distinct and
permanent, are steps in a regular gradation that leads through more
strongly-marked and more permanent varieties to sub-species, and thence
to species. The permanence of each race will thus depend on the perma-

nence of the conditions in which it is placed. So long as these remain
unchanged, the adapted form that has been once established as the best
will continue to hold the mastery; and all aberrations from it that unfit
the subject of them for maintaining its ground in the battle for life will
be borne down in the *mêlée*. But let a change take place in any of the
conditions, however trivial they may appear, that either affect the organism
directly (as is the case with temperature, hygrometric state, pressure of
the air or water), or do so by an alteration in its relation to other organisms
(as by affecting its supply of food, or its means of obtaining it, or by
subjecting it to attacks which require increased means of resistance or escape), the race must either be capable of adapting itself to that change,
or it must succumb. In the one case, the original form will give place
to some modification directly proceeding from it by genetic descent; in
the other, it will be superseded by some rival form derived from a different
ancestry, which presses in and occupies its place; just as we see, in the
social battles of life, that the families of our older aristocracy hold their
ground, or are displaced by the *parvenus* whom they regard as their natural
enemies, in proportion as they either adapt themselves to the spirit of the
age and take advantage of its requirements, or as they hold tenaciously
to their time-honoured customs, and refuse to profit by any thing that
shall lower them in their own artificial scale of dignity.

We are disposed to believe, then, that Mr. Darwin and Mr. Wallace
have assigned a *vera causa* for that diversification of original types of structure which has brought into existence vast multitudes of species, sub-species,
and varieties, referable to the same generic forms; and we think that the
weight of evidence is decidedly in favour of such an extension of this
doctrine from the present to the past, as will enable us to account for
the modification of specific types which is presented to us as we pass from
one geological formation to another. It will probably meet with less opposition among British than among continental palaeontologists; for with the
former it has come to be generally admitted that many species really do
range through a long succession of formations; whilst with some of the
latter it is sufficient for shells or corals to present themselves in two different
strata, to establish their specific diversity, however close may be their structural conformity. Are such the men, we would ask, whose opinion should
have any weight in a discussion like the present? They are of the same
class with the botanists who make new species of our commonest plants,
because they find them in remote parts of the globe; and who affirm that
an Australian or a New-Zealand species *cannot* be identical with a Euro-

pean, for no other reason, that we can discover, than that they have established in their own minds as a law of nature that the species of the northern and southern hemispheres *must* be dissimilar.

Now the tendency of all modern geological inquiry has been, as has been well stated by Professor Powell,[3] to substitute the idea of *continuous* for that of *interrupted* succession, of physical change; in breaking up large divisions into smaller; in obliterating sharp lines of demarcation by subordinate gradations; in tracing intermediate deposits in one locality which fill up breaks in another; and in thus connecting more and more closely with each other the successive members of the series of stratified deposits. And though we may be as yet very far from realising this is in all instances, yet no one who has followed the progress of geological research for the last quarter of a century, can refuse to admit that the general doctrines originally enunciated by Sir Charles Lyell on this subject have been progressively acquiring a firmer basis and a more extended application. The doctrine of continuous succession, once in physical geology, seems almost to necessitate the admission of the like doctrine as a corollary in regard to the succession of organic life; and such a deduction is fully borne out by our present knowledge of the relation of our existing fauna and flora to that of the later tertiary and post-tertiary epochs. "Throughout all the most recent formations," again to use the language of Professor Powell, "we find a continuous series of allied species, and a succession of organised structures, in a chain absolutely unbroken, and marked only by the minutest specific differences in its successive links, down to forms now existing; and as this is carried backwards through countless ages, by degrees we find fewer features of the present and more of the past, and even come to whole genera and orders of extinct races coexisting with some which have survived them." And through at intervals in the course of this series of close and continual connection, there are real or apparent interruptions of greater or less magnitude, in which the immediate affinity seems broken off between the species characterizing one formation and those most nearly allied to them in the next, yet the only fair inference from such a fact would be that no fossiliferous deposits took place *in that locality* during a long series of ages, through which a progressive change of organic forms was going on elsewhere, that marked itself in the entire change of specific and generic types presenting themselves in the next deposit in the first locality. A geologist who had formed his notions of the succession of strata

3. *Unity of Worlds* p. 335.

only from the study of those wide areas in the continent of America over which the palaeozoic are immediately overlaid by the tertiary formations, would be justly rebuked by his European brother for ignoring the whole series of secondary strata, and the varied forms of life which they contain; yet there are still geologists in this country who have the presumption to affirm, that because the continuity both of stratification and of organic life seems to have been completely interrupted at the end of the palaeozoic period in the limited areas hitherto explored, such interruptions must have prevailed universally over that vast proportion of the earth's surface of whose geological history we know absolutely nothing. As regards the supposed break between the latest secondary and the tertiary strata, the tendency of recent inquiries has most unequivocally been to remove the difficulties which have been urged on the strength of it; the study of the strata which intervene between the chalk and the nummulitic limestone of the south of Europe having revealed an unexpected continuity alike in stratification and in the succession of organic forms. And the more carefully the cretaceous, and even the oolitic, as well as the early tertiary fossils are compared with existing types, the more numerous are the instances that are found to present themselves, in which their conformity is so close as fully to sanction the idea of the continuous descent of the latter from the former, with more or less of intervening modification.

Of such modifications, occurring under circumstances which permit both their source and their continuity to be traced out, we may cite a characteristic example in the changes presented by certain univalve shells that occur in three ·successive beds in the tertiary formations of the Island of Cos, as described by the late Professor E. Forbes. The genera in question (*Paludina* and *Neritina*) are remarkable for their power of sustaining considerable alterations in the nature of the medium they inhabit: for although properly fresh-water mollusks, they are not limited to lakes and rivers, but are often found in estuaries, in which they are either subjected to alternations of salt water with fresh or live in water which is pretty constantly brackish. Now the lowest of the beds just referred to is obviously of purely freshwater formation; for there are embedded in it, with Paludinae and Neritinae of their ordinary smooth and unwrinkled type, various species of shells that could not have inhabited any other medium— amongst others, pulmoniferous water-snails. In the second bed, however, these last, with other forms most rigidly limited to fresh water, disappear; whilst the Paludinae and Neritinae have their shells belted by a strong fold or corrugation, such as is presented by existing shells of the same tribes inhabiting waters with a slight admixture of brine. And in the third all the fresh-water forms are wanting, save such as are known to be capable of living in brackish estuaries, and are replaced by marine shells, which have a like power of partial adaptation; and the shells of the Paludinae and Neritinae are deeply furrowed and surrounded by strong spiral ridges. The subsequent formations, which overlie these unconformably, are undoubtedly of purely marine origin. There seems no room for doubt, then, that the changes in the circumstances under which these successive beds were formed, were such as intermixed progressively increasing proportions of salt water with the stream of fresh water by which the materials of the lowest bed were deposited; and that by this increase was effected such a gradual change of type in the shells of the Paludinae and Neritinae, from the forms of the first to those of the third bed, as would, if the intermediate link had not been presented to us in the second bed, or the parallel modifications occurring in these genera at the present time had been unknown, have been held by the conchologist fully to justify him in referring those forms to different species.

From such a case as this, the transition is easy to that presented by the succession of deposits constituting the great chalk formation; through the whole of which there is a strong general resemblance in the organic remains, though the species are for the most part distinct in each stage. As the accumulation of each deposit has often been interrupted, and as long blank intervals have doubtless intervened between successive deposits, we have no right to expect to find in any one or two of them all the intermediate varieties between the species which appear at the commencement and at the close of these periods; but we ought to find after intervals, very long as measured by years, but only moderately long as measured geologically, closely-allied forms, or, as some authors term them, representative species. Now as this is just what we do find, the theory of descent, with modification, is so far conformable to positive facts, that it must be admitted to have at least as valid a foundation in a broad basis of phenomena as the theory of successive creations.

That we should meet with a similar gradational transition in all other cases, is assuredly what we have no right to expect, if we bear in mind *the extreme imperfection of the Geological Record,*—a consideration on which Mr. Darwin dwells very strongly, but not, in our estimation, one whit too strongly. "We are not only ignorant," he pithily says, "but we do not know how ignorant we are." To our minds the great wonder is, that palaeontological research should have already yielded so much information as to the past life of the globe, not that it should afford so little. The indications recently afforded in regard to the antiquity of the human race,[4] taken in connection with the progress of discovery of the air-breathing forms of vertebrata in the earlier formations,[5] teach a valuable lesson of caution in drawing inferences as to the *non-existence* of any particular type at any period whatever, from the mere *negative* fact that we have not hitherto met with its remains.

A considerable part of Mr. Darwin's treatise is occupied by a discussion of the principal scientific objections (he wisely refrains from taking notice of any others) that can be urged in opposition to his views. Having already noticed by anticipation the geological difficulty, we shall only say, that we think he has conclusively shown that no value whatever can be attached to the "breeding test," on which reliance is commonly placed as a means of discriminating species from varieties; and that the facts of geographical distribution are, when rightly viewed, rather in his favour than otherwise. A greater difficulty than either seems to us to be presented by those cases of extraordinary aberration, whether of structure or habit, whereby particular animals are distinguished from their kind; many of which it is difficult to imagine to have been acquired gradually by any process of consecutive modification.

The history of every science shows that the great epochs of its progress

4. We refer, of course, to the satisfactory evidence lately obtained by Mr. Prestwich, Sir. C. Lyell, and other geologists of the highest authority, as to the existence, in gravel-beds elevated a hundred feet above the level of the Somme, of large numbers of flint implements, obviously shaped by the hand of man, in association with the bones of large mammals now extinct. Notwithstanding the ingenious theories which have been invented, either to account for their production by the forces of nature rather than by human art, or, admitting them to be man's handiwork, to account for their presence in these gravel-beds on the hypothesis of the modern origin of the human race, we take upon ourselves to affirm, that no unprejudiced person can carefully examine a large series of these objects without coming to recognize them as the products of a rude handicraft directed by a definite purpose; and further, that it is an inevitable deduction from the circumstances under which they are found, that, whether or not the beings that made them were contemporaneous with the Mammoth, the Tichorhine Rhinoceros, and other great extinct mammals, with whose bones they arē associated, the gravel-beds containing them must have been first covered with layers of marl, clay, and sand, in some places forty feet thick, whose slowness of deposit is attested by the perfect preservation of the delicate land-shells they contain, and must have been afterwards upheaved at least a hundred feet; whilst, subsequently to this upheaval, the present valley of the Somme must have been excavated by its stream through the elevated land which now forms its high banks.

5. At the last meeting of the Geological Society, the discovery was announced by Dr. Dawson, of Canada, of remains of *six reptiles* in the trunk of one fossil tree in the celebrated section of Carboniferous strata at the "Joggins'" in Nova Scotia.

are those not so much of new discoveries of *facts,* as of those new *ideas* which have served for the colligation of facts previously known into general principles, and which have thenceforward given a new direction to inquiry. It is in this point of view that we attach the highest value to Mr. Darwin's work. Naturalists have gone on quite long enough on the doctrine of the "permanence of species." Their catalogues are becoming more and more encumbered with these hypothetical "distinct creations." And the difficulty of distinguishing between true species and varieties increases, instead of diminishing, with the extension of their researches. The doctrine of progressive modification by Natural Selection propounded by Mr. Darwin, will give a new direction to inquiry into the real genetic relationship of species, existing and extinct; and it has a claim to respectful consideration, not merely on account of the high value of Mr. Darwin's previous contributions to zoological science, and the thoroughly philosophical spirit in which it is put forth, but also because it brings into mutual reconciliation the antagonistic doctrines of two great schools—that of Unity of Type, as put forward by Geoffroy St. Hilaire and his followers of the Morphological School, and that of Adaptation to Conditions of Existence, which has been the leading principle of Cuvier and the Teleologists. Nor is it the least of its recommendations that it enables us to look at the War of Nature constantly going on around us as not marked only by suffering and death, but as inevitably tending towards the progressive exaltation of the races engaged in it; just as, in the world of mind, it is only by intellectual collision that Truth can become firmly established, and only by moral conflict, whether in the individual or in society, that Right can obtain an undisputed sway.

Comments on Carpenter

W. B. Carpenter studied medicine at the universities of Bristol, London, and Edinburgh, eventually becoming Fullerian professor of physiology at the Royal Institution of London in 1844. From 1847 to 1852 he was editor of the *British and Foreign Medico-Chirurgical Review.* His continuing influence on the editorial policy of this journal may well explain the extensive coverage it gave to the *Origin of Species* and subsequent writings on evolutionary theory. In 1856 Carpenter became registrar of the Univeristy of London, a post he held until 1879. During this period he continued his scientific researches. His publications tended to be either strictly descriptive or else broadly "theoretical." He published two reviews of the

Origin, one in the *National Review* (1860a), the other in the *British and Foreign Medico-Chirurgical Review* (1860b).

Darwin was pleased to have so prominent a physiologist come out in favor of evolution, but he was a little disappointed that Carpenter did not make greater use of his knowledge of physiology in defending evolutionary theory against its critics. As Carpenter's correct use of such terms as "deducible," "vera causa," and "colligation of facts" indicates, he had read some philosophy of science. He was especially impressed by Baden Powell's *Essays on the Spirit of the Inductive Philosophy, the Unity of Worlds, and the Philosophy of Creation* (1855). He agreed with Powell—as did almost every other author in the present anthology—that God's power was revealed to be more magnificent in creation by natural laws than by periodic, miraculous interventions. Like Powell, Carpenter could see nothing "abstractly impossible" about evolution, even in the essentially unscientific formulations set out in Chambers' *Vestiges of Creation* (1845). Theologians had adapted to a heliocentric universe, the nebular hypothesis, and the facts of paleontology. In good time they would reconcile themselves to the evolution of species.

Carpenter saw that the major stumbling block in the path of evolutionary theory was the explanation of how mind and soul could have evolved. Man could think and feel moral compunctions. How could such a thinking and moralizing being evolve from ancestors who could do neither? Carpenter attempted to answer this question, first, in his presidential address to the British Association for the Advancement of Science (1872a) and then in a paper entitled "On Mind and Will in Nature" (1872b). He thought he could resolve the conflict by means of the notion of force as a mediating concept between mind and matter. He felt that force (not to mention mind) could be conceived as existing independently of matter.

In the preceding review, Carpenter touches upon the problem which was one of the most frustrating for Darwin—the difficulty of finding criteria for the identification of species. Few naturalists doubted that at least some domestic varieties had been obtained by selection from wild varieties. The question was whether these productions and their ancestors were truly species or only varieties. Were they natural species or naturalists' species? In the traditional view, the former were real, immutable, and specially created. It was the task of the naturalist to make the species which he recognized coincide with those in nature. To accomplish this end, species criteria were necessary.

The common view among less serious naturalists was that natural species could always be distinguished both by morphological gaps and by sterility barriers. More serious naturalists, however, realized that these two criteria did not always co-vary. "The degrees of sterility do not coincide strictly with the degrees of difference between the parents in external structure or habits of life" (*Darwin,* 1871, p. 172). When these two criteria conflicted, biologists were at odds with each other about what to do. Some, like Huxley, gave precedence to physiological species. Others, like Carpenter, preferred the morphological criterion. In either case Darwin was anxious to show that species were not sharply defined as one would expect if they were specially created. To do this, he had to show that gradual variation occurred with respect to both morphological similarity and interfertility.

One of the most significant features of Carpenter's review is his statement that "no species can be fairly admitted as having a real existence in nature, until its *range of variation* has been determined both *over space* and *through time.*" He even goes so far as to use the term "population" in this connection. Surprisingly, even an essentialist like Louis Agassiz agreed with him, but for opposite reasons. Carpenter thought that variation over time and space had to be studied because it was of the essence of species to vary. Among contemporaneous species, gaps might well exist between certain species and their nearest neighbors, but since speciation is going on continuously, at any one time there must be a certain percentage of species in the midst of speciating—more than varieties but not quite species. And when species variation was followed in time, it was necessarily continuous. In spite of the evidence of variation in nature, Agassiz believed that morphological gaps could always be discovered among contemporaneous species. The extensive sampling which he advocated was intended to distinguish between the variable, inessential traits and the constant essential traits. Agassiz did not have to worry about continuous variation in time, since he believed that all species were created specially, albeit serially.

Darwin thought that by showing that the size of the morphological gaps between extant species varied from large differences to none at all, he could lend some plausibility to the claim that species could also vary morphologically in time. He wished to do the same for the sterility criterion. Early in his career, in his "Second Notebook," Darwin can be found saying, "As *species* is a real thing with regard to contemporaries—fertility must settle it" (DeBeer, 1960, p. 99). Later he argued: "With forms which must be ranked as undoubted species, a perfect series exists from those which are absolutely sterile, when crossed, to those which are almost completely fertile" (Darwin, 1871, p. 171).

In emphasizing that both the size of the morphological gaps and the degrees of sterility which existed between species varied continuously, Darwin did not wish to argue that these two species criteria were irrelevant in deciding species status; only that they were not infallible, and hence that species were not absolutely distinct. In later life, Darwin seemed to give greater weight to the criterion of constancy of character than to degree of sterility for species recognition. For example, he said that the mutual fertility of the various races of mankind had not been proved and "even if proved would not be an absolute proof of their specific identity." The strongest argument for the conspecificity of man was "that they graduate into each other, independently in many cases, as far as we can judge, of their having intercrossed" (Darwin, 1871, p. 174).

Vacillation over what criterion actually distingushed species was both a hindrance and a help in Darwin's attempt to establish the existence of evolution. It was a help in that it showed species were not unproblematic, clear-cut natural units as special creation implied they should be. It was a hindrance in that any attempt to show that one species had evolved into another could be put in jeopardy by raising doubts as to the status of the ancestral forms. The frustrations which Darwin felt over this vacillation was understandable, but under the circumstances, ambiguity over the nature of species could not be avoided, since the term "species" was the key theoretical term in the conceptual revolution which Darwin had begun. It was being transferred from one scientific theory to another, from natural theology and a metaphysics of essences to evolutionary theory. Even after evolutionary theory had been accepted, the meaning of the term "species" was to continue to change as evolutionary theory itself evolved. One of the consequences of this growth has been the increasing role which reproductive isolation plays in both genetics and population biology, until now it is the chief criterion for species status according to the advocates of the synthetic theory of evolution.*

* See also Carpenter (1839 and 1862), Oppenheimer (1959), Ghiselin (1969), and Vorzimmer (1970).

H. G. Bronn (1800–1862)

I have had this morning a letter from old Bronn (who, to my astonishment, seems slightly staggered by Natural Selection), and he says a publisher in Stuttgart is willing to publish a translation, and that he, Bronn, will to a certain extent superintend. Have you written to Kölliker? If not perhaps I had better close with this proposal—what do you think?— C. Darwin to T. H. Huxley, Down, February 2, 1860 (*More Letters*, 1:139).

I thank you sincerely for your most kind letter; I feared that you would much disapprove of the 'Origin,' and I sent it to you merely as a mark of my sincere respect. I shall read with much interest your work on the productions of Islands whenever I receive it. I thank you cordially for the notice in the 'Neues Jahrbuch für Mineralogie,' and still more for speaking to Schweitzerbart about a translation, for I am most anxious that the great and intellectual German people should know something about my book.

I have told my publisher to send immediately a copy of the *new* (second) edition to Schweitzerbart, and I have written to Schweitzerbart that I gave up all right to profit for myself, so that I hope a translation will appear. I fear that the book will be difficult to translate, and if you could advise Schweitzerbart about a *good* translator, it would be of very great service. Still more, if you would run your eye over the more difficult parts of the translation; but this is too great a favour to expect. I feel sure that it will be difficult to translate, from being so much condensed.— C. Darwin to H. G. Bronn, Down, February 4, 1860 (*Life and Letters*, 2:71–72).

On my return home, after an absence of some time, I found the translation of the third part of the 'Origin,' and I have been delighted to see a final chapter of criticisms by yourself. I have read the first few paragraphs and final paragraph, and am perfectly contented, indeed more than contented, with the generous and candid spirit with which you have considered my views.—C. Darwin to H. G. Bronn, Down, July 14, 1860 (*Life and Letters*, 2:73)

I ought to apologise for troubling you, but I have at last carefully read your excellent criticisms on my book. I agree with much of them, and wholly with your final sentence. The objections and difficulties which may be urged against my view are indeed heavy enough almost to break my back, but it is not yet broken! You put very well and very fairly that I can in no one instance explain the course of modification in any particular instance. I could make some sort of answer to your case of the two rats; and might I not turn round and ask him who believes in the separate creation of each species, why one rat has a longer tail or shorter ears than another? I presume that most people would say that these characters were of some use, or stood in some connection with other parts; and if so, Natural Selection would act on them. But as you put the case, it tells well against me. You argue most justly against my question, whether the many species were created as eggs or as mature, etc. I certainly had no right to ask that question. I fully agree that there might have been as well a hundred thousand creations as eight or ten, or only one. But then, on the view of eight or ten creations (*i.e.* as many as there are distinct types of structure) we can on my view understand the homological and embryological resemblance of all the organisms of each type, and on this ground almost alone I disbelieve in the innumerable acts of creation. There are only two points on which I think you have misunderstood me. I refer only to one Glacial period as affecting the distribution of organic beings; I did not wish even to allude to the doubtful evidence of glacial action in the Permian and Carboniferous periods. Secondly, I do not believe that the process of development has always been carried on at the same rate in all different parts of the world. Australia is opposed to such a belief. The nearly contemporaneous equal development in past periods I attribute to the slow migration of the higher and more dominant forms over the whole world, and not to independent acts of development in different parts. Lastly, permit me to add that I cannot see the force of your objection, that nothing is effected until the origin of life is explained: surely it is worth while to attempt to follow out the action of electricity, though we know not what electricity is.—C. Darwin to H. G. Bronn, Down, October 5, 1860 (*More Letters*, 1:172–173)

I send the translation of Bronn,* the first part of the chapter with generalities and praise is not translated. There are some good hits. He makes an apparently, and in part truly, telling case against me, says that I cannot explain why one rat has a longer tail and another longer ears, etc. But he seems to muddle in assuming that these parts did not all vary together, or one part so insensibly before the other, as to be in fact contemporaneous. I might ask the creationist whether he thinks these differences in the two rats of any use, or as standing in some relation from laws of growth; and if he admits this, selection might come into play. He who thinks that

* [A manuscript translation of Bronn's objections had been made into English.]

God created animals unlike for mere sport or variety, as man fashions his clothes, will not admit any force in my *argumentum ad hominem.*

Bronn blunders about my supposing several Glacial periods, whether or not such ever did occur.

He blunders about my supposing that development goes on at the same rate in all parts of the world. I presume that he has misunderstood this from the supposed migration into all regions of the more dominant forms.—C. Darwin to C. Lyell, 15 Marine Parade, Eastbourne, October 8, 1860 (*Life and Letters,* 2:139)

I thank you for your extremely kind letter. I cannot express too strongly my satisfaction that you have undertaken the revision of the new edition, and I feel the honour which you have conferred on me. I fear that you will find the labour considerable, not only on account of the additions, but I suspect that Bronn's translation is very defective, at least I have heard complaints on this head from quite a large number of persons. It would be a great gratification to me to know that the translation was a really good one, such as I have no doubt you will produce. According to our English practice, you will be fully justified in entirely omitting Bronn's Appendix, and I shall be very glad of its omission.—C. Darwin to Victor Carus, Down, November 10, 1866 (*Life and Letters,* 2:232–233)

If I think continuously on some half-dozen structures of which we can at present see no use, I can persuade myself that Natural Selection is of quite subordinate importance. On the other hand, when I reflect on the innumerable structures, especially in plants, which twenty years ago would have been called simply morphological and useless, and which are now known to be highly important, I can persuade myself that every structure may have been developed through Natural Selection. It is really curious how many out of a list of structures which Bronn enumerated, as not possibly due to Natural Selection because of no functional importance, can now be shown to be highly important.—C. Darwin to T. H. Huxley, Down, May 11, 1880 (*More Letters,* 1:387)

Review of the *Origin of Species**

[H. G. BRONN]

The fundamental ideas set out in this publication are causing even a greater stir in the scientific world than when they were first developed somewhat less extensively by Lyell in his *Principles;* however, since there is no way to provide either irrefutable proof or decisive disproof of this later theory, as there was for Lyell's, there still remains some doubt whether it is factually true.

* From *Neues Jahrbuch für Mineralogie* (1860), pp. 112–116. Translation by D. L. H.

Species can vary. This is common knowledge! Differences in nutrition, environment, climate, in addition to many currently unknown factors produce varieties.[1] However, the most productive and the commonest cause of the formation of varieties is the *Wahl der Lebens-Weise* (natural selection). That is to say, since animals and plants propagate themselves much too abundantly, a large part of the next generation is frequently forced to seek other sources of food and to choose a different way of life. This change in life-style in turn necessitates and gradually produces varying uses of the organs, varying functions and varying structures, so that when the same external conditions persist from generation to generation, permanent races develop. These races in turn pass on their newly developed characters even under changed conditions, until it is frequently impossible to distinguish species from varieties. It is well known how seldom descriptive botanists and zoologists agree among themselves in cases of this sort. These newly formed permanent varieties or races are all very prolific and often much more inclined to vary than their ancestors. Cultivated plants and domesticated animals demonstrate how much variation from the original type can occur in a short time. Inasmuch as man carefully selects those individuals which best serve his purpose of deviating most from the original type, he achieves in the comparatively short time of a dozen or a hundred years success so extraordinary that it could not be matched by nature in a period of ten or even a hundred times that long. It does show what can be done given the time. However, if we admit that, in hundreds or thousands of years, fortuitously appearing differences can bring about permanent races in this way and that these in turn develop into species, then we are forced to admit that a hundred thousand years is all that is needed to generate diverse tribes out of these species and a few million years to produce new orders and classes from these. If we have no dearth of time, then no objections can be raised to all of this, even though there may be considerable difficulty in producing explanations for individual cases and special occurrences. Of course, it is also possible that it was decided to propagate original plant and animal forms over the entire face of the earth. Changes in the temperature and the surface of the earth, ice ages and the like, have encouraged these forms to search for alternative ways of life in all directions and to utilize various land bridges and sea channels which may have opened up and then closed down again from time to

1. In our *History of Nature* a host of such cases are collected and traced back to their possible causes, as, for example, the consequences of interspecific hybridization; but the results are the same even if they are multiplied by 100,000,000.

time. On this view Darwin believes that all animal forms can be traced back to 4 or 5 *Stamm-Individuen* (progenitors) and all plant forms to the same number or maybe fewer. Indeed, it is even possible that all plants and animals originate from only a single prototype! This then is the author's line of reasoning.

As we have said above, proof or disproof is not immediately available. It is impossible to prove either that varieties, in the commonly accepted sense of the word, are limited and do not transgress certain boundaries or that they are unlimited. Since the latter is the positive formulation, proof of a sufficiently conclusive kind can be provided only for it, and to do that a century would have to be devoted to a series of systematic experiments. In the meantime, however, natural scientists will align themselves into two camps—the believers and the non-believers.

The author seems to have the same doubts over unlimited variability as he has over the previously mentioned assertion about the number of basic types. But there are really only two possibilities. Either his theory is false (and cannot be extended beyond the field of common varieties) or else it is true. If it is true, then variability is unlimited, which means that the organic world was not created, which means that the natural force through which the organic world originated has been found and the assumption of Creation is unnecessary. If there are 10, 5, 3 or even 2 different prototypes of plants and animals, then Creation is necessary. Otherwise, the only origin for the organic world would have to be something like Priestley's green matter, a substance which is not representative of any organic species. Since the author has already accepted so many bold ideas, why does he draw the line at this one? At the beginning of the preceding year, the French Academy (as did the Vienna Academy in earlier years) actually addressed itself to the question of whether lower organisms, plants and animals can arise out of water which contains organic matter but which has been boiled for a long time to destroy all life and then hermetically sealed and kept that way. A whole series of experiments were performed which seemed to affirm this possibility. Afterwards there were more forms of lower organisms in the water than had been identified in it before. In spite of all the precautionary measures which had been taken, all those present at the Academy and many others connected with it, the presiding jury of natural history and physiology, declared that there was still the possibility that the seeds of these organisms had not been completely destroyed by the searing heat of the boiling water. Even though we agree with this conclusion, we must admit that these experiments which had been executed with such extreme care are not completely invalidated by this possibility. Rather it should motivate us to repeat these experiments, this time avoiding all those factors which had been objected to before. If Priestley's or some such organic matter could be generated from inorganic matter and if a faultless proof could be provided to show that organic species can arise in the manner suggested by Darwin, then his theory would receive the strongest possible support in the shortest possible time. But as long as neither possibility is confirmed, we still need a creative force. It matters little to our exposition or to science in general whether the Personal Creator placed 200,000 or 10 species of plants and animals on the earth or whether man alone is specially created.

The recognition of a few dozen species of plants and invertebrate animals which stretch from the lower Silurian to the Cretaceous period leads us to suspect that the number of forms of Protozoa, Actinozoa, Malacazoa and Entomozoa was much larger at that time than our present knowledge would lead us to believe. Thus, if Darwin wanted organic life to start at this time, he would have to postulate many more basic types than he designated above. But in this matter, he aligns himself with Lyell, who believes that under no condition is the Silurian the oldest geological deposit [*neptunischen Gesteine*] but that a whole series of geological strata lying below were transformed into crystalline forms by certain metamorphic processes. That Lyell accepts an endless process of change of this kind is a well-known fact. Thus, we are amazed to discover that Lyell opposes those geologists who believe in a progressive development of the inorganic world with Darwin's writings. Of course, the means of progression which Darwin advocates is very different from that accepted up until now. In the accepted view, increasingly more perfect and higher species, genera and other forms are specially created in the continuous struggle to adjust to the external conditions of existence, whereas Darwin would have them emerge from the old. To the extent that one is inclined toward Darwin's hypothesis, then to that extent one is committed inevitably to the acceptance of progressive development and, therefore, to a *first creation!*

As might be expected of a work by Darwin, this one is filled with the most attractive ideas set out under the constant appeal to observation and experience. It is especially instructive for those who are not disposed to accept the author's theory at once. It is the fruit of twenty-years' labor on this question, although on the whole it presents only his conclusions, since the elaboration of all the individual observations and facts accumulated for this purpose would fill a voluminous work. The author is currently occupied with precisely this task. Its completion, however, will be delayed as much by the author's poor health as by the continuing accumulation of new material. The theory itself, however, is not new. It was proposed by Lamarck in his *Philosophie zoologique*, Geoffroy St. Hilaire and others, but it appears to be stated by Darwin with all the intelligence and knowledge which only the present state of science affords the talented investigator.

We would like to reiterate our own conviction with these words: make organic matter with a cellular structure out of inorganic matter, proceed then to create seeds and eggs of lower organic species out of this organic matter—a task which modern science is equal to if it is at all possible. Then with the additional help of Darwin's theory, we can conceive of a natural force which might have brought forth all organic species. Thus, we will no longer be forced to seek recourse in personal acts of creation which fall outside the scope of natural law.[2] Once we are in possession of this advantage, we need no longer doubt as before in the possibility of later discoveries gradually filling in the enormous gaps which now confront us in the series of plant and animal forms and impede our complete consent. However, as long as this is not possible, Darwin's theory is as improbable as ever, since it brings us no closer to the solution of the great problem of creation. Though it is conceivable, it still remains undecided whether all organisms from the simplest filament to those that are intricately constructed like the butterfly, snake, horse, etc. can be the production of just blind natural forces!

Comments on Bronn

H. G. Bronn (1800–1862) was a respected German zoologist, professor of geology and palaeontology at Freiburg and later at Heidelberg. In 1859 he published a defense of Special Creationism entitled "On the laws of evolution of the organic world during the formation of the crust of the earth" (Bronn, 1859). Needless to say, "evolution" had not taken on Darwinian connotations. Rather, it referred to progressive, successive special creations. Darwin would have preferred that Albrecht von Kölliker (1817–1905) supervise the German edition of the *Origin*, but he had to settle for Bronn. As the quotations which preceded this selection indicate, Darwin eventually found Bronn's translation unsatisfactory, not because of Bronn's appendix, but because the translation itself was clumsy and

2. The inconsistency of such assumptions as contrasted to the impossibility of another alternative has been shown in our "Investigations into the Developmental Laws of the Organic World" [1858], pp. 77ff and 227ff.

because Bronn had on occasion omitted passages that he took to be offensive—for example, "Light will be thrown on the origin of man and his history." Although Kölliker might have produced a truer translation than Bronn did, he would have been even less likely to endorse Darwin's ideas (see Kölliker, 1861, 1864). In 1866 Victor Carus (1823–1903) translated the fourth edition of the *Origin* and translated all of Darwin's subsequent works into German.

Many commentators noted Darwin's pentateuchal wording in his discussion of the possible first origin, or origins, of life, but few stressed the other alternative—the spontaneous generation of living creatures from inorganic matter. Bronn emphasized the role of spontaneous generation in evolution both to cast doubt on Darwin's theory in general and to point up its materialistic tendencies. Was life created by a supernatural force? Or was it generated spontaneously, like Priestley's green matter? Joseph Priestley (1733–1804) had discovered oxygen but had interpreted its action in the context of the phlogiston theory. According to this theory, phlogiston was a subtle fluid which was exhausted in animal respiration and released as heat in combustion. Priestley discovered that the green substance often found in ponds and stagnant water restored "vitiated" air and made it suitable again for respiration and combustion. Priestley was a gross materialist, going so far as to propose a materialist theory of psychology. According to Priestley (1779, 1:342), green matter was neither of an animal nor of a vegetable nature but was a thing *sui generis*. Was Darwin willing to go this far? In 1872 H. C. Bastian published a long, cumbersome attack on Pasteur, arguing that simple forms of life were spontaneously generated. Wallace was greatly impressed by this work and wrote a review praising its facts if not its mode of presentation (Wallace, 1872).

At stake was the vitalism-materialism issue in its most straightforward form. If Darwin could take purely inorganic matter and create a living creature without adding any vital substance or principle, then he would have refuted vitalism and added a necessary proof for his theory. It was no accident that evolutionary theory found so many vocal champions among the materialists in Germany, most notably Haeckel (1834–1919). For Haeckel, evolutionism was not just a scientific theory but also a metaphysical system. Haeckel's *Morphologie* (1866) was metaphysics at its murkiest. Instead of dismissing it as "rubbish," as he had with less metaphysical presentations, Darwin helped to get a translation published in England. It is unlikely, however, that Darwin read very much of this book in the original German and probably understood even less.

Thomas Vernon Wollaston (1821–1878)

I have been deeply interested by Wollaston's book [*On the Variation of Species,* 1856], though I differ *greatly* from many of his doctrines. Did you ever read anything so rich, considering how very far he goes, as his denunciations against those who go further; "most mischievous," "absurd," "unsound." Theology is at the bottom of some of this. I told him that he was like Calvin burning a heretic. It is a very valuable and clever book in my opinion. He has evidently read very little out of his own line. I urged him to read the New Zealand essay. His Geology is also rather *eocene,* as I told him. In fact I wrote most frankly; I fear too frankly; he says that he is sure that ultrahonesty is my characteristic: I do not know whether he meant it as a smear; I hope not.—C. Darwin to J. D. Hooker, Down, June 17, 1856 (*Life and Letters,* 1:431–432)

Have you seen Wollaston's attack in the 'Annals'? The stones are beginning to fly. But Theology has more to do with these two attacks than Science.—C. Darwin to J. D. Hooker, Down, February 1860 (*Life and Letters,* 2:69)

I am perfectly convinced (having read this morning) that the review in the *Annals* is by Wollaston; no one else in the world would have used so many parentheses. I have written to him, and told him that the 'pestilent' fellow thanks him for the kind manner of speaking about him. I have also told him that he would be pleased to hear that the Bishop of Oxford says it is the most unphilosophical work he ever read. The review seems to me clever, and only misinterprets me in a few places. Like all hostile men, he passes over the explanation given of Classification, Morphology, Embryology, and Rudimentary Organs, &c.—C. Darwin to C. Lyell, Down, February 15, 1860 (*Life and Letters,* 2:78)

As you say, I have been thoroughly well attacked and reviled (especially by entomologists—Westwood, Wollaston, and A. Murray have all reviewed

and sneered at me to their hearts' content), but I care nothing about
their attacks; several really good judges go a long way with me, and I
observe that all those who go some little way tend to go somewhat
further.—C. Darwin to H. W. Bates, Down, November 22, [1860] (*More
Letters*, 1:176)

Review of the *Origin of Species**

[THOMAS VERNON WOLLASTON]

To endeavour to understand the various "beginnings" of the organic
world is so essentially the part of an inductive, inquiring mind, like that
of the distinguished naturalist who has lately given us the remarkable volume bearing the above title, that no amount of failure in the attempt
to do so can check the inherent desire that we possess to renew our efforts,
again and again, to discover them. Yet, in spite of this, no trains of reasoning have ever yet brought us, and none with which we are acquainted,
we may safely add, ever *can* bring us, to the absolute origin of the present
order of things, and unfold to us (what perhaps it was never intended
that we should know) the mysteries of creation. "We cannot in any of
the palaeontological sciences," says Dr. Whewell, "ascend to a beginning
which is of the same nature as the existing cause of events, and which
depends upon causes that are still in operation. Philosophers never have
demonstrated, and probably never will be able to demonstrate, what was
the original condition of the solar system, of the earth, of the vegetable
and animal worlds, of languages, of arts. On all these subjects the course
of investigation, followed backwards as far as our materials allow us to
pursue it, ends at last in an impenetrable gloom. We strain our eyes in
vain when we try, by our natural faculties, to discern an origin."

When we look abroad into the world around us, we find ourselves in
the midst of a variety of phenomena, and an endless arrray of organic
forms, all circling onwards, yet never, so far as we can see, altering in
aspect; so that, from the light of mere nature alone, there seems no reason
why they should not go on for ever,–

> Still changing, yet unchanged, still doomed to feel
> Endless mutation in perpetual rest.

Neither, on the same grounds, would it appear necessary to believe that
they had ever *commenced,* did not geology inform us that there was a

* From *Annals and Magazine of Natural History* (1860), 5:132–143.

time in the world's history when they did not exist, but were replaced by another class of beings which occupied their places; and that this latter set was, at a far earlier age, represented by another; and this, again, by an older one still; and so on, until we seem to reach at length the primordial beings with which this planet was originally stocked. It is to discuss, and to *account for,* this succession of beings throughout time and space that Mr. Darwin's book has been compiled; and the great principle by which he believes them all to have been successively produced he terms "Natural Selection."

The opinion amongst naturalists that species were independently created, and have not been transmitted one from the other, has been hitherto so general that we might almost call it an axiom. True it is that we cannot prove this; but then, on the other hand, we cannot prove the converse; and, since of two unproveable propositions we have a right to take our choice, the former has been universally accepted, as most in accordance with the intelligible announcements of revelation, and as aiding us in the otherwise hopeless task of understanding what a species really is. This proposition Mr. Darwin boldly calls in question, and believes, on the contrary, that all species (man included) may have been derived, each in its turn, from those below them by the mere "selecting power of nature,"–which is supposed to have been continually at work, through countless ages, in rejecting (by inevitable annihilation) the weakest and most ill-developed individuals which everywhere existed, and in preserving every little modification which chanced from time to time (in the "great struggle for life" which has ever been going on amongst organic beings) to turn out *for the benefit* of its possessor, and transmitting it, by the law of inheritance, to the next generation, to be further increased in the same direction, until at length, in the course of centuries, the various races have each become so far modified in structure (and that, too, *intermittently,* or, as it were, *en route,* according to their position, or advancement, in the animal pedigree) as to have assumed the various forms, past and present, which naturalists have described under the name of "species." The fossils of each geological formation, on this view, "do not mark a new and complete act of creation, but only an occasional scene, taken almost at hazard, in a slowly-changing drama" (p. 315); and "the fact of the fossil remains of each formation being in some degree intermediate in character between the fossils in the formations above and below is simply explained by their intermediate position in the chain of descent" (p. 476).

Now, whether right or wrong in their assumption, and however much they may differ in their exact definitions, it is quite evident that there is an *idea* involved by naturalists in the term "species" which is altogether distinct from the fact (important though it be) of mere outward resemblance,–viz. the notion of blood-relationship acquired by all the individuals composing it, through a direct line of descent from a common ancestor; and therefore, it is no sign of metaphysical clearness when our author (p. 51) refuses to acknowledge any kind of difference between "genera," "species," and "varieties," except one of *degree*. *Practically*, no doubt, the differences, as we define them, are entirely, and must be, of this nature, for we are necessarily driven to form our judgment solely from outward characters (and must often trust, as it were, to chance, that our decision. thus arrived at, is correct); so that it is quite possible (nay, almost certain) that what one naturalist may rank as a species, another may perhaps, occasionally, believe to be only a variety: nevertheless the idea *involved* in the two terms is not invalidated on that account; and it is simply taking advantage of the imperfections of our discernment (whilst compelled to conjecture from the mere characters which are externally visible), to throw discredit on a distinction between essentially different *ideas*. Man may blunder (and we have but too clear evidence that he often does); but that cannot make nature inconsistent.

There is one point, however, according to Mr. Darwin's own confession, which has struck him much: viz. That all those persons who have most closely investigated particular groups of animals and plants, with whom he has ever conversed, or whose treatises he has read, are firmly convinced that each of the well-marked forms was at the first independently created. But, says he, the explanation of this is simple: from long-continued study they are thoroughly impressed with the distinctions between the several races, and they ignore all general arguments,–refusing "to sum up in their minds slight differences accumulated during many successive generations." But is this more, we may ask, than special pleading? If anybody is capable of forming an opinion on the origin of species, it surely must be those who have most closely studied them; for, if otherwise, we should arrive at the monstrous conclusion that, in order to generalize well, it is desirable to have only a superficial knowledge of the objects generalized upon!–a conclusion to which our learned and amiable author, we feel sure, would not subscribe. The true explanation seems to be this: not that the study of small details unfits an observer for wider areas of thought, but simply that a generalizing mind is of a higher stamp, and therefore less common, than one of an opposite tendency; so that there are more *collectors* in the world than generalizers. But to suppose the accurate study of minutiae to be detrimental to an enlarged interpretation of their results is certainly contrary to experience.

But let us briefly examine the argument of this volume, and see how it is sustained. In the first chapter, Mr. Darwin ably discusses the question of the variation of certain animals and plants under domestication; and few have paid greater attention to this subject than he has, or been more successful in their experiments. A close study of the varieties (acknowledged as such by all) of the domestic pigeon, the innumerable races of our common cattle, and also of what gardeners term "sporting plants," has long convinced him, as well indeed it might, of the almost endless phases which may be gradually shaped out by the selecting power of man. This will be admitted by all, and by none more readily than by those who believe in the distinct origin of species; for, as no two species are alike, it follows that the constitutions of all are different; and if their number, therefore, be infinite, so will, likewise, be the degree of their pliability. Hence, if it should have happened (whether from chance, or, which is more probable, from actual selection, after experiment) that the most plastic organisms have been operated upon, we cannot marvel at the results, however extraordinary. But that equal variations are never brought about in creatures of a less flexible temperament, is abundantly shown (by Mr. Darwin's own admission) in the case of such animals, as the cat, donkey, goose, peacock, guineafowl, &c., which apparently, although so universally bred and domesticated, have not altered in the slightest degree in the course of time. Mr. Darwin explains this fact by supposing (p. 42) that the principle of selection has not been brought to bear upon them. But, if selection, *"unconscious"* as well as "methodical," has been going on to the extent believed, we cannot see why it should not also have silently acted, at any rate to a certain extent, in such cases as these, no less than in the others. To our mind the answer is plain: viz. that the species in question are by nature unpliant (like the great mass of animals), and therefore have not made any progress from their original starting-points.

But let us admit, for the sake of argument, that man, as an active, living agent, and therefore as an intelligent, efficient cause, capable of directing his experiments, and bringing judgment, taste, energy and intellect to bear upon them, possesses the power of altering, in the course of time, the external features (even though they be usually unimportant ones) of nearly *all* the organism, animal and vegetable, on which he may systematically operate: let us admit this (for we do not wish to be unnecessarily

sceptical); and then let us discuss the question, whether there is any principle in nature analogous to this selecting power of man; for, if there is,
why should not similar modifications be produced even in the external
world? Mr. Darwin believes that there is such a principle; and his second
chapter is consequently devoted to what (as we have already stated) he
calls "Natural Selection."

The rate at which all organisms would naturally multiply, if unopposed
by external checks, is perfectly enormous. The elephant, the slowest breeder
of all known animals, would in 500 years, says Mr. Darwin, produce fifteen
million elephants, descended from a single pair. There is no exception
to the rule that every organic being naturally increases at so high a rate
that, if not destroyed, there literally, in a few centuries, would not be
standing-room on the earth for its progeny! Hence arises the certain fact
that more individuals must be destroyed annually than are born, and that
therefore there must be a constant warfare going on amongst living beings,
and, as a consequence, a general struggle for life: and in this battle it
is reasonable to suppose that the most gifted, or fully developed, individual,
each of its kind, would have the best chance of success and (through
having survived) of begetting offspring,–which offspring would probably
inherit, to some extent, the advantages of their parents, and would in their
turn increase these advantages, and give birth to a still more highly gifted
progeny; and so on (it is urged) *to an unlimited extent.*

Now, when not pressed too far, so as to become ridiculous, there is
a speciousness, nay even a probability, about this theory to which most
naturalists would readily give assent. Although unquestionably a mere theory,
and incapable of proof when applied to the greater portion of the feral
world, there is a reasonableness about it which at once commands our
respect. It enables us to account for many a trifling variation which, because
permanent, naturalists have usually regarded as of necessity aboriginally
distinct, and smooths down some of the minor controversies concerning
the value of minute modifications which may be properly referred to direct
agencies from without. Indeed we will go a step further, and affirm that
there is no reason why *varieties,* strictly so called (though too often, we
fear, mistaken for species), and also geographical "sub-species," may not
be gradually brought about, even *as a general rule,* by this process of
"natural selection:" but this, unfortunately, expresses the limits between
which we can imagine the law to operate, and which any evidence, fairly
deduced from facts, would seem to justify: it is Mr. Darwin's fault that
he presses his theory too far. The mere fact of any such varieties thus

matured (if they do indeed exist in nature) being apt to be at times mistaken by naturalists for true species, is surely no argument against the
genuineness of the latter: it merely shows the imperfection of our limited
judgment, and that the best observers are liable to err, and either not
to catch the true characters of a species intuitively (which, in point of
fact, they could scarcely be expected to do), or else to assign at times
undue importance to differences which they may afterwards detect not
to be in reality specific.

We must candidly admit, however, that Mr. Darwin is most consistent
to his principles; and for this we would give him every credit: for if
he objects to the inconsistency of "several eminent naturalists," in the
"strange conclusion which they have lately arrived at," that certain species
have been created independently, whilst they deny the fact that a multitude
of *formerly reputed* species are in the same category (p. 482), we might
fairly take him on his own grounds, and cavil at his conviction (p. 484)
"that all animals have descended from the, at most, only four or five progenitors, and plants from an equal or less number," and contend that
he is bound to advance even still further than this, seeing that he objects
to the existence of a limit simply *because we cannot* (by the nature of
the case, for, *in its entirety,* it is not a "truth of sense") *strictly define
it,* or (in our short-sightedness and stupidity) are apt to blunder and often-
times to mistake its position. But he cleverly anticipates this objection (and
a very serious one, for him, it would have been) by nipping it in the
bud: "Analogy," he says, "would lead me one step further, viz. to the
belief that all animals and plants have descended from some one prototype."
"Therefore I should infer, from analogy, that probably all the organic
beings [i.e. animals as well as plants] which have ever lived on this earth
have descended from some one primordial form into which life was first
breathed" (p. 484). This is plain language, at any rate!

But, having said a few words on the narrowness of the limits within
which we can honestly conceive this ingenious fancy to be applicable, we
might call attention to many other considerations arising out of it, did
space permit. To our mind indeed the whole theory of "natural selection"
is far too utilitarian, and its importance immensely overrated. "An extraordinary amount of modification," says Mr. Darwin, "implies an unusually
large and long-continued amount of variability, which has been continually
accumulated, by natural selection, *for the benefit of the species* " (p. 153);
but surely every naturalist must, in his own province, have observed that
a vast number of "modifications" have apparently no reference whatsoever

to the "good," or advancement, of the species (a fact indeed which has not altogether escaped, *teste* p. 90, our author's sagacious ken), but are often merely, as it were, fantastic, or grotesque, having no connexion with either its well-being or mode of life, and the final cause of which it is utterly hopeless to discuss. Moreover, some of these "developments" (so called) seem merely given for the adornment or elegance of the creature, and frequently display an arrangement of colouring which nothing but an actual intelligence could have planned, and which therefore no amount of mere chance "selection" by an imaginary agent called "nature" can be supposed to have effected. Nor can such characters be referred to what our author would call "sexual selection," seeing that, in the majority of instances, they pertain to both males and females. Neither can they be due to "correlation of growth;" for we cannot conceive that such marvellous perfection of painting as, for instance, the tints of certain butterflies (which are blended together with such nicety and consummate skill, in accordance with the laws of colouring, as to surpass an artist's touch) could have been brought about through mere correlation with a change in some other part of the organism. Such cases bespeak thought, imagination, and judgment, all and each of the highest stamp, and are utterly inexplicable on any of the three principles above alluded to.

Besides, to make "nature" accomplish anything requiring intelligence and foresight, and other attributes of mind, is nothing more or less than to personify an abstraction, and must be regarded therefore as in the highest degree uphilosophical. We believe it was Coleridge who first called attention to this fact, that to treat a mere abstraction as an efficient cause is simply absurd. But that this is the plain and undoubted tendency of our modern materialists, the following sentence, taken at random from the present volume, will certainly go far to corroborate: "As man can produce, and certainly has produced, a great result by his methodical and unconscious means of selection, what may not *nature* effect? Man can act only on external and visible characters: *nature* cares nothing for appearances, except in so far as they may be useful to any being. *She* can act on every internal organ, on every shade of constitutional difference, on the whole machinery of life. Man selects only for his own good; *Nature* only for that of the being which *she* tends. Every selected character is fully exercised by *her;* and the being is placed under well-suited conditions of life" (p. 83).

But who is this "Nature," we have a right to ask, who has such tremendous power, and to whose efficiency such marvellous performances are ascribed? What are her image and attributes, when dragged from her wordy lurking-place? Is she aught but a pestilent abstraction, like dust cast into our eyes to obscure the workings of an Intelligent First Cause of all?

Although it is quite possible that there may be a final cause for every thing, and every character of a thing, in nature (in the same sense as one of our acutest metaphysicians has contended that religion is the final cause of the human mind), we should nevertheless be exceedingly reluctant to press this doctrine too far, for all experience warns us that it may become an impediment, rather than a help, to the progress of scientific discovery. Yet it is one thing to give it more than its due, another to reject it altogether: and those who, like our author, prefer being shipwrecked bodily on the rocks of Scylla to running the slightest risk from the opposite Charybdis, need but to be reminded that a proper use of it has been as fruitful in guiding the researches of our greatest physiologists as the abuse of it has been instrumental in perverting them. And we may confidently affirm that Bacon's famous censure on the "barreness" of these "vestal virgins" (which was applied, be it remembered, to *physics only,* and which has been made so much of by the advocates for the sufficiency of secondary causes in the organic world) would have been less severe "could he have prophetically anticipated," as Sedgwick has well remarked, "the modern discoveries in physiology."

But, before dismissing these immediate considerations, we must say a word or two on the fact of "individual variability," which we cannot but think has been made too much of throughout the volume before us. Without it, "natural selection" would be of course impossible—that is evident; but is its *presence* sufficiently significant to render the theory in any degree *probable?* This is the question with which we are now concerned. Mr. Darwin says that it is only necessary for an individual to vary, be it ever so little, for the principle of natural selection to be established; but to us it seems almost incredible that the general "struggle for existence," or even the extreme pressure of peculiar circumstances from without, should find in mere "individual variability" a sufficient *primum mobile* to lay the foundation of a series of after-divergences (in a given, undeviating direction) destined, each, to accumulate, by infinitesimal degrees, into such successive, intermittent, well-marked forms as to merit, at each stage, the rank of "species." For "individual variability" (so called) is scarcely more, after all, than one of the many proofs, or indices, of individuality; so that to assert its *existence* is simply to state a truism. Amongst the millions of people who have been born into the world, we are certain that no two have ever been *precisely* alike in *every* respect; and, in a similar manner, it is not too much to affirm the same of all living creatures (however alike some of them may seem to our uneducated eyes) that have ever existed. We cannot demonstrate this, undoubtedly, for it is not a truth of *sense;* but it is a truth, nevertheless, of the highest reason (founded on a limited experience), which a reflecting mind will at once receive without evidence; and it may therefore be almost regarded as an axiom. But what does this fact (self-evident as it is) indicate, except this: that, whilst "individual variation" (in each species) is literally endless, it is at the same time strictly prescribed within its proper morphotic limits (as regulated by its specific range), even though *we may be totally unable to define their bounds?* For, if otherwise, how could it happen that, whilst individually different *ad infinitum,* they are nevertheless (in many species) so alike in the mass as to appear to our rough judgment absolutely identical? Hence, we cannot regard "individual variability" as a phenomenon of any real importance or signification, but simply as a *fact* almost involved, as it were, in our very notions of individuality; for, if ever there was a truth more certain than another, it is this: that "there is no similitude in nature that owneth not also to a difference."

But, although we cannot honestly believe, except to a very limited extent, in this "natural selection" theory, as being directly opposed to the doctrine of Efficient Causation (which involves the conception of intelligence, free-agency, and will), as excluding even the idea of creative foresight from the natural world, and so rendering final causes both absurd and impossible, and, moreover, as built chiefly upon negative evidence, and unsupported by the majority of *facts,* still we by no means wish to imply that Mr. Darwin's volume (so full, as it is, of bold hypotheses and philosophical suggestions) is not a most valuable and important fund of knowledge, but, on the contrary, that it will doubtless prove a solid and lasting contribution to science, as one which will inevitably direct a mass of future observations into a new channel; for to leave an opposite impression would be the deepest act of ingratitude on our part for the great profit that we have derived from the careful perusal of its contents. His remarks on geographical distribution (a subject which he has so long and so carefully studied) are most instructive and admirable, and will supply an explanation for many an obscure and puzzling fact which has so often perplexed observers, concerning the appearance of similar and closely allied forms in regions far removed from each other. Especially interesting, too, is the whole of his Section on "Dispersal during the Glacial Period," from which we should be tempted to quote largely did space permit. But as such is unfortunately wanting, we must leave this subject, as well as the entire portion concerning the geological succession and the imperfections of its record, altogether unglanced at. He states his difficulties with honesty, precision, and clearness, and sometimes (as it appears to us) even exposes them more than is necessary—to his own disadvantage: but we wish that we could add that, in spite of this candour on his part (a candour which is so manly and outspoken as almost to "cover a multitude of sins"), we thought *any* of them satisfactorily replied to. There is a clever and ingenious pleading for them all; but, if we look back into the volume, we find (to use the mildest expression) that each, in its turn, has been left in doubt—awaiting further evidence. So that, until this is forthcoming, we cannot but feel that, whilst the theories are in one direction (and made to dovetail into each other), the great body of facts is unquestionably on the opposite side. More especially will this apply to that gravest of all objections (as Mr. Darwin frankly admits), *the thorough and complete absence* (both in geological collections, imperfect though they be, and those, extensive and endless as they are, of the Recent Period) of that countless host of transitional links which, on the "natural selection" theory, must certainly have existed at one period or another of the world's history. They *may* be forthcoming some day; we cannot tell (and so, truly, may many other things, after the same fashion of reasoning!) : but at present it is absolutely certain that we have not so much as a shadow of evidence either that they do exist, or have ever existed. On whichever side we turn we find order and symmetry to be the law of creation, instead of confusion and disorder. To an uneducated eye, which views things only in the mass, this may not appear *prima facie* evident; but those who have worked closest and longest at details, in the open face of nature, know that it is true. Naturalists may quarrel and blunder about the relative importance of minute differences, and therefore about the limits of their "species"—and perhaps nearly all of them have erred in drawing too tightly the boundaries between which "varieties" are supposed to occur; but nevertheless the plain fact remains, that, on a broad scale, more or less abrupt and well-defined forms alone have as yet been discovered, and that they do *not* shade off into each other by that legion of osculant infinitesimal links on which the very life, as it were, of this ingenious theory mainly depends.

As to the evidence to be gathered from the endless phases which have been gradually matured in our domestic cattle and pigeons by the long and systematic efforts of man, we deny that any parallel can be drawn from them, on a general scale, in the feral world; for everything tends to prove that the whole system of certain species (though *not,* as it is admitted, of all), when under domestication, tends to become plastic; whilst, moreover, we cannot ascribe to the operation of a doubtful, unproved (and perhaps altogether imaginary) natural "law" effects in any way analogous to those produced by an active, living agent (and therefore an intelligent efficient cause) who has been capable for centuries of concentrating his efforts with judgment, caution, discernment, and skill, and of carefully selecting, by a direct action of *mind,* all the various divergences that were favourable for his purpose, and so of "adding them up" (as Mr. Darwin happily expresses it), one by one, in a given direction, beforehand decided on, until he has at last succeeded (though at times, even then, with the greatest difficulty) in accomplishing the purpose which he had in view. And, besides all this, it is admitted that there are, after all, *some* forms which he cannot succeed in modifying: which certainly would tend to prove that *even his* most persevering efforts can only avail with certain more or less *naturally elastic* organisms. And that some undoubtedly *are* "naturally elastic," as compared with others, every naturalist who has worked six months under the open skies (instead of in his closet) is absolutely certain. Indeed, whilst we see hosts of species scarcely alter at all, in whatever circumstances and regions they are placed, and therefore whilst exposed to *innumerable conditions of surrounding organic forms*—so that (in a broad sense) they may be regarded as almost independent of the various influences alongside them,–we see, on the contrary, that other species are by constitution so unstable and shifting, in their external details, as scarcely to present *two phases alike* in even the several localities and altitudes of a continuous, unbroken tract. Nor is this mere assertion, for we are prepared to support it by the plainest facts; whilst, at the same time, we could point to a country in which nearly *all* the land-shells now existing (upwards of one hundred species) are found in a fossil state, conglomerated together in beds of indurated mud often twenty feet in thickness, and which have not altered, apparently, *so much as a puncture or a granule* during the enormous period (even though it be geologically recent) which has elapsed since they were first deposited,–a period, moreover, in which there is every reason to believe that the various physical conditions (and perhaps extent) of the whole region have most materially changed: which, at any rate, does not tally with that steady movement towards perfection, that certain progress, of some kind or other (even though slow), of organic forms, which a reception of this "natural selection" idea so loudly and positively demands.

As to the theological difficulties of this question, we must decline entering into them; for we believe that science and theology are best discussed apart, and that neither of them was ever intended to teach us the other. Nevertheless we fear it must be admitted that they are exceedingly grave, if not absolutely insurmountable; and, although as yet they have been altogether, and studiously, kept out of view, the time will assuredly come when, like all other objections, they must be fairly stated, the arguments on both sides candidly examined by competent judges, and each of them impartially weighed, on its own merits. Although it is obviously desirable, for more reasons than one, not to bring revelation and science into unnecessary contact (for the evils which have resulted from injudicious attempts to do so have usually been but too evident), still no man who loves truth, in all its phases, for its own sake, will long rest contented in accepting *as such* a zoological creed which is in direct antagonism with his theological one; for, since two opposite sets of statements cannot be both true, one or the other of them must eventually fall. The question simply is: which, in this case, shall it be? Although we might hazard a hasty reply, we nevertheless will not do so; though we can anticipate the feelings of our more learned theologians, were a bouquet of some of the leading conclusions culled for their special contemplation. What, for instance, would they think, when told that, in spite of their honest convictions (convictions which they had supposed to be coeval with our race), it has been lately discovered that man, with all his lofty endowments and future hopes, was, in point of fact, never "created" at all, but was merely, in the fullness of time, a development from an ape; and not merely from an ape, but that he was originally derived from the same source as bears, cats, rats, mice, geese, mussels, periwinkles, beetles, worms, and sponges (nay, even, perhaps, from the same as the very plants themselves); that, in all probability, he will at length beget some higher creature still, and will himself "become utterly extinct," like each of the beasts (throughout time and space) before him; and that, moreover, "as all the living forms of life are the lineal descendants of those which lived long before the Silurian epoch,"–"*hence* [–a cold shuddering comes over us at what we are compelled to regard as a glorious *non sequitur,* and that, too, from premises which we cannot admit!]–*hence* (we repeat) we may look with some confidence to a secure future of equally inappreciable length!" Hard doctrine, this, for "unphilosophical" minds like ours! And, were we inclined to be sceptical as to the *data* on which this sweeping conclusion is built, we might naturally ask, how is it, *if the above premises be true* (*i.e.* if it indeed be a fact that man has been gradually qualified, by self-improvement, for his advanced post, after passing through an endless array of lower forms),–how is it that *no traditions whatsoever* bearing on the previous and more simple conditions of the human structure (immediately before it attained its climax of perfection) have ever been extant; for it is quite inconceivable that so radical an organic change could have been slowly brought about without, at the least, *some vague tradition of it* having become *a fact of the human mind.* When probed by such-like inquiries, the entire theory (*to the extent that it is pushed*) fairly crumples up.

But we must conclude this notice. Did space permit, we might have offered many remarks on the general tendencies of the Selection theory, *when carried out to its full extent.* We might have dived below the surface, to ascertain the main object of Mr. Darwin's clever and ingenious volume; and have asked, what it was that first prompted him to undertake it. If it was the marvel of creation (and a real marvel it assuredly is) that offered the primary stumbling-block to a philosophical mind, we might have asked whether the marvel would have been got rid of had we been able to reduce the number of the separate, independent acts. To our mind, the wonder consists in the act *at all,* and not in the number of times that it may have been repeated: for a Being that *can create* may surely do so as often as (He) pleases; and we have no right therefore to limit that act,–at any rate on the question of its *probability;* for, if we admit that it has been exerted so much as once, there is no *a priori* reason why it should not have been a million times repeated, or why, if He had so willed it, it might not, at some period or other, have been in even constant operation. Such an idea is difficult to conceive, we admit; but (be it remembered) *it is not one atom more so* than the process of creation *at all:* and with respect to the *marvel* of it (so difficult, and impossible, to understand), it may be well to recollect that it has been contended by some of our greatest minds that even the *sustaining* power of Nature is, in point of fact, as much of a miracle as the creative power.

Although we have felt compelled to say thus much against the theory so ably pleaded for in Mr. Darwin's book, we repeat that, *in a very limited sense indeed,* there seems no reason why the theory may not be a sound one; but at present, even to that extent, it remains to be substantiated. The volume is eloquently written, and its immense array of facts most carefully collected. But we are bound to add, that many an equivocal idea is shrouded under the fairest garb; and we find that we have sometimes swallowed a dose unconsciously, on account of the pleasant medium through which it was administered. And, as an instance of this, we will quote the concluding sentence of the whole work, which is certainly very beautiful, though we can scarcely believe that our author was in earnest when he wrote it. Here it is, without comment (the italics are our own):

> It is interesting to contemplate an entangled bank, clothed with many plants of many kinds, with birds singing on the bushes, with various insects flitting about, and with worms crawling through the damp earth, and to reflect that these elaborately constructed forms, so different from each other, and dependent on each other in so complex a manner, have all been produced by laws acting around us: these laws, taken in the largest sense, being Growth with Reproduction; inheritance, which is almost implied by reproduction; variability from the indirect and direct action of the external conditions of life, and from use and disuse; a ratio of increase so high as to lead to a struggle for life, and as a consequence to Natural Selection, entailing divergence of character and the extinction of less-improved forms. Thus, from the war of nature, from famine and death, *the most exalted object which we are capable of conceiving, namely, the production of the higher animals,* directly follows. There is a grandeur in this view of life, with its several powers, having been originally breathed into a few forms or into one; and that, whilst this planet has gone cycling on according to the fixed laws of gravity, from so simple a beginning, endless forms most beautiful and most wonderful have been and are being evolved.

Would not one step more plunge us headlong into the Nebular Hypothesis, and the whole theory of Spontaneous Generation?

Comments on Wollaston

Wollaston's review in many respects is typical: He advises caution and moderation in all things, although he himself had argued for the existence of the lost continent of Atlantis in an earlier publication! He admits that varieties may well arise by natural selection but contends that Darwin goes too far in claiming that variation ever proceeds past the limits of natural species, even though he was totally unable to define these limits! The action of natural selection is to be tempered with a little divine guidance, but this doctrine in turn must not be pushed too far. Science and theology are best kept separate. The absolute origin of the present order of things is too far in the distant past for us ever to know the actual state of affairs with any certainty. Since of two unprovable propositions we have a right to take our choice, Wollaston chooses special creation. The fallaciousness of these theses is only too apparent. The views which

Wollaston is contrasting are incompatible. Once reference to miracles is permitted in scientific explanations, there is no principle by which their intrusion can be limited. Only on the supposition that the phenomena under investigations are not supernatural do they allow of scientific explanation. Of two equally likely *natural* explanations, we have the right to take our choice. Special creation by divine fiat is not a natural explanation.

Carpenter objected to precisely the position taken by Wollaston on species definition. Wollaston complains that Darwin lacked "metaphysical clearness" because he failed to recognize that by "species" naturalists *mean* those forms related by blood through a direct line of descent from a common ancestor. If a formerly acknowledged species can be shown to be descended from another formerly acknowledged species, then these two forms were not actually species but varieties. Certain paleontologists recognized morphologically identical forms in successive strata as separate species just because they were in separate strata. Other naturalists refused to include morphologically identical synchronous interfertile forms in the same species if they were geographically separate. Carpenter complained, "They have established in their minds as a law of nature that species of the northern and southern hemispheres *must* be dissimilar." (See Darwin's comments on Agassiz as well as the papers by Fawcett and Hopkins.) Darwin took this difficulty to be hopeless, "for if two hitherto reputed varieties be found in any degree sterile together, they are at once ranked by most naturalists as species . . . If we thus argue in a circle, the fertility of all varieties produced under nature will assuredly have to be granted" (Darwin, 1859, p. 267).

Although Carpenter's and Darwin's irritation is understandable, their objections are not justified. A naturalist can define "species" any way he chooses, *if* he is willing to accept the consequences. Species can be defined as those forms which have been specially created without precluding evolution, but if it can be shown that no formerly acknowledged species was specially created, but all evolved, then it has been shown that *there are no species.* As Hopkins in a later paper will say, "Every natural species must by definition have had a separate and independent origin, so that all theories—like those of Lamarck and Mr. Darwin—which assert the derivation of all classes of animals from one origin, do, in fact, deny the existence of natural species at all." But the only thing that has been lost is a word. On a different definition of "species," natural species might well exist. Derivation from a common ancestry does not preclude the existence of species on all definitions.*

* See also Wollaston (1856).

François Jules Pictet (1809–1872)

There has been one prodigy of a review, namely, an *opposed* one (by Pictet, the palaeontologist, in the *Bib. Universelle* of Geneva) which is *perfectly* fair and just, and I agree to every word he says; our only difference being that he attaches less weight to arguments in my favour, and more to arguments opposed, than I do. Of all the opposed reviews, I think this the only quite fair one, and I never expected to see one.—Charles Darwin to Asa Gray, Down, April 3, [1860], *Autobiography*, p. 244

On the Origin of Species by Charles Darwin*

FRANÇOIS JULES PICTET

The book on species by Mr. C. Darwin, which appeared in London a few months ago, has made quite a sensation in England and deservedly so. It has been a long time since we have read anything more comprehensive and more interesting on this difficult and controversial question. The facts are set out with clarity and in a lively manner, in a novel form, and in a way freed from the ordinary routine. It is impossible for his study not to engender reflection or to force one to see certain questions in a new light, even when one can not accept all the theoretical consequences to which the learned author tries to lead the mind of his reader.

I am undertaking a difficult task in attempting to render an account of this work because I am afraid that I will weaken the force of his argument by compressing it too much. Moreover, Mr. Darwin's book is an extract and an abridgment of a more extensive book on which he is laboring. This book will bring together in greater detail all the facts that contribute to his demonstration. Accordingly I will discuss the important questions which it raises, showing the spirit and the method of the work rather than trying so much to give an incomplete analysis of it.

* From *Archives des Sciences de la Bibliotheque Universelle* (1860), 3:231–255. Translation by D. L. H.

I have to say at the outset what my general impression has been. In
the first few chapters I have followed the author step by step with the
greatest of pleasure. As long as he concerned himself with discussing those
facts which are currently familiar, I have been disposed to accept all his
reasoning; I felt that he was prudent and wise, and I had hoped that
this same prudence would permit me to follow him to the very end without
objection. But a point came at which his imagination advanced more quickly
than mine and drew conclusions from the accepted facts which seem incompatible with these same facts. There seemed to me to be, as it were, a
sort of disparity between his premises and his conclusion, the premises
being so prudent, so just, and so limited, and the conclusion on the contrary
appearing so extremely speculative.

My purpose in this article is to draw the attention of naturalists to
this general impression and to the host of facts noted by Mr. Darwin.
This learned writer, in reviving the celebrated theory of Lamarck but in
a more judicious and acceptable form, thinks like him that the diverse
zoological characters are the product of gradual modifications. He believes,
for example, that the ancestors of birds lacked wings and that this important
organ was formed and developed gradually over a long series of generations.
He thinks that the earth worm and the butterfly have a common ancestor
from which they were produced by successive modifications, etc. There
are, it is true, profound differences between his theory and that of Lamarck.
Thus, in an ingenious manner, he brings in a new agent which he calls
natural selection and explains the probable action of it very well, but I
believe that he accords it too much importance. In order to make it possible
for the reader to appreciate the arguments which one can produce for
and against this idea, I will extract the empirical facts from the first chapters
and then I will discuss the theoretical consequences which one can draw
from them.*

But I am obliged to stop here and not multiply these details further
lest I surpass the limits which I have imposed upon myself. In the preceding pages I have provided a summary analysis of the first part of Mr.
Darwin's work which contains the exposition of his theory and the evidence
which he provides from the present day. One might find perhaps that
I have weakened the author's argument, for this is inevitable, but I hasten
to add that I have never done so intentionally, since I am quite willing
to accept most of the facts and the ideas which I have set forth. I see
no serious objection to the view that varieties can be formed in the present

* [Pictet's description of Darwin's theory has been omitted.]

day by natural selection and that, in ascending from previous epochs to our own, one can explain the origin of neighboring species by this law if one is willing to admit a very long period of time for this. Because he accepts the theory of gradual causation of Sir Charles Lyell, Mr. Darwin is not niggardly in the thousands of years needed to permit the accumulation of such variations. I do not have the same need for all this which Mr. Darwin so willingly grants. If ten thousand generations are not enough, he says, take twenty thousand, or a hundred thousand. In order to show how his imagination does not recoil from such numbers, all we need cite is the calculation which he himself makes of the time which a single geological phenomenon has lasted. He does not fear to estimate the denudation of the Wealdian terrain, an epoch which is already quite near our own relative to the entire history of the earth, to be at least three hundred million years. It is clear that before these numbers the proofs which we have been in the habit of giving for the permanence of species lose much of their force. If the 4000 years which separates us from the mummies of Egypt have been insufficient to modify the crocodile and the ibis, then Mr. Darwin can always reply that this period of time is really trifling. I dare not argue with such weapons whose range I cannot appreciate. But to return to simple varieties and to closely related species, I believe that Darwin's theory can explain numerous things and illuminate many questions. Moreover I have allowed myself to be won over by his ideas, as I have said repeatedly, because there must be a material bond between very similar species of two successive faunas and that in all probability the innumerable species about whose limits we are often so seldom in accord were not created without exception with all their distinctive characters.

But as I said at the beginning of this article, in Mr. Darwin's book there is a sudden leap by which the reader is asked to pass from the prudent study of facts to the most extreme theoretical consequences. We have arrived at this juncture and it is here that I regret that I must part company with the author and can no longer follow him on his new path whose tendencies and perils I am going to indicate.

As we have seen, Mr. Darwin accepts on the one hand the possibility of slight variations and on the other an immense series of centuries. He then multiplies these two factors by each other and concludes that there are powerful and profound variations, not only in the external form but also in the most essential organs. Thus, he accepts the successive modification of specific characters, next generic characters, then the limits of families, of orders and of classes, and finally, forced by an inflexible logic, he is led to have all contemporary animals as well as all those of previous faunas derived from a very few primitive types and perhaps only one.

Because these bold deductions do not appear to be justified by the facts, it is necessary to have a more powerful argument before accepting them. Immediately a general objection emerges. Nothing which he has said proves to me that slight and superficial variations can in the long run change nature and degenerate in such grave modifications. I find nothing in the examples chosen by Mr. Darwin which authorizes me to believe that it is all a matter of degree, that just because one has demonstrated that in the course of some thousands of generations, the size, color, and the form of the beak can be modified, that the proportions of the limbs can be changed a little, etc., I am now forced to conclude that additional thousands of generations or of years will change gills into lungs, sprout wings, create an eye, or change an oviparous into a viviparous animal.

All known facts demonstrate on the contrary that the prolonged influence of modifying causes has an action which is constantly restrained within sufficiently confined limits. If we study the modifications brought about by domestication, remembering that these modifications are probably much more intense than those of natural variation, we still cannot find a single example of an influence which has been able to modify the essential nature of an organ. Thus dogs, which have deviated more profoundly from their original state than any other domesticated animal and which are so diverse in their external form, retain an astonishing constancy of characters. If one may be permitted a slightly trivial observation, after all, dogs themselves are never confused. They always mount another dog, as different as the other dog may be from itself, whereas they would never mount a cat or a fox.

If the variations in all domesticated animals have been superficial and slight, then I must say that I know of nothing which proves nor any example which forces me to believe that the contrary took place in nature. Before I can accept Mr. Darwin's deductions, I must see for myself a known case of an important organ beginning to form or of a modification of some value in essential characters. As long as there is no proof that grave changes can be introduced regularly in the order of direct generation, then I am drawn to the daily observation which leads me to believe the contrary. Everything in living nature seems to proclaim this tendency of the conservation of specific forms. When for thousands of years, we see acorns constantly reproducing oak trees, which are alike in their carriage and appearance down to the accidental details, when we reflect on the powerful and mysterious force which acts on this little grain in order to bring about so constant a development, when we see the same phenomena repeated in all organisms, would not the force of the induction be that permanence of form is the rule and variation is only an exception? This permanence seems to be demonstrated by proofs much more direct and positive than those for natural selection and the divergence of characters.

It is equally impossible to attribute much importance to the accidental anomalies of internal organs, and in particular to the insufficient example which Mr. Darwin gives of the variations in the branchings of the arterial system, the nervous system, etc. These modifications are without importance. How can one see in the application of natural selection and the struggle for existence to these modifications the formation of new types? Similar examples have been described in man, but none of them have resulted in the perfecting of an organism. Where is there to be found a difference of this kind which has resulted in the formation of a more perfect and robust race? These little anomalies, sometimes transmitted for two or three generations, are soon lost because of the law of the reversion to type.

At the same time I must acknowledge that even if Mr. Darwin has not found sufficient direct proof to justify the possibility of his hypothesis he has been able to support it with some indirect proofs whose relevance is real and incontestable. If one passes over the absence of any direct proof, one still must acknowledge that his theory squares quite well with the great facts of comparative anatomy and zoology. It provides particularly admirable explanations for the unity of organic composition, the representative or rudimentary organs, and the species plus genera which form natural series. It corresponds equally well with many palaeontological facts. It accords well with the specific resemblances which exist between two consecutive faunas, with the parallelism which one observes sometimes between the sequences of palaeontological and embryological development, etc.

Thus we find ourselves in a singular position. We are presented with a theory which on the one hand seems to be impossible because it is inconsistent with the observed facts and on the other hand appears to be the best explanation of how organized beings have been developed in the epochs previous to ours.

Thus we return quite naturally to the question of what we can substitute for it. Here my intellect fails me and I am quite near to saying *I do not know,* the usual answer to these mysterious questions. Nevertheless, I will say a few words about the hypothesis with which I have accounted for the facts up until now.

I have always represented the succession of organized beings as under the influence of two forces. The one, which I have named *normal generation,* acts under our very eyes, creating the resemblances between children and their parents, assuming the permanence of species during numerous successive generations, and which, nevertheless, assumes and permits the variations which the contemporary world exhibts in superabundance. I believe that the long sequence of geological time could have augmented its force and permitted it by the accumulation of similar results to give birth to a few neighboring species from a single species. The other force, which I call *creative force,* acted in the beginning to produce immediately a varied and abundant fauna. Its action manifesting itself during long periods of time has given birth successively to distinct types whose existence is attested to by palaeontology. Even though I am well aware that the nature and mode of action of this mysterious force escapes us, I believe that everything points to the fact that the action of the one must be augmented by the other.

I will even go a little further and say that Mr. Darwin is also obliged to accept these two forces, only he limits the action of the latter as much as possible. When he attempts to bring his principal proofs into accordance with the laws of comparative anatomy, he must go all the way. It is evident that the classes whose members are interconnected by all the analogies of unity in composition and subordination of parts must have proceeded from the same author. Also, Mr. Darwin readily acknowledges that originally there must have existed four or five progenitors for the animal kingdom and somewhat fewer for the plant kingdom. But it is impossible to deny that the embranchements have real analogies between each other, and Mr. Darwin, with a little repugnance, admits that a rigorous logic forces him to acknowledge only one type for each kingdom. But everything points equally to the fact that there is no sharp dividing line between the animal and plant kingdoms; therefore, all organized beings including man would have to have proceeded from a single type. Man, the sponge, and the oak tree would be descended from a common ancestor by way of normal generation and by a continuous succession of parents and off-spring, just as it occurs today!

Even after Mr. Darwin is forced to this extreme position, he still must explain the formation of this common ancestor. We congratulate him for frankly rejecting the doctrine of *spontaneous generation* which is ordinarily (and none too illogically) the basis of similar theories. He says that life has been breathed into the first primordial form. It is our creative force that has done it. Consequently, both theories acknowledge the existence of the two *forces* and differ only to the degree that each is employed. Arguments advanced to the effect that it is absurd to admit an occult and mysterious force fall of themselves. The entire question seems to be in the balancing of these two forces and in the role which each plays. Mr. Darwin believes that the creative force has acted only once and thereafter the force of normal generation has taken over. We think that the role of the creative force, though stronger in the beginning, has continued for a long time, and on this count we have to go into some details.

I don't see why the creative force, in its mode of action as completely unknown as it may be, could not have brought about results analogous to those which the law of variation in normal generation has created. I must remark on this subject that all those who have maintained its existence have been victims of the unlucky answer of *successive creations,* an expression ordinarily applied to temporal sequences. The advocates of such a view have been too ready to see in the appearance of each new type the direct intervention of a Supreme Will which has no connection except some general *plan* with all the rest of creation. Or, when they attempt to state this plan, it was easily explained in terms of the physiological concordance of the harmony of the organs; but then it was rather more difficult to understand why there were so many rudimentary organs which are physiologically useless but similar in appearance to necessary organs in other types.

It *is much more probable,* as I have said elsewhere,[1] that the creative force is under control of a general law established in the beginning by the Creator, a law whose nature we have not been able to discover and whose mode of action probably cannot ever be completely understood. Why couldn't we say that this law has a certain analogy to normal generation and that this same analogy supplies the explanation of the unity of organic composition? But I hasten to descend from this sphere of overly elevated hypothesis to the solid ground of facts.

The general objection which I have emphasized above and which arises from the total absence of proof in favor of the doctrine of profound modifications is my principal objection and the one which prevents me from accepting Mr. Darwin's extreme conclusions. But even if we accept Mr. Darwin's basic premises, there are still more detailed objections to be made. Mr. Darwin has been much too wise not to have anticipated nearly all of these and has endeavored to refute them in advance. I believe, neverthe-

1. *Traite de paleontologie,* [vol.] I, p. 87.

less, that he has not completely destroyed their value. Even though I consider them of only secondary importance, I must say a few words to show that Mr. Darwin's alleged indirect proofs are far from being general and that, on this same score, his theory faces immense difficulties.

In order to fulfill the obligation of limiting an article which is already too long, I will not enter into a discussion of all these objections. For example, I will not enter into a discussion of the inferences which could be made from the sterility of hybrids, since I recognize with Mr. Darwin that it could result *a posteriori* from acquired differences between species and not be a barrier preventing crossings.

Similarly I will not treat the question of instinct, not that I have been convinced by Mr. Darwin's arguments, and admit, for example, that the marvelous instinct of the bee could be gradually developed in the descendants from a common ancestor. But since we cannot perceive instinct in geological time, we do not have any documentation in this matter and in consequence the discussion would be incomplete.

I will discuss especially those objections which are raised by the study of animal fossils, and I will attempt to show that a host of important facts are all but inexplicable just by natural selection and that these facts make the admission of a creative force indispensable. Here, again, for the sake of brevity, I will limit myself to setting out just five of these objections.

First objection. The fauna of the lower Silurian period, the most ancient of which we have any knowledge, is admirably rich in a variety of forms which are nearly as diversified as in nature today. But in Mr. Darwin's theory, there were only a very few types in the beginning and these types required thousands of years to vary. Mr. Darwin responds to this powerful objection by supposing that the Silurian period may be as distant in time from the origin of things as we are from it and that millions of centuries saw the development of more and more varied beings of which we have no knowledge. Perhaps, it is added, the formations which contain the remains are at the bottom of today's seas, an hypothesis which is impossible for me to discuss but which appears to need support by some proof.

Second objection. The theory of Mr. Darwin supposes that species were modified during thousands of generations by very minimal changes in their characters. Thus, all species must be connected with their common ancestors and, in consequence, with each other in all possible degrees. Why don't we find these gradations in the fossil record, and why, instead of collecting thousands of identical individuals, do we not find more intermediary forms? To this Mr. Darwin replies that we have knowledge of such a small proportion of fossils that one cannot construct proofs and that contiguous geological deposits are often separated by long periods of time because each sea has not always been able to form layers of sediment. Consequently, according to him, we have only a few incomplete pages in the great book of nature and the transitions have been in the pages which we lack. But why then and by what peculiar rules of probability does it happen that the species which we find most frequently and most abundantly in all the newly discovered beds are in the immense majority of the cases species which we already have in our collections?

Third objection. Mr. Darwin's theory accords poorly with the history of clearcut and well-defined forms which appear to have survived for only a limited time. Hundreds of cases could be cited, such as the flying reptiles, ichthyosaurs, belemnites, ammonites, etc. One sees these numerous types appear, so to speak, in one fell swoop and suddenly they disappear. Mr. Darwin's response to this objection would be nearly the same as that which he made to the preceding objection. These appearances and disappearances seem to be sudden because we know nearly nothing of the history of the globe and that the intermediary forms have escaped us. I avow for my part that I do not believe in these intermediaries, and I have considerable difficulty connecting this possibility with natural selection. Take birds, for example. Admit, for instance, that they sprang from a common progenitor with mammals and reptiles. The wing then must have been formed by successive alterations in the anterior limb of the prototype. But I do not see how natural selection could act for the conservation of future birds, since this modified member, this future wing, being neither a real arm nor a real wing, could not possibly be of physiological value. The transitional forms which must have been endowed with it for thousands of generations do not appear to me *a priori* to have a good chance in the struggle for existence.

Fourth objection. It frequently happens that from the moment of their appearance, genera are represented by a great variety of specific forms. But, according to Mr. Darwin's theory, these varied forms could be produced only by descent from a common progenitor. Thus, the belemnites are unknown before the Jurassic period and had acquired all their numerical development whether in species or in individuals in the lowest strata of this formation. The mentioning of this objection is sufficient to indicate its significance; it would have to be discussed elsewhere in a manner analogous to the preceding.

Fifth objection. The zoological forms are modified in the same manner all over the face of the earth. The same genera and families were formed together and disappeared together, or at least we can establish with certainty that diverse types have succeeded each other in the same order. Thus, the Paleozoic era has exactly the same characteristic genera in America as in Europe. It is followed by deposits which have the same overall traits in their fauna as our Jurassic and Cretaceous terrains. The most superficial deposits correspond quite well with our tertiary divisions. One finds there the same thing as in the East Indies, and geologists have established there the same for the Jurassic, Cretaceous, Nummulitic, and other deposits. How could natural selection bring about these parallelisms?

In summary, I have said that everything proves the existence of two forces, one normal generation and the other the creative force. I can neither deny this latter force nor limit its role to any great extent. I differ in opinion with Mr. Darwin in the sense that I believe that he exaggerates the effects of variation in normal generation; but I hasten to add that his book provides quite an interesting study and may well succeed in partially modifying the pre-existing opinions of the creative limits of normal generation. These limits are perhaps *a little more* extended than was generally believed, and the variations produced in the normal course of generation can probably explain some of the specific differences in the most natural groups.

This brings us back before we terminate this review to respond to a question which the practicing naturalists, who use species either to characterize present-day geographic zones or to distinguish the layers of the earth and the geological strata, are likely to ask. Does the increase in the limit of species variability occasioned by Mr. Darwin's research have any effect on their methods?

I do not think so, and I hold fast to this point of view in complete assurance. No matter what theory one might adopt, in a given period of time, species are fixed; they are fixed nearly as much for Mr. Darwin as they are for the most convinced partisan of absolute permanence. There must in effect be thousands of generations for a nearly imperceptible modification to be introduced. Species therefore are constant in the present. If four thousand years has not been sufficient to change the smallest character in the crocodile, ibis or ichneumon fly of Egypt, then it is quite probable that an additional four thousand years will not do more, and our descendants in forty centuries will find themselves in the same situation. The same thing holds for each geologic period and for the same reasons.

With respect to the accumulation of specific modifications due to the accumulation of previous centuries, the acceptance of one or the other theories cannot in practice influence these applications. Typically, all forms are well-defined and precise; this is as true for a geographic zone as for a geologic stage. The beautiful work of Mr. Darwin does lead us to a new point of view. We no longer need to place great importance on knowing whether such and such a form is properly called a species, an incipient species, or a constant variety, because its origin can be of the same nature; but we must exercise much greater care in the study of variations, which need to be defined more carefully. We are easily convinced that these variations are of two sorts. Some variations are contemporaneous, and in palaeontology indicate the same epoch. Depending on whether they are fixed or are connected by continuous series, they merit or do not merit a specific name. Other variations appear successively through the centuries and are of a palaeontological character if they are at a stage in specialization. Whether they are very close neighbors to the species which have preceded them or those that follow, or whether they are quite distinct, they have the same significance if they are susceptible to a rigorous definition; in the latter case they must be designated by names. We also are learning to take into account the significant relationships which exist between species of the same natural genus, and we have a good deal more freedom in searching for the expression of these preceding relationships without occasioning the discouragement which so often attacks those who concern themselves with the study of species.

In other words, the beautiful book of Mr. Darwin will open up a new field of reconciliation and ideas to zoologists and palaeontologists. In order to summarize our opinion in a few words, let me say that reading Darwin's book would be truly useful to all zoologists, especially if they exercise constraint in following the author in his seducing argument and stop before they are led from the certainty of slight variations to the dangerous doctrine that over long periods of time these modifications have no limit.

Comments on Pictet

Darwin was extremely pleased with Pictet's review because in it Pictet set out a fair and full summary of evolutionary theory for French-speaking scientists, and Darwin anticipated that they would be the hardest to convert. In England, Lyell had already established uniformitarianism among geologists and paleontologists—with a few major exceptions. For example, Hooker, in his review of the *Origin*, says that "geologists are now pretty unanimous in believing that the host of changes of the globe were the same in kind and degree as those now going on." On the continent, especially in France, catastrophism still reigned. It was enough that a fossil appeared in two different strata to warrant its inclusion in two different species. As Carpenter observed in his review of the *Origin*: "It will probably meet with less opposition among British than among continental palaeontologists; for with the former it has come to be generally admitted that many species really do range through a long succession of formations.

Furthermore, the idea of the transmutation of species was even more strongly identified in France with the name of Lamarck than it was in England and consequently was held in even greater disrepute. Part of Lamarck's reputation was engendered by his own personality, part by the great influence of Cuvier. Cuvier viewed himself as a hard-nosed empiricist, though the existence of eternal immutable forms played a central role in his philosophy. Strangely enough, when evolution was accepted in France, it was Lamarckian evolution. To this day, French biologists prefer some sort of directed evolution to the synthetic theory of English- and German-speaking biologists.

Darwin had expected French reaction to be more in the spirit of the review published by Marie Jean Pierre Flourens, the influential Perpetual Secretary of the French Academy of Science (Flourens, 1864). In one place, for example, Flourens declaimed, "At last Mr. Darwin's work has appeared. One cannot help being struck by the talent of the author. But what unclear ideas, what false ideas! What metaphysical jargon clumsily hurled into natural history! What pretentious and empty language! What childish and out-of-date personifications! Oh lucidity! Oh French stability of mind, where art thou?"

Pictet goes even further than Wollaston in his tentative acceptance of the possibility of evolution: Not only varieties but also species may have arisen by natural selection, but no group could have evolved which necessitated the development of a new organ. Pictet admits that there is considerable indirect evidence for a more extended view of evolution but no direct proof. On the contrary, he asserts, all observations demonstrate that the prolonged influence of modifying causes is constantly restrained within sufficiently confined limits. Look at the permanence which we see all around us. Would not the force of induction be that permanence of form is the rule and variation the exception? Pictet's line of reasoning is a good example of the dangers inherent in over-reliance on superficial appearances. Aristotle and Bacon believed in the spontaneous generation of multicellular organisms because these creatures certainly seemed to spring up in the absence of copulation and the production of eggs. After all, our every sense proclaims that the earth does not move. Would not the force of induction be that it neither rotates nor revolves?

Pictet concludes: "Before I can accept Mr. Darwin's deduction, I must see for myself a known case of an important organ beginning to form or of a modification of some value in essential characters." This was in 1860. By 1864 Pictet had become a convert (Darwin, ed., 1903, 1:258), and two years later published a paper (Pictet, 1866) supporting evolutionary theory. His conversion, we might add, was accomplished without the direct proof he earlier required.

The most striking feature of Pictet's review is the candor with which he introduces his notion of creative force, a candor reminiscent of Darwin himself. Normal generation and natural selection just did not seem adequate to explain the origin and evolution of species. There must be something else, and Pictet freely admits that he knows nothing whatsoever about this something else, this creative force. As Pictet sees the situation, it is either creative force or spontaneous generation, and spontaneous generation is too abhorrent even to consider. Pictet justifies his position by seizing upon Darwin's pentateuchal phrase to accuse him of introducing an unknown factor in the origin of life and of rejecting spontaneous generation. Pictet's justification for his own acceptance of occult qualities is even more peremptory. Arguments against occult qualities in science "fall of themselves." As the Introduction to this volume should have amply proven, the situation with respect to occult qualities is not nearly so cut and dried. Pictet's creative force, mysterious and unknown as it might be, was not necessarily supernatural, being under the control of a general law established in the beginning by the Creator.*

* See also the reviews by Carpenter, Owen, and Mivart.

Adam Sedgwick (1785–1873)

I have had a kind yet slashing letter against me from poor dear Old Sedgwick, 'who has laughed till his sides ached at my book.'—C. Darwin to T. H. Huxley, Down, November 25 [1859] (*More Letters,* 1:130)

By the way, I hear from Murray that all the attacks heaped on my book do not seem to have at all injured the sale, which will make poor dear old Sedgwick groan.—C. Darwin to W. H. Miller, Down, December 1 [1859] (*More Letters,* 1:123–124)

I now feel certain that Sedgwick is the author of the article in the *Spectator*. No one else could use such abusive terms. And what a misrepresentation of my notions! Any ignoramus would suppose that I had *first* broached the doctrine, that the breaks between successive formations marked long intervals of time. It is very unfair. But poor dear old Sedgwick seems rabid on the question. 'Demoralised understanding!' If ever I talk with him I will tell him that I never could believe that an inquisitor could be a good man; but now I know that a man may roast another, and yet have as kind and noble a heart as Sedgwick's.—C. Darwin to C. Lyell, Down, March 24, 1860 (*Life and Letters,* 2:91)

You may like to hear about reviews on my book. Sedgwick (as I and Lyell feel *certain* from internal evidence) has reviewed me savagely and unfairly in the *Spectator*. The notice includes much abuse, and is hardly fair in several respects. He would actually lead any one, who was ignorant of geology, to suppose that I had invented the great gaps between successive geological formations, instead of its being an almost universally admitted dogma. But my dear old friend Sedgwick, with his noble heart, is old, and is rabid with indignation.—C. Darwin to Asa Gray, Down, April 3, [1860], (*Life and Letters,* 1887, p. 90)

I wonder whether Sedgwick noticed in the *Edinburgh Review* about the 'Sacerdotal Revilers,'* so the revilers are tearing each other to pieces. I suppose Sedgwick will be very fierce against me at the Philosophical Society. Judging from his notice in the *Spectator,* he will misrepresent me, but it will certainly be unintentionally done. In a letter to me, and in the above notice, he talks much about my departing from the spirit of inductive philosophy. I wish, if you ever talk on the subject to him, you would ask him whether it was not allowable (and a great step) to invent the undulatory theory of light, i.e., hypothetical undulations, in a hypothetical substance, the ether. And if this be so, why may I not invent the hypothesis of Natural Selection (which from the analogy of domestic productions, and from what we know of the struggle of existence and of the variability of organic beings, is, in some very slight degree, in itself probable) and try whether this hypothesis of Natural Selection does not explain (as I think it does) a large number of facts in geographical distribution—geological succession, classification, morphology, embryology, etc. I should really much like to know why such an hypothesis as the undulation of the ether may be invented, and why I may not invent (not that I did invent it, for I was led to it by studying domestic varieties) any hypothesis, such as Natural Selection . . .

I can perfectly understand Sedgwick or anyone saying that Natural Selection does not explain large classes of facts; but that is very different from saying that I depart from right principles of scientific investigation.—C. Darwin to J. S. Henslow, Down, May 8 [1860], (*More Letters,* 1:149)

I have been greatly interested by your letter to Hooker, and I must thank you from my heart for so generously defending me, as far as you could, against my powerful attackers. Nothing which persons say hurts me for long, for I have entire conviction that I have not been influenced by bad feelings in the conclusions at which I have arrived. Nor have I published my conclusions without long deliberations and they were arrived at after far more study than the public will ever know of or believe in.—I am certain to have erred in many points, but I do not believe so much as Sedgwick and Co. think."—C. Darwin to J. S. Henslow, Down, May 14, 1860 (*More Letters,* 1:150)

Last Friday we all went to the Bull Hotel at Cambridge to see the boys, and for a little rest and enjoyment. The backs of the Colleges are simply paradisaical. On Monday I saw Sedgwick, who was most cordial and kind; in the morning I thought his brain was enfeebled; in the evening he was brilliant and quite himself. His affection and kindness charmed us all. My visit to him was in one way unfortunate; for after a long

* [In his anonymous review in the *Edinburgh Review,* Richard Owen condemned sacerdotal revilers such as Sedgwick.]

sit he proposed to take me to the museum, and I could not refuse, and in consequence he utterly prostrated me; so that we left Cambridge next morning, and I have not recovered from the exhaustion yet. Is it not humiliating to be thus killed by a man of eighty-six, who evidently never dreamed that he was killing me? As he said to me, 'Oh, I consider you as a mere baby to me!'—C. Darwin to J. D. Hooker, Down, May 25, [1870], (*Life and Letters*, 2:305–306)

Letter to Charles Darwin from
Adam Sedgwick, December 1859*

I write to thank you for your work on the *Origin of Species*. It came, I think, in the latter part of last week; but it *may* have come a few days sooner, and been overlooked among by book-parcels, which often remain unopened when I am lazy or busy with any work before me. So soon as I opened it I began to read it, and I finished it, after many interruptions, on Tuesday. Yesterday I was employed—1st in preparing my lecture; 2ndly, in attending a meeting of my brother Fellows to discuss the final propositions of the Parliamentary Commissioners; 3rdly, in lecturing; 4thly, in hearing the conclusion of the discussion and the College reply, whereby, in conformity with my own wishes, we accepted the scheme of the Commissioners; 5thly, in dining with an old friend at Clare College; 6thly, in adjourning to the weekly meeting of the Ray Club, from which I returned at 10 P.M., dog-tired, and hardly able to climb my staircase. Lastly, in looking through the *Times* to see what was going on in the busy world.

I do not state this to fill space (though I believe that Nature does abhor a vacuum), but to prove that my reply and my thanks are sent to you by the earliest leisure I have, though that is but a very contracted opportunity. If I did not think you a good-tempered and truth-loving man, I should not tell you that (spite of the great knowledge, store of facts, capital views of the correlation of the various parts of organic nature, admirable hints about diffusion, through wide regions, of many related organic beings &c. &c.) I have read your book with more pain than pleasure. Parts of it I admired greatly, parts I laughed at till my sides were almost sore; other parts I read with absolute sorrow, because I think them utterly false and grievously mischievous. You have *deserted*—after a start in that tram-road of all solid physical truth—the true method of induction, and started us in machinery as wild, I think, as Bishop Wilkins's locomotive that was to sail with us to the moon. Many of your wide conclusions are based upon assumptions which can neither be proved nor disproved, why then express them in the language and arrangement of philosophical induction? As to your grand principle—*natural selection*—what is it but a secondary consequence of supposed, or known, primary facts? Development is a better word, because more close to the cause of the

* *Life and Letters* (1887), pp. 42–45.

fact? For you do not deny causation. I call (in the abstract) causation
the will of God; and I can prove that He acts for the good of His creations
by laws which we can study and comprehend. Acting by law, and under
what is called final causes, comprehends, I think, your whole principle. You
write of "natural selection" as if it were done consciously by the selecting
agent. 'Tis but a consequence of the pre-supposed development, and the
subsequent battle for life. This view or nature you have stated admirably,
though admitted by all naturalists and denied by no one of common-sense.
We all admit development as a fact of history: but how came it about?
Here, in language, and still more in logic, we are point-blank at issue. There
is a moral or metaphysical part of nature as well as a physical. A man who
denies this is deep in the mire of folly. 'Tis the crown and glory of organic
science that it *does* through *final cause,* link material and moral; and yet
does not allow us to mingle them in our first conception of laws, and our
classification of such laws, whether we consider one side of nature or the
other. You have ignored this link; and, if I do not mistake your meaning,
you have done your best in one or two pregnant cases to break it. Were it
possible (which, thank God, it is not) to break it, humanity, in my mind,
would suffer a damage that might brutalize it, and sink the human race into
a lower grade of degradation than any into which it has fallen since its written records tell us of its history. Take the case of the bee-cells. If your development produced the successive modification of the bee and its cells (which
no mortal can prove), final cause would stand good as the directing cause
under which the successive generations acted and gradually improved. Passages in your book, like that to which I have alluded (and there are others
almost as bad), greatly shocked my moral taste. I think, in speculating on
organic descent, you *over*-state the evidence of geology; and that you *under*-
state it while you are talking of the broken links of your natural pedigree:
but my paper is nearly done, and I must go to my lecture room. Lastly,
then, I greatly disliked the concluding chapter—not as a summary, for
in that light it appears good—but I disliked it from the tone of triumph
and confidence in which you appeal to the rising generation (in a tone
I condemned in the author of the *Vestiges*)* and prophesy of things not
yet in the womb of time, nor (if we are to trust the accumulated experience
of human sense and the inferences of its logic) ever likely to be found
anywhere but in the fertile womb of man's imagination. And now to say
a word about a son of a monkey and an old friend of yours; I am better, far
better, than I was last year. I have been lecturing three days a week (formerly I gave six a week) without much fatigue, but I find by the loss of
activity and memory, and of all productive powers, that my bodily frame is
sinking slowly towards the earth. But I have visions of the future. They are
as much a part of myself as my stomach and my heart, and these visions
are to have their antitype in solid fruition of what is best and greatest.
But on one condition only—that I humbly accept God's revelation of Himself both in His works and in His word, and do my best to act in conformity

* [See Sedgwick, (1845).]

with that knowledge which He only can give me, and He only can sustain me in doing. If you and I do all this, we shall meet in heaven.

I have written in a hurry, and in a spirit of brotherly love, therefore forgive any sentence you happen to dislike; and believe me, spite of any disagreement in some points of the deepest moral interest, your true-hearted old friend,

A. Sedgwick.

Objections to Mr. Darwins Theory of the Origin of Species*
[ADAM SEDGWICK]

Before writing about the transmutation theory, I must give you a skeleton of what the theory is:–

1st. *Species* are *not permanent; varieties* are the beginning of new species.

2nd. Nature began from the simplest forms—probably from one form—the primaeval *monad,* the parent of all organic life.

3d. There has been a continual ascent on the organic scale, till organic nature became what it is, by one continued and unbroken stream of onward movement.

4th. The organic ascent is secured by a Malthusian principle through nature,–by a battle of life, in which the best in organization (the best varieties of plants and animals) encroach upon and drive off the less perfect. This is called the theory of *natural selection.*

It is admirably worked up, and contains a great body of important truth; and it is eminently amusing. But it gives no element of strength to the fundamental theory of transmutation; and without specific transmutations natural selection can do nothing for the general theory.[1] The flora and

---

* [Sedgwick's paper was originally published in *The Spectator* on March 24, 1860, but since Sedgwick had not seen the proof sheets and the published version contained two errors, the paper was republished on April 7, 1860. The paper presented here is the corrected and revised version. Sedgwick's corrections were minimal. The name "Owen" had been substituted for "Oken" in one place, and "Pachyderm" had been misspelled. The revisions were confined to the concluding paragraphs, which Sedgwick expanded. Both versions of the concluding paragraphs are presented here.]

1. It is worth remarking that though no species of the *horse* genus was found in America when discovered, two or three *fossil* species have been found there. Now, if these horses had (through some influence of climate) been transmuted into tapirs or buffaloes, one might expect to see the *tendency* at least towards such a change in the numerous herds of wild horses—the descendants of those brought from Europe—which are now found in both South and North America.

fauna of North America are very different from what they were when the Pilgrim Fathers were driven out from old England; but changed as they are, they do not one jot change the collective fauna and flora of the actual world.

5th. We do not mark any great organic changes *now*, because they are so slow that even a few thousand years may produce no changes that have fixed the notice of naturalists.

6th. But *time is the agent*, and we can mark the effects of time by the organic changes on the great geological scale. And on every part of that scale, where the organic changes are great in two contiguous deposits of the scale, there must have been a corresponding lapse of time between the periods of their deposition—perhaps millions of years.

I think the foregoing heads give the substance of Darwin's theory; and I think that the great broad facts of geology are directly opposed to it.

Some of these facts I shall presently refer to. But I must in the first place observe that Darwin's theory is not *inductive*,–not based on a series of acknowledged facts pointing to a *general conclusion*,–not a proposition evolved out of the facts, logically, and of course including them. To use an old figure, I look on the theory as a vast pyramid resting on its apex, and that apex a mathematical point. The only facts he pretends to adduce, as true elements of proof, are the *varieties* produced by domestication, or the *human artifice* of cross-breeding. We all admit the varieties, and the very wide limits of variation, among domestic animals. How very unlike are poodles and greyhounds! Yet they are of one species. And how nearly alike are many animals,–allowed to be of distinct species, on any acknowledged views of species. Hence there may have been very many blunders among naturalists in the discrimination and enumeration of species. But this does not undermine the grand truth of nature, and the continuity of true species. Again, the varieties, built upon by Mr. Darwin, are varieties of domestication and human *design*. Such varieties could have no existence in the old world. Something may be done by cross-breeding; but mules are generally sterile, or the progeny (in some rare instances) passes into one of the original crossed forms. The Author of Nature will not permit His work to be spoiled by the wanton curiosity of Man. And in a state of nature (such as that of the old world before Man came upon it) wild animals of different species do not desire to cross and unite.*

* [In his review of the *Origin* included in this volume, F. W. Hutton (1860) quotes the three paragraphs that follow as if they were by three separate authors. The numerous trivial changes in word order suggest that perhaps Hutton had seen a prepublication copy of Sedgwick's paper and was quoting from it.]

Species have been constant for thousands of years; and time (so far as I see my way) though multiplied by millions and billions would never change them, so long as the conditions remained constant. Change the conditions, and old species would disappear; and new species *might* have room to come in and flourish. But how, and by what causation? I say by *creation*. But, what do I mean by creation? I reply, the operation of a power quite beyond the powers of a pigeon-fancier, a cross-breeder, or hybridizer; a power I cannot imitiate or comprehend; but in which I can believe, by a legitimate conclusion of sound reason drawn from the laws and harmonies of Nature. For I can see in all around me a design and purpose, and a mutual adaptation of parts which I *can* comprehend,–and which prove that there is exterior to, and above, the mere phenomena of Nature a great prescient and designing cause. Believing this, I have no difficulty in the repetition of new species during successive epochs in the history of the earth.

But Darwin would say I am introducing a *miracle* by the supposition. In one sense, I am; in another, I am not. The hypothesis does not suspend or interrupt an established law of Nature. It does suppose the introduction of a new phenomenon unaccounted for by the operation of any *known* law of Nature; and it appeals to a power above established laws, and yet acting in harmony and conformity with them.

The pretended physical philosophy of modern days strips Man of all his moral attributes, or holds them of no account in the estimate of his origin and place in the created world. A cold atheistical materialism is the tendency of the so-called material philosophy of the present day. Not that I believe that Darwin is an atheist; though I cannot but regard his materialism as atheistical; because it ignores all rational conception of a final cause. I think it untrue because opposed to the obvious course of Nature, and the very opposite of inductive truth. I therefore think it intensely mischievous.

Let no one say that it is held together by a cumulative argument. Each series of facts is laced together by a series of assumptions, which are mere repetititons of the one false principle. You cannot make a good rope out of a string of air-bubbles.

I proceed now to notice the manner in which Darwin tries to fit his principles to the facts of geology.

I will take for granted that the known series of fossil-bearing rocks or deposits may be divided into the Palaeozoic; the Mesozoic; the Tertiary or Neozoic; and the Modern—the Fens, Deltas, &c., &c., with the spoils of the actual flora and fauna of the world, and with wrecks of the works of Man.

To begin then, with the Palaeozoic rocks. Surely we ought on the transmutation theory, to find near their base great depositis with *none but the lowest forms of organic life*. I know of no such deposits. Oken contends that life began with the infusorial forms. They are at any rate well fitted for fossil preservation; but we do not find them. Neither do we find beds exclusively of hard corals and other humble organisms, which ought, on the theory, to mark a period of vast duration while the primaeval monads were working up into the higher types of life. Our evidence is, no doubt, very scanty; but let not our opponents dare to say that it makes *for them*. So far as it is positive, it seems to me pointblank *against them*. If *we* build upon imperfect evidence, they commence without any evidence whatsoever, and against the evidence of actual nature. As we ascend in the great stages of the Palaeozoic series (through Cambrian, Silurian, Devonian, and Carboniferous rocks) we have in each a *characteristic* fauna; we have no wavering of species,—we have the noblest cephalopods and brachiopods that ever existed; and they preserve their typical forms till they disappear. And a few of the types have endured, with specific modifications, through all succeeding ages of the earth. It is during these old periods that we have some of the noblest icthyc forms that ever were created. The same may be said, I think, of the carboniferous flora. As a whole, indeed, it is lower than the living flora of our own period; but many of the old types were grander and of higher organization than the corresponding families of the living flora; and there is no wavering, no wanting of organic definition, in the old types. We have some land reptiles (batrachians), in the higher Palaeozoic periods, but not of a very low type; and the reptiles of the permian groups (at the very top of the Palaeozoic rocks), are of a high type. If all this be true, (and I think it is), it gives but a sturdy grist for the transmutation-mill, and may soon break it cogs.[2]

We know the complicated organic phenomena of the Mesozoic (or Oolitic) period. It defies the transmutationist at every step. Oh! but the document, says Darwin, is a fragment. I will interpolate long periods to account for all the changes. I say, in reply, if you deny my conclusion

2. I forebear to mention the Stagonolepis, a very highly organized reptile, the remains of which were found, by Sir R. I. Murchison, in a rock near Elgin, supposed to belong to the old red sandstone. Some doubts have been expressed about the age of the deposit. Should the first opinion prove true (and I think it will), we shall then have one of the oldest reptiles of the world exhibiting, not a very low, but a very high organic type.

grounded on positive evidence, I toss back your conclusions, derived from negative evidence—the inflated cushion on which you try to bolster up the defects of your hypothesis. The reptile fauna of the Mesozoic period is the grandest and highest that ever lived. How came these reptiles to die off, or to degenerate? And how came the Dinosaurs to disappear from the face of Nature, and leave no descendants like themselves, or of a corresponding nobility? By what process of *natural selection* did they disappear? Did they tire of the land, and become Whales, casting off their hind-legs? And, after they had lasted millions of years as whales, did they tire of the water, and leap out again as Pachyderms? I have heard of both hypotheses; and I cannot put them into words without seeming to use the terms of mockery. This I do affirm, that if the transmutation theory were proved true in the actual world, and we could hatch rats out of eggs of geese, it would still be difficult to account for the successive forms of organic life in the old world. They appear to me to give the lie to the theory of transmutation at every turn of the pages of Dame Nature's old book.

The limits of this letter compel me to omit any long discussion of the Tertiary Mammals, of course including man at their head. On physical grounds, the transmutation theory is untrue, if we reason (as we ought to do) from the known to the unknown. To this rule, the Tertiary Mammals offer us no exception. Nor is there any proof, either ethnographical or physical, of the bestial origin of man.

And now for a few words upon Darwin's long inter*polated periods* of geological ages. He has an eternity of past time to draw upon; and I am willing to give him ample measure; only let him use it logically, and in some probable accordance with facts and phenomena.

1st. I place the theory against facts viewed collectively. I see no proofs of enormous *gaps* of geological time, (I say nothing of years or centuries,) in those cases where there is a sudden change in the ancient fauna and flora. I am willing, out of the stock of past time, to lavish millions or billions upon each epoch, if thereby we can gain rational results from the operation of *true causes*. But time and "natural selection" can do nothing if there be not a *vera causa* working with them.[3] I must confine myself to a very small number of the collective instances.

2d. Towards the end of the carboniferous period, there was a vast extinction of animal and vegetable life. We can, I think, account for this extinction mechanically. The old crust was broken up. The sea bottom underwent

3. See reference on *Time,* in the *Annotations of Bacon's Essays.*

a great change. The old flora and fauna went out; and a new flora and fauna appeared, in the ground, now called permian, at the base of the new red sandstone, which overlies the carboniferous rocks. I take the fact as it *is,* and I have no difficulty. The time in which all this was brought about *may* have been very long, even upon a geological scale of time. But where do the *intervening* and connecting types exist, which are to mark the *work of natural selection?* We do not find them. Therefore, the step onwards gives no true resting-place to a baseless theory; and is, in fact, a stumbling-block in its way.

3d. Before we rise through the new red sandstone, we find the muschel-kalk (wanting in England, though its place on the scale is well-known) with an *entirely new* fauna: where have we a proof of any enormous lapse of geological time to account for the change? We have no proof in the desposits themselves: the presumption they offer to our senses is of a contrary kind.

4th. If we rise from the muschel-kalk to the Lias, we find again a new fauna. All the anterior species are gone. Yet the passage through the upper members of the new red sandstone to the Lias is by insensible gradations, and it is no easy matter to fix the physical line of their demarcation. I think it would be a very rash assertion to affirm that a great geological interval took place between the formation of the upper part of the new red sandstone and the Lias. Physical evidence is against it. To support a baseless theory, Darwin would require a countless lapse of ages of which we have no commensurate physical monuments; and he is unable to supply any of the connecting organic links that ought to bind together the older fauna with that of the Lias.

I cannot go on any further with these objections. But I will not conclude without expressing my deep aversion to the theory; because of its unflinching materialism;–because it has deserted the inductive track,–the only track that leads to physical truth;–because it utterly repudiates final causes, and therby indicates a demoralized understanding on the part of its advocates. By the word, demoralized, I mean a want of capacity for comprehending the force of moral evidence, which is dependent on the highest faculties of our nature. What is it that gives us the sense of right and wrong, of law, of duty, of cause and effect? What is it that enables us to construct true theories on good inductive evidence? Theories which enable us, whether in the material or the moral world, to link together the past and the present. What is it that enables us to anticipate the future, to act wisely with reference to future good, to believe in a future state, to acknowledge the

being of a God? These faculties, and many others of like kind, are a part
of ourselves quite as much so as our organs of sense. All nature is subordinate to law. Every organ of every sentient being has its purpose bound up in
the very law of its existence. Are the highest conceptions of man, to which
he is led by the necessities of his moral nature, to have no counterpart
or fruition? I say *no,* to all such questions; and fearlessly affirm that we
cannot speculate on man's position in the actual world of nature, on his
destinies, or on *his origin,* while we keep his highest faculties out of our
sight. Strip him of these faculties, and he becomes entirely bestial; and
he may well be (under such a false and narrow view) nothing better
than the natural progeny of a beast, which has to live, to beget its likeness,
and then die for ever.

By gazing only on material nature, a man may easily have his very
senses bewildered (like one under the cheatery of an electro-biologist);
he may become so frozen up, by a too long continued and exclusively
material study, as to lose his relish for moral truth, and his vivacity in
apprehending it. I think I can see traces of this effect, both in the origin
and in the details of certain portions of Darwin's theory; and, in confirmation of what I now write, I would appeal to all that he states about those
marvellous structures,–the comb of a common honey-bee, and the eye of
a mammal. His explanations make demands on our credulity, that are
utterly beyond endurance, and do not give us one true natural step towards
and explanation of the phenomena—viz., the perfection of the structures,
and their adaptation to their office. There *is* a light by which a man may
see and comprehend facts and truths such as these. But Darwin wilfully
shuts it out from our senses; either because he does not apprehend its
power, or because he disbelieves in its existence. This is the grand blemish
of his work. Separated from his sterile and contracted theory, it contains
very admirable details and beautiful views of nature,–especially in those
chapters which relate to the battle of life, the variations of species, and
their diffusion through wide regions of the earth.

In some rare instances, Darwin shows a wonderful credulity. He seems
to believe that a white bear, by being confined to the slops floating in
the Polar basin, might in time be turned into a whale; that a lemur might
easily be turned into a bat; that a three-toed tapir might be the great
grandfather of a horse; or that the progeny of a horse may (in America)
have gone back into the tapir.

But any startling and (supposed) novel paradox—maintained very boldly
and with an imposing plausibility, derived from a great array of facts

all interpreted hypothetically—produces, in some minds, a kind of pleasing
excitement, which predisposes them in its favour; and if they are unused
to careful reflection, and averse to the labour of accurate investigation,
they will be likely to conclude that what is (apparently) *original,* must
be a production of original *genius,* and that anything very much opposed
to prevailing notions must be a grand *discovery,*–in short, that whatever
comes from "the bottom of a well" must be the "truth" which has been
long hidden there.[4]

I need hardly go on any further with these objections. But I cannot
conclude without expressing my detestation of the theory, because of its
unflinching materialism;—because it has deserted the inductive track, the
only track that leads to physical truth;—because it utterly repudiates final
causes, and therby indicates a demoralized understanding on the part of
its advocates. In some rare instances it shows a wonderful credulity. Darwin
seems to believe that a white bear, by being confined to the slops floating
in the Polar basin, might be turned into a whale; that a Lemur might
easily be turned into a bat; that a three-toed Tapir might be a great
grandfather of a horse! or the progeny of a horse (in America) have
gone back to the tapir.

But any startling and (supposed) novel paradox, maintained very boldly
and with something of imposing plausibility, produces, in some minds, a
kind of pleasing excitement, which predisposes them in its favour; and
if they are unused to careful reflection, and averse to the labour of accurate investigation, they will be likely to conclude that what is (apparently) *original* must be a production of original *genius,* and that anything
very much opposed to prevailing notions must be a grand *discovery,*–in
short, that whatever comes from "the bottom of a well" must be the "truth"
supposed to be hidden there.

Comments on Sedgwick

Adam Sedgwick was one of the founders of the science of geology in
England. He was Woodwardian Professor of Geology at Trinity College,
Cambridge, from 1818 until his death in 1873. Although Darwin did not
attend Sedgwick's geology lectures during his short stay at Cambridge,
he did accompany Sedgwick on a three-week walking tour to North Wales.
This excursion was almost the only preparation which Darwin was to receive

4. [The two versions of Sedgwick's paper are nearly identical up to this point.
The original version, however, concluded with the two paragraphs that follow.]

for his voyage on the *Beagle*. Throughout his long tenure at Cambridge, Sedgwick remained active in university affairs, publishing a slim volume in 1834, a *Discourse on the Studies of Cambridge*. In spite of its title, Sedwick's *Discourse* was largely an attack on the associationist psychology of James Mill, John Stuart Mill's father, and the utilitarian ethics of Jeremy Bentham. William Whewell had so high an opinion of Sedgwick that he dedicated to him his *Philosophy of the Inductive Science* (Whewell, 1840), commenting that Sedgwick could have been his fellow-laborer or even master in philosophy if only he had not devoted his life to geology. As might be expected, John Stuart Mill held quite a different opinion of Sedgwick's philosophical abilities:

> Professor Sedgwick, a man of eminence in a particular walk of natural science, but who should not have trespassed into philosophy, had lately published his *Discourse on the Studies of Cambridge,* which had as its most prominent feature an intemperate assault on analytic psychology and utilitarian ethics, in the form of an attack on Locke and Paley [Mill, 1924, p. 140.]

Sedgwick used later editions of his *Discourse* to mount a continuing attack on Chambers' *Vestiges of Creation* (1845). Numerous scientists, including Huxley (1845), had objected strongly to the unscientific character of the *Vestiges,* but no scholar of Sedgwick's stature had devoted so much energy to attacking a theory which so few scientists took seriously. In one edition after another, his *Discourse* swelled with prefaces and appendixes until the original text was all but lost. The reason for Sedgwick's preoccupation was the *Vestiges* is easy to discover. He looked upon science as a powerful force to aid in the task of keeping mankind faithful to the word of God. Like many naturalists of the day, Sedgwick was a minister, but unlike many, he took his calling seriously. The chief goal of science was to "teach us to see the finger of God in all things animate and inanimate."

After abandoning Wernerian geology, Sedgwick remained highly conservative throughout the rest of his life. He came to accept the glacier theory of Agassiz and Buckland, but only slowly and grudgingly. He could never quite bring himself to adopt Lyell's uniformitarian principles, objecting to his a priori reasoning (Lyell, 1881, 2:3). There was not the slightest chance that Sedgwick would be able to look favorably upon the *Origin of Species*. While Darwin was publishing his uncontroversial journals on the voyage of the *Beagle,* Sedgwick predicted that Darwin would have "A great name among the naturalists of Europe" (Darwin, 1958, p. 145). Sedgwick's opinion changed markedly when he read the *Origin*. In it, Darwin had deserted the true method of induction, the tramroad to all solid physical truth. In Sedgwick's view, Darwin's theory was damnable on two counts: it was radically new and it openly conflicted with Scripture. Early in his career, Sedgwick was able to make the necessary reinterpretations of Scripture which the facts of geology required. In fact, he went so far as to defend William Buckland's liberal interpretation of the Bible set out in Buckland's Bridgewater Treatise (1836) against the attacks of a Reverend Cockburn. Thereafter, the pious clergyman harassed Sedgwick with long letters and in publications.

In spite of his experience with the Reverend Cockburn, Sedgwick proceeded to treat Darwin in much the same manner as Cockburn had misused him. Sedgwick's colleagues without exception attested to his gaiety of spirit, kindliness of heart, and goodness of character. Yet this same man was capable of lacerating another without mercy if he threatened the foundations of natural theology. Sedgwick's polemics were always biting, but never more so than when he thought that science was on a collision course with theology. Sedgwick's blatant appeal to theological standards in his attacks on the *Vestiges* elicited no reproach from the scientific community, perhaps because scientists felt that Chambers' exposition was so uncritical that it deserved whatever treatment it might receive. The *Origin of Species,* however, was quite another matter. Nearly every author included in this volume inveighed against sacerdotal revilers (see especially Hutton's review).

A truer measure of the sentiment among scientists on the relation between science and theology can be gathered from the response given to a declaration circulated by a group of scientists asking their colleagues to affirm the proposition that the results of science could never conflict with the Written Word, correctly interpreted. The statement went on to declare that no scientist had the right to presume that his own conclusions must be right and the statements of Scripture wrong. On the one hand, two hundred and ten names had been appended to the document at the time of publication. On the other hand, only two of these names were especially prominent in science—J. P. Joule and Adam Sedgwick. Sir John Herschel was so irritated by this resolution that he published a personal repudiation of it as an infringement upon the freedom of religious opinion (Herschel, 1864, p. 375).

Sedgwick had found the *Vestiges* "ignorant, superficial and pernicious" (Clark and Hughes, 1890, 2:85). The book was dangerous because it was written in a style that appealed to the masses. The *Origin* was even more dangerous because it was neither ignorant nor superficial. It lent scientific support to the popularity of the argument of the *Vestiges*. In reply to an inquiry, Sedgwick summarized his position of the *Origin* as follows:

I have read Darwin's book. It is clever, and calmly written; and therefore, the more mischievous, if its principles be false; and I believe then *utterly false*. It is the system of the author of the *Vestiges* stripped of his ignorant absurdities. It repudiates all reasoning from final causes; and seems to shut the door upon any view (however feeble) of the God of Nature as manifested in His works. From first to last it is a dish of rank materialism cleverly cooked and served up. As a system of philosophy it is not unlike the Tower of Babel, so daring in its high aim as to seek a shelter against God's anger; but it is like a pyramid poised on its apex. It is a system embracing all living nature, vegetable and animal; yet contradicting—point blank—the vast treasure of facts that the Author of Nature has, during the past two or three thousand years, revealed to our senses. And why is this done? For no other solid reason, I am sure, except to make us independent of a Creator (Mill 1924, p. 140).

As is often the case when proper scientific method is invoked, Sedgwick was highly inconsistent in its application. Darwin's reasoning was just a string of bubbles, but his own alternative explanation in terms of creative powers which he could not comprehend was "a legitimate conclusion of sound reason drawn from the laws and harmonies of nature." Sedgwick lauded all but the shoddiest study if it supported his views on creation and condemned any that were opposed as not being properly inductive. As Lyell observed in his own disputes with Sedgwick, the man was incapable of disinterested inquiry. Strangely enough, the relationship between Sedgwick and Darwin remained cordial throughout their lives in spite of Darwin's heretical ideas and Sedgwick's savage attacks. One presumes that they discussed neither evolution nor religion in their private conversations.

In his style of writing and his opposition to evolutionary theory, Sedgwick had much in common with St. George Jackson Mivart. Both men were religiously motivated. Mivart was a convert to Catholicism and took his religion seriously. Both men seemed to lose their sense of proportion in such controversies, but Sedgwick somehow always managed to stay just this side of personal calumny. On occasion, Mivart overstepped the bounds of Victorian good taste. In spite of his polemics, Sedgwick always enjoyed an honored place in the scientific community. At best, Mivart maintained a peripheral position, both because of the quality of his scientific achievements and because of the hatred which he had engendered among the Darwinians (see Gruber, 1960).

Sedgwick may not have been the greatest of philosophers, but he accurately diagnosed the danger which evolutionary theory posed for natural theology. Since Sedgwick was one of the most prominent advocates of natural theology in Victorian England, the danger was very real to him. Evolutionary theory *was* a dish of rank materialism, and it *did* repudiate the kind of reasoning from final causes that had been used to support belief in God. Sedgwick refused to eliminate completely God's direct intervention in natural phenomena. A universe governed completely by law—even divinely instituted law—was devoid of significance for theology.

Richard Owen (1804–1892)

How curious I shall be to know what line Owen will take; dead against us, I fear; but he wrote me a most liberal note on the reception of my book, and said he was quite prepared to consider fairly and without prejudice my line of argument.—C. Darwin to C. Lyell, Down, December 5, 1859 (*Life and Letters*, 2:35)

I had very long interviews with Owen, which perhaps you would like to hear about . . . I infer from several expressions that, at bottom, he goes an immense way with us. . .

He said to the effect that my explanation was the best ever published of the manner of formation of species. I said I was very glad to hear it. He took me up short: 'You must not at all suppose that I agree with you in all respects.' I said I thought it no more likely that I should be right in nearly all points, than that I should toss up a penny and get heads twenty times running. I asked him what he thought the weakest part. He said he had no particular objection to any part. He added:–

'If I must criticise, I should say, "we do not want to know what Darwin believes and is convinced of, but what he can prove!" I agreed most fully and truly that I have probably greatly sinned in this line, and defended *deduction* my general line of argument of inventing a theory and seeing how many classes of facts the theory would explain. I added that I would endeavor to modify the 'believes' and 'convinceds.' He took me up short: 'You will then spoil your book, the charm of (!) it is that it is Darwin himself.' He added another objection, that the book was too *teres atque rotundus*—that it explained everything, and that it was improbable in the highest degree that I should succeed in this. I quite agree with this rather queer objection, and it comes to this that my book must be very bad or very good.—C. Darwin to C. Lyell, December 12, 1859 (*Life and Letters*, 2:36–37)

I have just read the 'Edinburgh,' which without doubt is by Owen. It is extremely malignant, clever, and I fear will be very damaging. He is atrociously severe on Huxley's lecture, and very bitter against Hooker. So we three *enjoyed* it together. Not that I really enjoyed it, for it made me uncomfortable for one night; but I have got quite over it to-day. It requires much study to appreciate all the bitter spite of many of the remarks against me; indeed I did not discover all myself. It scandalously misrepresents many parts. He misquotes some passages, altering words within inverted commas . . .

It is painful to be hated in the intense degree with which Owen hates me.—C. Darwin to C. Lyell, April 10, 1860 (*Life and Letters,* 2:94)

I have gone over Owen's review again, and compared passages, and I am astonished at the misrepresentations. But I am glad that I resolved not to answer. Perhaps it is selfish, but to answer and think more on the subject is too unpleasant. I am so sorry that Huxley by my means has been thus atrociously attacked. I do not suppose you much care about the gratuitous attack on you.—C. Darwin to J. D. Hooker, April 13, 1860 (*Life and Letters,* 2:96)

Owen is indeed very spiteful. He misrepresents and alters what I say very unfairly. But I think his conduct towards Hooker most ungenerous: viz., to allude to his essay (Australian Flora), and not to notice the magnificent results on geographical distribution. The Londoners say he is mad with envy because my book has been talked about; what a strange man to be envious of a naturalist like myself, immeasurably his inferior! From one conversation with him I really suspect he goes at the bottom of his hidden soul as far as I do!—C. Darwin to J. S. Henslow, Down, May 8, 1860 (*More Letters,* 1:149)

What a quibble to pretend that he did not understand what I meant by *inhabitants* of South America; and any one would suppose that I had not throughout my volume touched on Geographical Distribution. He ignores also everything which I have said on Classification, Geological Succession, Homologies, Embryology, and Rudimentary Organs–p. 496.

He falsely applies what I said (too rudely) about 'blindness of preconceived opinions' to those who believe in creation, whereas I exclusively apply the remark to those who give up multitudes of species as true species, but believe in the remainder–p. 500.

He slightly alters what I say,–I *ask* whether creationists really believe that elemental atoms have flashed into life. He says that I describe them as so believing, and this, surely, is a difference–p. 501.

He speaks of my 'clamouring against' all who believe in creation, and this seems to me an unjust accusation–p. 501.

He makes me say that the dorsal vertebrae vary; this is simply false: I nowhere say a word about dorsal vertebrae–p. 522.

What an illiberal sentence that is about my pretension to candour, and about my rushing through barriers which stopped Cuvier: such an argument would stop any progress in science–p. 525.

How disingenuous to quote from my remark to you about my *brief* letter, as if it applied to the whole subject–p. 530.

How disingenuous to say that we are called on to accept the theory, from the imperfection of the geological record, when I over and over again [say] how grave a difficulty the imperfection offers–p. 530.—C. Darwin to Asa Gray, May 22, 1860 (*Life and Letters,* 2 : 106–107)

[In an earlier letter to T. H. Huxley (April 9, 1860, *More Letters,* 1 : 146–147), Darwin had presented a shorter list of objections to Owen's review. It did contain, however, the following additional objection:

"The quotation on P. 512 of the *Review* about 'young and rising naturalists with plastic minds,' attributed to 'nature of limbs,' is a false quotation, as I do not use the words 'plastic minds.' "

The quotation, however, was not cited as being from the *Origin of Species* but from Owen's 'On the nature of the limbs.' Later in his review Owen does correctly quote Darwin's comment about the few naturalists, endowed with much "flexibility of mind" who might be influenced by Darwin's work.]

In the *Saturday Review* (one of our cleverest periodicals) of May 5th, p. 573, there is a nice article on Owen's review, defending Huxley, but not Hooker; and the latter, I think, Owen treats most ungenerously. But surely you will get sick unto death of me and my reviewers—C. Darwin to Asa Gray, Down, June 8, 1860, (*More Letters,* 1 : 153)

Have you seen the last *Saturday Review?* I am very glad of the defence of you and of myself. I wish the reviewer had noticed Hooker. The reviewer, whoever he is, is a jolly good fellow, as this review and the last on me showed. He writes capitally, and understands well his subject. I wish he had slapped Owen a little bit harder.—C. Darwin to T. H. Huxley, Down, May, 1860 (*Life and Letters,* 2 : 105)

I anticipated your request by making a few remarks on Owen's review . . . he grossly misrepresents and is very unfair to Huxley. You say that you think the article must be by a pupil of Owen; but no one fact tells so strongly against Owen, considering his former position at the College of Surgeons, as that he has never reared one pupil or follower.—C. Darwin to Asa Gray, Down, June 8, 1860, *More Letters,* 1 : 153

I liked Rolleston's paper, but I never read anything so obscure and not self-evident as his 'canons.' I had a dim perception of the truth of your profound remarks—that he wrote in fear and trembling 'of God, man, and monkeys,' but I would alter it into 'God, man, Owen, and monkeys.' Huxley's letter was truculent, and I see that every one thinks it too truculent; but in simple truth I am become quite demoniacal about

Owen—worse than Huxley; and I told Huxley that I should put myself under his care to be rendered milder. But I mean to try and get more angelic in my feelings; yet I never shall forget his cordial shake of the hand, when he was writing as spitefully as he possibly could against me. But I have always thought that you have more cause than I to be demoniacally inclined towards him. Bell told me that Owen says that the editor mutilated his article in the *Edinburgh Review,* and Bell seemed to think it was rendered more spiteful by the Editor; perhaps the opposite view is as probable. Oh, dear! this does not look like becoming more angelic in my temper!—C. Darwin to J. D. Hooker, Down, April 23, 1861 (*More Letters,* 1:185)

I have been atrociously abused by my religious countrymen; but as I live an independent life in the country, it does not in the least hurt me in any way, except indeed when the abuse comes from an old friend like Professor Owen, who abuses me and then advances the doctrine that all birds are probably descended from one parent.—C. Darwin to J. L. A. de Quatrefages, Down, July 11, 1862 (*More Letters,* 1:202)

By the way, one of my chief enemies (the sole one who has annoyed me), namely Owen, I hear has been lecturing on birds; and admits that all are descended from one, and advances as his own idea that the oceanic wingless birds have lost their wings by gradual disuse. He never alludes to me, or only with bitter sneers, and coupled with Buffon and the *Vestiges.*—C. Darwin to Asa Gray, Down, July 23, 1862 (*More Letters,* 1:203)

You would laugh if you could see how indignant all Owen's mean conduct about *E. Columbi* made me. I did not get to sleep till past 3 o'clock. How all you lash him, firmly and severely, with unruffled temper, as if you were performing a simple duty. The case is come to such a pass, that I think every man of science is bound to show his feelings by some overt act, and I shall watch for a fitting opportunity.—C. Darwin to T. H. Huxley, Down, December 28, 1862 (*More Letters,* 1:232)

How quiet Owen seems! I do at last begin to believe that he will ultimately fall in public estimation.—C. Darwin to T. H. Huxley, Down, June 27, 1863 (*More Letters,* 1:242)

I often saw Owen, whilst living in London, and admired him greatly, but was never able to understand his character and never became intimate with him. After the publication of the *Origin of Species* he became my bitter enemy, not owing to any quarrel between us, but as far as I could judge out of jealousy at its success. Poor Dear Falconer, who was a charming man, had a very bad opinion of him, being convinced that he was not only ambitious, very envious and arrogant, but untruthful and dishonest. His power of hatred was certainly unsurpassed. When in former days I used to defend Owen, Falconer often said, 'You will find him out some day,' and so it has passed.—C. Darwin, *Autobiography*, p. 104

Darwin on the Origin of Species*
[RICHARD OWEN]

In the works above cited† the question of the origin, succession, and extinction of species is more or less treated of, but most fully and systematically by the accomplished Naturalist who heads the list. Mr. Charles Darwin has long been favourably known, not merely to the Zoological but to the Literary World, by the charming style in which his original observations on a variety of natural phenomena are recorded in the volume assigned to him in the narrative of the circumnavigatory voyage of H.M.S. Beagle, by Capt. (now Admiral) Fitz Roy, F.R.S. Mr. Darwin earned the good opinion of geologists by the happy applications of his observations on coral reefs,‡ made during that voyage, to the explanation of some of the phenomena of the changes of level of the earth's crust. He took high rank amongst the original explorers of the minute organisation of the invertebrate animals, upon the appearance of his monographs, in the publications by the Ray Society, on the Cirripedia, Sub-Classes Lepadidae (1851), and Balanidae (1854). Of independent means, he has full command of his time for the prosecution of original research: his tastes have led him to devote himself to Natural History; and those who enjoy his friendship and confidence are aware that the favourite subject of his observations and experiments for some years past has been the nature and origin of the so-called *species* of plants and animals. The octavo volume, of upwards of 500 pages, which made its appearance towards the end of last year, has been received and perused with avidity, not only by the professed naturalist, but by that far wider intellectual class which now takes interest in the higher generalisations of all the sciences. The same pleasing style which marked Mr. Darwin's earliest work, and a certain artistic disposition and sequence of his principal arguments, have more closely recalled the

* From the *Edinburgh Review*, April 1860, 11:487–532.
† [By Darwin (1859), Wallace (1859), Flourenz (1864), Agassiz (1857), Hooker (1859), Powell (1855), Pouchet (1859), and Owen (1855, 1858, 1860).]
‡ *On the Structure and Distribution of Coral Reefs*, 8vo. 1842.

attention of thinking men to the hypothesis of the inconstancy and transmutation of species, than had been done by the writings of previous advocates
of similar views. Thus, several, and perhaps the majority, of our younger
naturalists have been seduced into the acceptance of the homoeopathic
form of the transmutative hypothesis now presented to them by Mr. Darwin,
under the phrase of 'natural selection.'

[Owen cites Hooker and possibly Lyell as converts to Darwin's theory.
He then presents a long series of quotations from the *Origin* which are
supposedly the only important original observations which Darwin puts
forth as evidence for his theory. They include Darwin's discovery of the
homology between the ovigerous frena in the pedunculated cirripedes and
the branchiae of the sessile cirripedes (Darwin, 1958:191), the slave-making
habits of certain ants (Darwin, 1958:221–223), the hive-making instincts
of bees (Darwin, 1958:228–230), the transference of freshwater flora and
fauna on the bodies of freshwater migratory birds (Darwin, 1958:385–386,
397), the derivation of all domestic pigeons from the rock pigeon (Darwin,
1958:445), and Darwin's amusing description of the interconnections between clover, humble-bees, mice, and cats (Darwin, 1958:73). After quoting Darwin's brief description of the evolution of species by natural selection, Owen launches his attack.]

The scientific world has looked forward with great interest to the facts
which Mr. Darwin might finally deem adequate to the support of his theory
on this supreme question in biology, and to the course of inductive original
research which might issue in throwing light on 'that mystery of mysteries.'
But having now cited the chief, if not the whole, of the original observations
adduced by its author in the volume now before us, our disappointment
may be conceived. Failing the adequacy of such observations, not merely
to carry conviction, but to give a colour to the hypothesis, we were then
left to confide in the superior grasp of mind, strength of intellect, clearness
and precision of thought and expression, which might raise one man so
far above his contemporaries, as to enable him to discern in the common
stock of facts, of coincidences, correlations and analogies in Natural History,
deeper and truer conclusions than his fellow-labourers had been able to
reach.

These expectations, we must confess, received a check on perusing the
first sentence in the book.

'When on board H.M.S. "Beagle," as naturalist, I was much struck with
certain facts in the distribution of the inhabitants of South America, and
in the geological relations of the present to the past inhabitants of that

continent. These facts seemed to me to throw some light ·on the origin
of species—that mystery of mysteries, as it has been called by some of
our greatest philosophers' (p. 1).

What is there, we asked ourselves, as we closed the volume to ponder
on this paragraph,–what can there possibly be in the inhabitants, we suppose he means aboriginal inhabitants, of South America, or in their distribution on that continent, to have suggested to any mind that man might
be a transmuted ape, or to throw any light on the origin of the human
or other species? Mr. Darwin must be aware of what is commonly understood by an 'uninhabited island;' he may, however, mean by the inhabitants
of South America, not the human kind only, whether aboriginal or otherwise, but all the lower animals. Yet again, why are the fresh-water polypes
or sponges to be called 'inhabitants' more than the plants? Perhaps what
was meant might be, that the distribution and geological relations of the
organised beings generally in South America, had suggested transmutational
views. They have commonly suggested ideas as to the independent origin
of such localized kinds of plants and animals. But what the 'certain facts'
were, and what may be the nature of the light which they threw upon
the mysterious beginning of species, is not mentioned or further alluded
to in the present work.

The origin of species is the question of questions in Zoology; the supreme
problem which the most untiring of our original labourers, the clearest
zoological thinkers, and the most successful generalizers, have never lost
sight of, whilst they have approached it with due reverence. We have
a right to expect that the mind proposing to treat of, and assuming to
have solved, the problem, should show its equality to the task. The signs
of such intellectual power we look for in a clearness of expression, and
in the absence of all ambiguous or unmeaning terms. Now, the present
work is occupied by arguments, beliefs, and speculations on the origin of
species, in which, as it seems to us, the fundamental mistake is committed,
of confounding the questions, of species being the result of a secondary
cause of law, and of the nature of that creative law. Various have been
the ideas promulgated respecting its mode of operation; such as the reciprocal action of an impulse from within, and an influence from without, upon
the organisation (Demaillet, Lamarck); premature birth of an embryo
at a phase of development, so distinct from that of the parent, as, with
the power of life and growth, under that abortive phase, to manifest differences equivalent to specific (*Vestiges of Creation*); the hereditary transmission of what are called 'accidental monstrosities;' the principle of gradual

transmutation by 'degeneration' (Buffon) as contrasted with the 'progressional' view.

In reference to the definition of species, Lamarck,[1] in 1809, cited, as the most exact, that of 'a collection of like (semblables) individuals produced by other individuals equally like them (pareils a eux). But the progress of discovery, especially, perhaps, in palaeontology, led him to affirm that species were not as ancient as Nature herself, nor all of the same antiquity; that this alleged constancy was relative to the circumstances and influences to which every individual was subject, and that as certain individuals, subjected to certain influences, varied so as to constitute races, such variations might and do graduate (s'avancent) towards the assumption of characters which the naturalist would arbitrarily regard, some as varieties, others as species. He comments in almost the words of Mr. Darwin, on the embarrassment and confusion which the different interpretation of the nature and value of such observed differences, in the works of different naturalists, had occasioned.[2] The true method of surveying the diversities of organisation is from the simple to the compound forms, which course Lamarck affirms to be graduated and regularly progressive, save where local circumstances, and others influencing the mode of life, have occasioned anomalous diversities.

Cuvier had preceded Lamarck in specifying the kinds and degrees of variation, which his own observations and critical judgment of the reports of others led him to admit. 'Although organisms produce only bodies similar to themselves, there are circumstances which, in the succession of generations, alter to a certain point their primitive form.'[3] Here it may be remarked, that the whole question at issue hinges upon the proof of the determination of that limit of variety. Cuvier gives no proof that the alteration stops 'at a certain point.' It merely appears from what follows, that his means of knowing by his own and others' observations had not carried him beyond the point in question, and he was not the man to draw conclusions beyond his premises.

'Less abundant food,' he goes on to say, 'makes the young acquire less size and force. Climate more or less cold, air more or less moist, exposure to light more or less continuous, produce analogous effects; but, above all, the pains bestowed by man on the animal and vegetable productions which he raises for his uses, the consecutive attention with which he restricts them in regard to exercise, or to certain kinds of food, or to influences

1. *Philosophie zoologique*, 8vo. (1809), vol. I, p. 54.
2. Ibid., p. 55.
3. Cuvier, *Tableau Elementaire de l'Histoire Naturelle*, 8vo. (1798), p. 9.

other than those to which they would be subject in a state of nature, all tend to alter more quickly and sensibly their properties.'

Cuvier admits that the determination by experiment of these variable properties, of the precise causes to which they are due, of the degree of variability and of the powers of the modifying influences, is still very imperfect ('mais ce travail est encore tres-imparfait.') The most variable properties in organisms are, according to Cuvier, *size* and *colour.*

'The first mainly depends on abundance of food; the second on light and many other causes so obscure that it seems to vary by chance. The length and strength of the hairs are very variable. A villous plant, for example, transported to a moist place, becomes smooth. Beasts lose hair in hot countries, but gain hair in cold. Certain external parts, such as stamens, thorns, digits, teeth, spines, are subject to variations of number both in the more and the less; parts of minor importance, such as barbs of wheat, &c., vary as to their proportions; homologous parts ('des parties de nature analogue') change one into another, *i.e.,* stamens into petals as in double flowers, wings into fins, feet into jaws, and we might add, adhesive into breathing organs (as in the case of the barnacles cited by Mr. Darwin).'

As to the alleged test of difference between a species and a variety by the infecundity of the hybrid of two parents which may differ in a doubtful degree, Cuvier, in reference to this being the case when the parents are of distinct species, and not mere varieties, emphatically affirms, 'Cette assertion ne repose sur aucune preuve' (p. 11); it is at least constant that individuals of the same species, however different, produce together; 'quelque differens qu'ils soient, peuvent toujours produire ensemble.' But Cuvier warns us not to conclude, when individuals of two different races produce an intermediate and fecund offspring, that they must be of the same species, and that they have not been originally distinct (p. 13).

' "The number of varieties, or amount of variation," says Cuvier, "relates to geographical circumstances." At the present day, many such varieties appear to have been confined around their primitive centre, either by seas which they could neither traverse by swimming or by flight, or by temperatures which they were not able to support, or by mountains which they could not cross, &c.'[4]

Daily observation, comparison, and reflection, on recent and extinct or-

4. 'Les varietes de chacune ont du etre d'autant plus fortes et plus nombreuses, que les circonstances des lieux ou de sa nature, lui ont permis de s'etendre plus loin; c'est ce qui peut faire croire que les grandes differences que se trouvent parmi les hommes, les chiens, et les autres etres repandues partout le monde, ne sont que des effets des causes accidentelles, en un mot, des *varietes*' (p. 14).

ganisms, pursued from the date of these remarks (1798) to the close of his career (1832) failed to bring the requisite proof, or to impress the mind of Cuvier with any amount of belief worth mentioning, as to the nature of the cause operative in the production of the species of which he was the first to demonstrate the succession.

Lamarck, without contributing additional results from observation and experience, affirms that the changes defined by Cuvier do not 'stop at a certain point,' but progress with the continued operation of the causes producing them. That, moreover, such changes of form and structure induce corresponding changes in actions, and that a change of actions, growing to a habit, becomes another cause of altered structure; that the more frequent employment of certain parts or organs leads to a proportional increase of development of such parts; and that, as the increased exercise of one part is usually accompanied by a corresponding disuse of another part, this very disuse, by inducing a proportional degree of atrophy, becomes another element in the progressive mutation of organic forms.[5]

These principles seem entitled to be regarded as of the nature of those called 'verae causae' by Bacon, and they are agreeable with known powers and properties of animated beings; only observation has not disclosed more than a very limited extent of their operation,–limited both as to the time in which that operation has been watched, and limited consequently as to the amount of the change produced.

When Cuvier affirms that such capacity to vary proceeds only to a certain point, he may mean that it has not been watched and traced beyond such point. Cuvier admits the tendency to hereditary transmission of characters of variation. Neither he nor any other physiologist has demonstrated the organic condition or principle that should operate so as absolutely to prevent the progress of modification of form and structure correlatively with the operation of modifying influences, in successive generations. But those who hastily or prematurely assume an indefinite capacity to deviate from a specific form are as likely to obstruct as to promote the solution of the question.

The principles, based on rigorous and extensive observation, which have been established since the time of Cuvier, and have tended to impress upon the minds of the most exact reasoners in biology the conviction of a constantly operating secondary creational law, are the following:–The law of irrelative or vegetative repetition, referred to at p. 437 of Mr. Darwin's work; the law of unity of plan or relations to an archetype; the

5. *Philosophie zoologique,* 8vo. (1809), vol. I., chaps. iii, vi, vii.

analogies of transitory embryonal stages in a higher animal to the matured forms of lower animals; the phenomena of parthenogenesis; a certain parallelism in the laws governing the succession of forms throughout time and space; the progressive departure from type, or from the more generalised to more specialised structures exemplified in the series of species from their first introduction to the existing forms.[6] In his last published work[7] Professor Owen does not hesitate to state 'that perhaps the most important and significant result of palaeontological research has been the establishment of the axiom of *the continuous operation of the ordained becoming of living things.*' The italics are the author's. As to his own opinions regarding the nature or mode of that 'continuous creative operation,' the Professor is silent. He gives a brief summary of the hypotheses of others, and as briefly touches upon the defects in their inductive bases.[8] Elsewhere he has restricted himself to testing the idea of progressive transmutation by such subjects of Natural History as he might have specially in hand: as, e.g. the characters of the chimpanzee, gorilla, and some other animals.

All who have brought the transmutative speculations to the test of observed facts and ascertained powers in organic life, and have published the results, usually adverse to such speculations, are set down by Mr. Darwin as 'curiously illustrating the blindness of preconceived opinion;' and whosoever may withhold assent to his own or other transmutationists' views, is described as 'really believing that at innumerable periods of the earth's history certain elemental atoms suddenly flashed into living tissues' (p. 483). Which, by the way, is but another notion of the mode of becoming of a species as little in harmony with observation as the hypothesis of natural selection by external influence, or of exceptional birth or development. Nay, Mr. Darwin goes so far as to affirm—

> All the most eminent palaeontologists, namely, Cuvier, Owen, Agassiz, Barrande, Falconer, E. Forbes, &c., and all our greatest geologists, as Lyell, Murchison, Sedgwick, &c., have unanimously, often vehemently, maintained the immutability of species (p. 310).

But if by this is meant that they as unanimously reject the evidences of a constantly operative secondary cause of law in the production of the

---

6. The most numerous illustrations of this principle are to be found in Owen's palaeontological works and memoirs: but he refrains from announcing it as a general law, probably regarding the induction as being yet incomplete.

7. *Palaeontology, or a Systematic Summary of Extinct Animals, and their Geological Relations,* 8vo. (1860), p. 3; and President's Address to the British Association at Leeds (1858), p. 3.

8. *Palaeontology,* p. 404.

succession of specifically differing organisms, made known by Palaeontology, it betrays not only the confusion of ideas as to the fact and the nature of the law, but an ignorance or indifference to the matured thoughts, and expressions of some of those eminent authorities on this supreme question in Biology.

One of the disciples would seem to be as short-sighted as the master in regard to this distinction.

'It has been urged,' writes Dr. Hooker, 'against the theory that existing species have arisen through the variation of pre-existing ones and the destruction of intermediate varieties, that it is a hasty inference from a few facts in the life of a few variable plants, and is therefore unworthy of confidence; but it appears to me that the opposite theory, which demands an independent creative act for each species, is an equally hasty inference' (Hooker, p. xxv).

Here it is assumed, as by Mr. Darwin, that no other mode of operation of a secondary law in the foundation of a form with distinct specific characters, can have been adopted by the Author of all creative laws than the one which the transmutationists have imagined. Any physiologist who may find the Lamarckian, or the more diffused and attenuated Darwinian, exposition of the law inapplicable to a species, such as the gorilla, considered as a step in the transmutative production of man, is forthwith clamoured against as one who swallows up every fact and every phenomenon regarding the origin and continuance of species 'in the gigantic conception of a power intermittently exercised in the development, out of inorganic elements, of organisms the most bulky and complex as well as the most minute and simple.' Significantly characteristic of the partial view of organic phenomena taken by the transmutationists, and of their inadequacy to grapple with the working out and discovery of a great natural law, is their incompetency to discern the indications of any other origin of one specific form out of another preceding it, save by their way of gradual change through a series of varieties assumed to have become extinct.

But has the free-swimming medusa, which bursts its way out of the ovicapsule of a campanularia, been developed out of inorganic particles? Or have certain elemental atoms suddenly flashed up into acalephal form? Has the polype-parent of the acalephe necessarily become extinct by virtue of such anomalous birth? May it not, and does it not proceed to propagate its own lower species, in regard to form and organisation, notwithstanding its occasional production of another very different and higher kind. Is the fact of one animal giving birth to another not merely specifically, but generically and ordinally, distinct, a solitary one? Has not Cuvier, in a score or more of instances, placed the parent in one class, and the fruitful offspring in another class, of animals? Are the entire series of parthenogenetic phenomena to be of no account in the consideration of the supreme problem of the introduction of fresh specific forms into this planet? Are the transmutationists to monopolise the privilege of conceiving the possibility of the occurrence of unknown phenomena, to be the exclusive propounders of beliefs and surmises, to cry down every kindred barren speculation, and to allow no indulgence in any mere hypothesis save their own? Is it to be endured that every observer who points out a case to which transmutation, under whatever term disguised, is inapplicable, is to be set down by the refuted theorist as a believer in a mode of manufacturing a species, which he never did believe in, and which may be inconceivable?

We would ask Mr. Darwin and Dr. Hooker to give some thought to these queries, and if they should see the smallest meaning in them, to reconsider their future awards of the alternative which they may be pleased to grant to a fellow-labourer, hesitating to accept the proposition, either that life commenced under other than actually operating laws, or that 'all the beings that ever lived on this earth have descended,' by the way of 'natural selection,' from a hypothetical unique instance of a miraculously created primordial form.

We are aware that Professor Owen and others, who have more especially studied the recently discovered astounding phenomena of generation summed up under the terms Parthenogenesis and Alternation of Generations, have pronounced against those phenomena having, as yet, helped us 'to penetrate the mystery of the origin of different species of animals,' and have affirmed, at least so far as observation has yet extended, that 'the cycle of changes is definitely closed;' that is, that when the ciliated 'monad' had given birth to the 'gregarina,' and this to the 'cercaria,' and the 'cercaria' to the 'distoma'—that the fertilised egg of the flukeworm again excludes the progeny under the infusorial or monadic form, and the cycle again recommences.[9] But circumstances are conceivable,–changes of surrounding influences, the operation of some intermittent law at long intervals, like that of the calculating-machine quoted by the author of 'Vestiges,'–under which the monad might go on splitting up into monads, the gregarina might go on breeding gregarinae, the cercaria cercariae, &c., and thus four or five not merely different specific, but different generic, and ordinal forms, zoologically viewed, might all diverge from an antecedent

9. President's Address to the British Association at Leeds, p. 27.

quite distinct form. For how many years, and by how many generations, did the captive polype-progeny of the *Medusa aurita* go on breeding polypes of their species (*Hydra tuba*), without resolving themselves into any higher form, in Sir John Dalyell's aquarium![10] The natural phenomena already possessed by science are far from being exhausted on which hypotheses, other than transmutative, of the production of species by law might be based, and on a foundation at least as broad as that which Mr. Darwin has exposed in this essay.

We do not advocate any of these hypotheses in preference to the one of 'natural selection,' we merely affirm that this at present rests on as purely a conjectural basis. The exceptions to that and earlier forms of transmutationism which rise up in the mind of the working naturalist and original observer, are so many and so strong, as to have left the promulgation and advocacy of the hypothesis, under any modification, at all times to individuals of more imaginative temperament; such as Demaillet in the last century, Lamarck in the first half of the present, Darwin in the second half. The great names to which the steady inductive advance of zoology has been due during those periods, have kept aloof from any hypothesis on the origin of species. One only, in connexion with his palaeontological discoveries, with his development of the law of irrelative repetition and of homologies, including the relation of the latter to an archetype, has pronounced in favour of the view of the origin of species by a continuously operative creational law; but he, at the same time, has set forth some of the strongest objections or exceptions to the hypothesis of the nature of that law as a progressively and gradually transmutational one.

Mr. Darwin rarely refers to the writings of his predecessors, from whom, rather than from the phenomena of the distribution of the inhabitants of South America, he might be supposed to have derived his ideas as to the origin of species. When he does allude to them, their expositions on the subject are inadequately represented. Every one studying the pages of Lamarck's original chapters (iii. vi. vii., vol. I., and the supplemental chapter of 'additions' to vol. II, of the 'Philosophie Zoologique'), will see how much weight he gives to inherent constitutional adaptability, to hereditary influences, and to the operation of long lapses of time on successive generations, in the course of transmuting a species. The common notion of Lamarck's philosophy, drawn from the tirades which a too figurative style of illustrating the reciprocal influence of innate tendencies and outward

10. See the beautiful work entitled 'Rare and Remarkable Animals of Scotland,' 4to. vol. i. 1847, by Sir. J. G. Dalyell.

influences have drawn upon the blind philosopher, is incorrect and unjust.
Darwin writes:–

Naturalists continually refer to external conditions, such as climate,
food, &c., as the only possible cause of variation. In one very limited
sense, as we shall hereafter see, this may be true; but it is preposterous
to attribute to mere external conditions, the structure, for instance, of
the woodpecker, with its feet, tail, beak, and tongue, so admirably adapted
to catch insects under the bark of trees. In the case of the misseltoe,
which draws its nourishment from certain trees, which has seeds that
must be transported by certain birds, and which has flowers with separate
sexes absolutely requiring the agency of certain insects to bring pollen
from one flower to the other; it is equally preposterous to account for
the structure of this parasite, with its relations to several distinct organic
beings, by the effects of external conditions, or of habit, or of the volition
of the plant itself.

The author of the "Vestiges of Creation" would, I presume, say that,
after a certain unknown number of generations, some bird had given
birth to a woodpecker, and some plant to the misseltoe, and that these
had been produced perfect as we now see them; but this assumption
seems to me to be no explanation, for it leaves the case of the coadapta-
tions of organic beings to each other and to their physical conditions
of life untouched and unexplained (p. 3).

The last cited ingenious writer came to the task of attempting to unravel
the 'mystery of mysteries,' when a grand series of embryological researches
had brought to light the extreme phases of form that the higher animals
passed through in the course of foetal development, and the striking analogies which transitory embryonal phases of a higher species presented to
series of lower species in their permanent or completely developed state.
He also instances the abrupt departure from the specific type manifested
by a malformed or monstrous offspring, and called to mind the cases in
which such malformations had lived and propagated the deviating structure.
The author of 'Vestiges,' therefore, speculates—and we think not more
rashly or unlawfully than his critic has done—on other possibilities, other
conditions of change, than the Lamarckian ones; as, for example, on the
influence of premature birth and of prolonged foetation in establishing
the beginning of a specific form different from that of the parent. And
does not the known history of certain varieties, such as that of M. Graux's
cachemir-wooled sheep, which began suddenly by malformation, show the
feasibility of this view?[11] 'The whole train of animated beings,' writes the

11. *Reports of the Juries Exhibition of the Works of All Nations,* 8vo. (1852),
p. 70.

author of 'Vestiges of Creation,' 'are the results *first,* of an inherent impulse in the forms of life to advance, in definite times, through grades of organisation terminating in the highest dicotyledons and mammals; *second,* of external physical circumstances, operating reactively upon the central impulse to produce the requisite peculiarities of exterior organisation,–the adaptation of the natural theologian.' But he, likewise, requires the same additional element which Mr. Darwin so freely invokes.

The gestation of a single organism is the work of but a few days, weeks, or months; but the gestation (so to speak) of a whole creation is a matter involving enormous spaces of time . . . Though distinctions admitted as specific are not now, to ordinary observation, superable, time may have a power over these . . . Geology shows successions of forms, and grants enormous spaces of time within which we may believe them to have changed from each other by the means which we see producing varieties. Brief spaces of time admittedly sufficing to produce these so-called varieties, is it unreasonable to suppose that large spaces of time would effect mutations somewhat more decided, but of the same character?[12]

Unquestionably not, replies Mr. Darwin:–

To give an imaginary example from changes in progress on an island: let the organisation of a canine animal which preyed chiefly on rabbits, but sometimes on hares, become slightly plastic; let these same changes cause the number of rabbits very slowly to decrease, and the number of hares to increase; the effect of this would be that the fox or dog would be driven to try to catch more hares; his organisation, however, being slightly plastic, those individuals with the lightest forms, longest limbs, and best eyesight, let the difference be ever so small, would be slightly favoured, and would tend to live longer, and to survive during that time of the year when food was scarcest; they would also rear more young, which would tend to inherit these slight peculiarities. The less fleet ones would be rigidly destroyed. I can see no more reason to doubt that these causes in a thousand generations would produce a marked effect, and adapt the form of the fox or dog to the catching of hares instead of rabbits, than that greyhounds can be improved by selection and careful breeding![13]

Of course, prosaic minds are apt to bore one by asking for our proofs, and one feels almost provoked, when seduced to the brink of such a draught of forbidden knowledge as the transmutationists offer, to have the Circean

12. *Vestiges of Creation,* 8vo. 1846, p. 231.
13. 'On the Tendency of Species to form Varieties,' &c., in *Proceedings of the Linnaean Society,* 1858, p. 49.

cup dashed away by the dry remark of a President of the British Association:–

> Observation of animals in a state of nature is required to show their degreee of plasticity, or the extent to which varieties do arise: whereby grounds may be had for judging of the probability of the elastic ligaments and joint-structures of a feline foot, for example, being superinduced upon the more simple structure of the toe with the nonretractile claw, according to the principle of a succession of varieties in time.'[14]

This very writer has, however, himself suggested an operative cause in the development of organised beings of a different and opposite character to that conceived by 'Vestiges,' to produce the teleological adaptations. Professor Owen has pointed out the numerous instances in the animal kingdom of a principle of structure prevalent throughout the vegetable kingdom, exemplified by the multiplication of organs in one animal performing the same function, and not related to each other by combination of powers for the performance of a higher function. The Invertebrate animals, according to the Professor, afford the most numerous and striking illustrations of the principles which he has generalised as the 'Law of Irrelative Repetition.'

'We perceive,' says he,

> in the fact of the endoskeleton consisting of a succession of segments similarly composed—in the very power of enunciating special, general, and serial homologies—an illustration of that law of vegetative or irrelative repetition, which is so much more conspicuously manifest by the segments of the exoskeleton of the Invertebrate: as, for example, in the rings of the centipede and worm, and in the more multiplied parts of the skeleton of the Echinoderus. The repetition of similar segments in the spinal column, and of similar elements in a vertebral segment, is analogous to the repetition of similar crystals, as the result of the polarising force in the growth of an inorganic body. Not only does the principle of vegetative repetition prevail more and more as we descend in the scale of animal life, but the forms of the repeated parts of the skeleton approach more and more to geometrical figures; as we see, for example, in the external skeletons of the echini and star fishes: nay, the calcifying salt assumes the same crystalline figures which characterise it, when deposited and subject to the general polarising force out of the organised body. Here, therefore, we have direct proof of the concur-

14. [President's Address to the British Association at Leeds], p. 44. [The President in this instance was Owen himself; reprinted in Basalla, Coleman, and Kargon (1970).]

rence of such general all pervading polarising force, with the adaptive or special organising force, in the development of an animal body.

In addition, therefore, to the organising principle, however explained, producing the special 'adaptations,' and admitted as the 'second' power in the production of species by 'Vestiges,' Professor Owen states—

'There appears also to be in counter-operation during the building up of such bodies, a general polarising force, to the operation of which the similarity of forms, the repetition of parts, the signs of the unity of organisation may be mainly ascribed; the platonic *idea* or specific organising principle would seem,' he adds, 'to be in antagonism with the general polarising force, and to subdue and mould it in subserviency to the exigencies of the resulting specific form.'[15]

An index of the degree in which the polaric or irrelative repetitive force has operated is given by that character of the animal's organisation which is expressed by the term of 'a more generalised structure.' V[on] Baer pointed out that the structure was 'more generalised,' in the ratio of the proximity of the individual to the starting point of its existence. In proportion as the individual is subject to the action and reaction of surrounding influences, in other words, as it advances in life, does it acquire a more specialised structure—more decided specific and individual characters.[16] Owen has shown that the more generalised structure is, in a very significant degree, a characteristic of many extinct as compared with recent animals; and it may be readily conceived that specialisation of structure would be the result of the progressive modification of any organ applied to a special purpose in the animal economy.

We have cited these attempts to elucidate the nature of the organising forces, to show the prevalent condition of the most advanced physiological minds in regard to the cause of the successive introduction of distinct species of plants and animals. Demaillet invoked the operation of the external influences or conditions of life, with consentaneous volitional efforts, in order to raise species in the scale, as the fish, e.g., into the bird.[17] Buffon called in the same agency to lower the species by way of degeneration, as the bear, e.g., into the seal, and this into the whale.[18] Lamarck added

15. *Archetype of the Vertebrate Skeleton,* 8vo. (1840), p. 171.

16. 'The extent to which the resemblance, expressed by the term, "Unity of Organisation," may be traced between the higher and lower organised animals, bears an inverse ratio to their approximation to maturity' (*Owen, Lectures on Invertebrata* [1843], p. 645).

17. *Telliamed, ou Entretiens d'un Philosophe Indien avec un missionaire François,* Amsterdam, 8vo. (1748).

18. *Histoire Naturelle,* &c., 4to., [vol.] xiv, 1766.

to these outward influences the effects of increased or decreased use or action of parts. The Author of 'Vestiges,' availing himself of the ingenious illustration of a pre-ordained exception, occurring at remote intervals, to the ordinary course, derived by Babbage from the working of his Calculating Engine, threw out the suggestion of a like rare exception in the character of the offspring of a known species, and he cites the results of embryological studies, to show how such 'monster,' either by excess or defect, by arrest or prolongation of development, might be no monster in fact, but one of the preordained exceptions in the long series of natural operations, giving rise to the introduction of a new species. Owen has not failed to apply the more recent discoveries of Parthenogenesis to the same mysterious problem. A polype, e.g., breaks up into a pile of medusae; 'the indirect or direct action of the conditions of life' might tend to harden the integument and change the medusa into a starfish. But he resists the seduction of possibilities, and governed by the extent of actual observation, says:–'The first acquaintance with these marvels excited the hope that we were about to penetrate the mystery of the origin of species; but, as far as observation has yet extended, the cycle of changes is definitely closed.'[19]

Mr. Wallace calls attention to the 'tremendous rate of increase in a few years from a single pair of birds producing two young ones each year, and this only four times in their life; in fifteen years such pair would have increased to nearly ten millions!'[20] The passenger-pigeon of the United States exemplifies such rate of increase, where congenial food abounds. But, as a general rule, the animal population of a country is stationary, being kept down by a periodical deficiency of food and other checks. Hence the struggle for existence; and the successful result of adapted organisation and powers in a well developed variety, which Mr. Darwin generalises as 'Natural Selection,' and which Mr. Wallace[21] illustrates as follows:–

'An antelope with shorter or weaker legs must necessarily suffer more from the attacks of the feline carnivora; the passenger-pigeon with less powerful wings, would sooner or later be affected in its powers of procuring a regular supply of food.'[22] If, on the other hand, 'any species should produce a variety having slightly increased powers of preserving existence, that variety must inevitably in time acquire a superiority in numbers.' 'During any change tending to render existence more difficult to a species,

19. Address to the British Association at Leeds (1858), p. 27.
20. *Proceedings of the Linnaean Society*, 1858, p. 55.
21. *Proceedings of the Linnaean Society* (dated from 'Ternate,' February 1858), vol. III, p. 58.
22. Wallace, [*Proceedings of the Linnaean Society*, 1858, vol. III], p. 85.

tasking its utmost powers to avoid complete extermination, those individuals forming the most feebly organised variety would suffer first; the same causes continuing, the parent species would next suffer, would gradually diminish in numbers, and with a recurrence of similar unfavourable conditions, must soon become extinct. The superior variety would then alone remain, and on a return to favourable circumstances would rapidly increase in numbers and occupy the place of the extinct species and variety. The *variety* would now have replaced the *species,* of which it would be a more perfectly developed and a more highly organised form.'[23]

Buffon regarded varieties as particular alterations of species, as supporting and illustrating a most important principle—the mutability of species themselves. The so-called varieties of a species, species of a genus, genera of a family, &c., were, with him, so many evidences of the progressive amount or degrees of change which had been superinduced by time and generations upon a primordial type of animal. Applying this principle to the two hundred mammalian species of which he had given a history in his great work, he believed himself able to reduce them to a very small number of primitive stocks or families.[24] Of these he enumerates fifteen: besides which, Buffon specifies certain isolated forms, which represent, as he forcibly and truly expresses it, both species and genus:[25] such are the elephant, rhinoceros, hippopotamus, giraffe, camel, lion, bear, and mole.[26] Palaeontology has since revealed the evidences of the true nature and causes of the present seeming isolation of some of these forms.

Such evidences have been mainly operative with the later adopters and diffusers of Buffon's principle in the reduction of the number of primitive sources of existing species, and the contraction of the sphere of direct creative acts. Thus Lamarck[27] reduces the primordial forms or prototypes of animals to two, viz. the worm (vers), and the monad (infusoires); the principles which in the course of illimited time operated, on his hypothesis, to produce the present groups of animals led from the vibrio, through the annelids, cirripeds, and molluscs to fishes, and there met the other developmental route by way of rotifers, polypes, radiaries, insects, arachnides, and crustacea. The class of fishes, deriving its several forms from combinations of transmuted squids and crabs, then proceeded through the

23. Ibid., p. 58.
24. *Histoire naturelle,* vol. xiv, p. 338.
25. "Quelques especes isolees, qui, comme celle de Phomme, fassent en meme temps espece et genre" (ibid., p. 335).
26. Ibid., p. 360.
27. *Philosophie zoologique,* vol. II, p. 463.

well-defined vertebrate pattern up to man. With a philosophic consistency, wanting in his latest follower, Lamarck sums up: 'Cette serie d'animaux commencant par deux branches ou se trouvent les plus imparfaits, les premiers de chacune de ces branches ne recoivent l'existence que par generation directe ou spontanee.'[28]

Mr. Darwin, availing himself of the more exact ideas of the affinities and relationships of animal groups obtained by subsequent induction, says, 'I believe that animals have descended from at most only four or five progenitors.' evidently meaning, or answering to, the type-forms of the four or five 'sub-kingdoms' in modern zoology, 'and plants from an equal or lesser number.'

But if the means which produce varieties have operated 'through the enormous species of time, within which species are changed,'[29] the minor modifications which produce, under our brief scope of observation, so-called varieties, might well amount to differences equivalent to those now separating sub-kingdoms; and, accordingly, 'analogy,' Mr. Darwin logically admits, 'would lead us one step further, namely to the belief that all animals and plants have descended from some one prototype;[30] and, summing up the conditions which all living things have in common, this writer infers from that analogy, 'that probably all the organic beings which have ever lived on this earth, have descended from some one primordial form, into which life was first breathed.'[31]

By the latter scriptual phrase, it may be inferred that Mr. Darwin formally recognises, in the so-limited beginning, a direct creative act, something like that supernatural or miraculous one which, in the preceding page, he defines, as 'certain elemental atoms which have been commanded suddenly to flash into living tissues.' He has, doubtless, framed in his imagination some idea of the common organic prototype; but he refrains from submitting it to criticism. He leaves us to imagine our globe, void, but so advanced as to be under the conditions which render life possible; and he then restricts the Divine power of breathing life into organic form to its minimum of direct operation. All subsequent organisms henceforward result from properties imparted to the organic elements at the moment of their creation, pre-adapting them to the infinity of complications and their morphological results, which now try to the utmost the naturalist's faculties to comprehend and classify. And we admit with Buckland, that

28. Ibid. p. 463.
29. *Vestiges of Creation*, p. 231.
30. [*On the Origin of Species*], p. 484.
31. Ibid.

such an aboriginal constitution, 'far from superseding an intelligent agent, would only exalt our conceptions of the consummate skill and power, that could comprehend such an infinity of future uses, under future systems, in the original groundwork of his creation.' We would accordingly assure Professor Owen that he 'may conceive the existence of such ministers, personified as Nature, without derogation of the Divine power;' and that he, with other inductive naturalists, may confidently advance in the investigation of those 'natural laws or secondary causes, to which the orderly succession and progression of organic phenomena have been committed.'[32] We have no sympathy whatever with Biblical objectors to creation by law, or with the sacerdotal revilers of those who would explain such law. Literal scripturalism in the time of Lactantius, opposed and reviled the demonstrations of the shape of the earth; in the time of Galileo it reviled and persecuted the demonstrations of the movements of the earth; in the time of Dean Cockburn of York, it anathematised the demonstrations of the antiquity of the earth; and the eminent geologist who then personified the alleged antiscriptural heresy, has been hardly less emphatic than his theological assailant, in his denunciations of some of the upholders of the 'becoming and succession of species by natural law,' or by 'a continuously operating creative force.' What we have here to do, is to express our views of the hypothesis as to the nature and mode of operation of the creative law, which has been promulgated by Messrs. Wallace and Darwin.

The author of the volume 'On the Origin of Species,' starts from a single supernaturally created form. He does not define it; it may have been beyond his power of conception. It is, however, eminently plastic, is modified by the influence of external circumstances, and propagates such modification by generation. Where such modified descendants find favourable conditions of existence, there they thrive; where otherwise they perish. In the first state of things, the result is so analogous to that which man brings about, in establishing a breed of domestic animals from a selected stock, that it suggested the phrase of 'Natural Selection;' and we are appealed to, or at least 'the young and rising naturalists with plastic minds,'[33] are adjured, to believe that the reciprocal influences so defined have operated, through divergence of character and extinction, on the descendants of a common parent, so as to produce all the organic beings that live, or have ever lived, on our planet.

Now we may suppose that the primeval prototype began by producing,

32. *On the Nature of Limbs*, p. 86.
33. [*On the Origin of Species*], p. 482.

in the legal generative way, creatures like itself, or so slightly affected by external influences, as at first to be scarcely distinguishable from their parent. When, as the progeny multiplied and diverged, they came more and more under the influence of 'Natural Selection,' so, through countless ages of this law's operation, they finally rose to man. But, we may ask, could any of the prototype's descendants utterly escape the surrounding influences? To us such immunity, in the illimitable period during which the hypothesis of Natural Selection requires it to have operated, is inconceivable. No living being, therefore, can now manifest the mysterious primeval form to which Darwin restricts the direct creative act; and we may presume that this inevitable consequence of his hypothesis, became to him an insuperable bar to the definition of that form.

But do the facts of actual organic nature square with the Darwinian hypothesis? Are all the recognised organic forms of the present date, so differentiated, so complex, so superior to conceivable primordial simplicity of form and structure, as to testify to the effects of Natural Selection continuously operating through untold time? Unquestionably not. The most numerous living beings now on the globe are precisely those which offer such a simplicity of form and structure, as best agrees, and we take leave to affirm can only agree, with that ideal prototype from which, by any hypothesis of natural law, the series of vegetable and animal life might have diverged.

If by the patient and honest study and comparison of plants and animals, under their manifold diversities of matured form, and under every step of development by which such form is attained, any idea may be gained of a hypothetical primitive organism,—if its nature is not to be left wholly to the unregulated fancies of dreamy speculation—we should say that the form and condition of life which are common, at one period of existence, to every known kind and grade of organism, would be the only conceivable form and condition of the one primordial being from which 'Natural Selection' infers that all the organisms which have ever lived on this earth have descended.

Now the form in question is the nucleated cell, having the powers of receiving nutritive matter from without, of assimilating such nutriment, and of propagating its kinds by spontaneous fission. These powers are called 'vital,' because as long as they are continued the organism is said to live. The most numerous and most widely diffused of living being present this primitive grade of structure and vital force, which grade is inferior to that of the truly definable 'plant' or 'animal,' but is a grade represented and passed through by the germ of every, even the highest, class of animals, in the course of embryonic development. The next stages of differentiated or advanced organisation are defined as follows in Professor Owen's last publication:–

> When the organism is rooted, has neither mouth nor stomach, exhales oxygen, and has tissues composed of "cellulose" or of binary or ternary compounds, it is called a "plant." When the organism can move, when it receives the nutritive matter by a mouth, inhales oxygen, and exhales carbonic acid, and developes tissues, the proximate principles of which are quaternary compounds of carbon, hydrogen, oxygen, and nitrogen, it is called an "animal." But the two divisions of organisms called "plants" and "animals" are specialised members of the great natural groups of living things; and there are numerous organisms, mostly of minute size and retaining the form of nucleated cells, which manifest the common organic characters, but without the distinctive superadditions of true plants or animals. Such organisms are called "Protozoa," and include the sponges or *Amorphozoa,* the *Foraminifera* or Rhizopods, the *Polycystinea,* the *Diatomacea, Desmidia, Gregaerina,* and most of the so-called *Polygastria* of Ehrenberg, or infusorial animalcules of older authors.[34]

All these would be interpreted as the earliest evidences of the modifying and species-changing influences, according to the hypothesis of Lamarck. They are the organisms respecting which the first living physiologists hesitate to apply the Harveian axiom *omne vivum ab ovo,* believing the possibility of their spontaneous origin to be by no means experimentally disproved. The prevalence of the essential first step in the production of all higher organisms, viz. through the combined matter of the 'germ-cell' and 'sperm-cell,' has no doubt strongly inclined physiologists to believe impregnation to be an absolute condition of the beginning of all existing organisms. But, as the President of the British Association stated, in his 'Address' at Leeds:–

> In regard to lower living things, analogy is but hazardous ground for conclusions. The single-celled organisms, such as many of the so-called animalcules of infusions, which are at a stage of organisation too low for a definite transfer to either the vegetable or animal kingdom, offer a field of observation and experiment which may yet issue in giving us a clearer insight into the development of the organic living cell.—Whether an independent free-moving and assimilating organism, of a grade of structure similar to, and scarcely higher than, the "germ cell" may not arise by a collocation of particles, through the operation of a force analogous to that which originally formed the germ-cell in

34. Owen's *Palaeontology,* p. 4.

the ovarian stroma, is a question which cannot be answered until every possible care and pains have been applied to its solution (p. 28).

Professor Pouchet believes that he is authorised by the results of his experiments to answer that question in the affirmative. It is one of supreme importance, and which has, hitherto, never received such an amount of painstaking experimental research as it merits; and the best observations, the most carefully conducted and ingeniously devised arrangements for insuring success, are undoubtedly those of the patiently observant Professor of Zoology in the 'Ecole de Medecine,' and 'Ecole superieure des Sciences,' at Rouen.[35] This, at least, may be affirmed, that the inductive groundwork of his opponents is by no means such as can justify any dogmatic negation of Heterogeny as applicable to the simplest Protozoa.

On the basis, therefore, of analogical probability, it may be inferred:– that the primordial as well as all other forms of organic beings, originate, and have ever originated, from the operation of secondary and continuously operating creative laws: and that the various grades of organisms now in being, from the microscopic monad upwards, indicate the various periods in time at which the first step of the series they respectively terminate began. The monad that by 'natural selection' has ultimately become man, dates from the farthest point in the remote past, upon which our feigners of developmental hypotheses can draw with unlimited credit: the monad which by its superficial vibratile cilia darted across the field of the microscope we were looking through this morning, is the result of the collocation of particles which, without 'sudden flash,' took place under the operation of the heterogeneous organising force of yesterday.

Accordingly we find that every grade of structure, from the lowest to the highest, from the most simple to the most complex, is now in being—a result which it is impossible to reconcile with the Darwinian hypothesis of the one and once only created primordial form, the parent of all subsequent living things. The changes which our planet has undergone in the course of geological time have been accompanied by the loss of many minor links which connected together the existing evidences of gradational struc-. ture; but the general laws regulating the progress and diversity of organic forms, having been the same throughout all time, so it happens, according to the testimony of the most experienced palaeontologists, that—

'Every known fossil belongs to some one or other of the existing classes, and that the organic remains of the most ancient fossiliferous strata do

35. Pouchet [Félix Archimède], *Heterogenie, ou Traite de la Generation spontanée base sur des nouvelles Experiences*, 8vo., 1859.

not indicate or suggest that any earlier and different group of beings remains to be discovered, or has been irretrievably lost in the universal metamorphism of the oldest rocks.'[36]

That forms, recognised as species by their distinctive characters and the power of propagating them, have ceased to exist, and have successively passed away, is a fact now unquestioned; that they have been exterminated by exceptional cataclysmal changes of the earth's surface, as was surmised at the first acceptance of the fact of extinction, has not been proved; that their limitation in time may, in some instances, or in some degree, be due to constitutional changes, accumulating by slow degrees in the long course of generations, is possible: but all the traceable and observed causes of extirpation point either to continuous slowly operating geological changes, or to no greater sudden cause than the apparition of mankind on a limited tract of land not before inhabited. It is now, therefore, generally inferred that the extinction of species, prior to man's existence, has been due to ordinary causes—ordinary in the sense of agreement with the great laws of never-ending mutation of geographical and climatal conditions on the earth's surface. The individuals of species least adapted to bear such influences and incapable of modifying their organisation in harmony therewith, have perished. Extinction, therefore, on this hypothesis, is due to the want of self-adjusting, self-modifying power in the individuals of the species.

In the joint paper on the tendency of varieties to form species by natural means of selection,[37] one of the authors writes:–

'Any minute variation in structure, habits, or instincts, adapting the individual better to the new conditions, would tell upon its vigour and health. In the struggle it would have a better chance of surviving, and those of its offspring which inherited the variation would also have a better chance. Let this work go on for a thousand generations, and who will pretend to affirm,' asks Mr. Darwin, 'that a new species might not be the result?'

Thereupon is adduced the imaginary example of dogs and rabbits on an island, which we have already cited.

Now this, we take leave to say, is no very profound or recondite surmise; it is just one of those obvious possibilities that might float through the imagination of any speculative naturalist; only, the sober searcher after truth would prefer a blameless silence to sending the proposition forth as explanatory of the origin of species, without its inductive foundation.

36. Owen's *Palaeontology*, p. 18.
37. By Darwin and Wallace, 'Proceedings of the Linnaean Society,' August, 1858, p. 45.

In the degeneration-theory of Buffon, man is one of the primitive types,–the created apes and monkeys are derivatives. He might have illustrated it as follows:–

To give an imaginary example from changes in progress on an island: let the organisation of a wild man feeding chiefly on fruits become slightly plastic; let corresponding changes cause the sources of food on the ground very slowly to decrease, and those on the trees to increase: the effect of this would be that the man would try to climb more for food. Suppose also that a tiger or like destructive carnivore should swim over and settle in the island, which happened to be destitute of flints for weapons. The human organisation being slightly plastic, those individuals with the longest and strongest arms, and with the most prehensile use of the great toe, let the difference be ever so small, would be slightly favoured, would survive during that time of the year when food was scarcest on the ground, but ripe and ready on certain trees; they would also rear more young which would tend to inherit these slight peculiarities. The best climbers would escape the tigers, the worst would be rigidly destroyed.

Buffon would have seen no more reason to doubt that these causes, in a thousand generations, would produce a marked effect, and adapt the form of the wild man to obtain fruits rather than grains, than Darwin now believes that man can be improved by selection and careful interbreeding into a higher, more heroic, more angelic form! The advocate of Buffon's hypothesis might point out that it is on islands, as Borneo and Sumatra, for example, where the orang-utan—the obvious result of such 'degradation by natural selection'—is exclusively found. And is it not there also, and in some other islands of the Malayan Archipelago, where the next step in the scale of 'degeneration' is exhibited in the still longer-armed Ungkas and other tail-less *Hylobates?* And though we call them 'tail-less' yet they have the 'os coccygis;' and this being a terminal appendage of stunted vertebrae, offers the very condition for the manifestation of an occasional developmental variety. If cats, after accidental mutilation or malformation, can propagate a tail-less breed, why may not apes produce a tailed variety, and by natural selection in a long course of ages, degenerate into endless incipient species of 'baboons and monkeys'?

But Mr. Darwin, it may be said, repudiates the coarse transmutational conditions and operations of Buffon and Lamarck: or, if there be any parallel between his and Buffon's illustration of the changing of species, at all events such parallels must run in opposite directions. Mr. Darwin starts from a single created prototype, from which it is difficult to conceive he can mean any other course of organic progress than an ascensive one. But of this in the absence of a definition of the starting point, we cannot be perfectly sure. 'Natural selection' may operate in both directions. The following, for example, would have been cordially welcomed by Buffon as a testimony in favour of his 'degeneration' hypothesis:—

> In North America the black bear was seen by Hearne swimming for hours with widely opened mouth, thus catching, like a whale, insects in the water. Even in so extreme a case as this, if the supply of insects were constant, and if better adapted competitors did not already exist in the country, I can see no difficulty in a race of bears being rendered, by natural selection, more and more aquatic in their structure and habits, with larger and larger mouths, till a creature was produced as monstrous as a whale.[38]

If the ursine species had not been restricted to northern latitudes, we might have surmised this to have been one of the facts connected with 'the distribution of the inhabitants of South America,' which seemed to Mr. Darwin, when naturalist on board H.M.S. Beagle, 'to throw some light on the origin of species.' But the close resemblance of the style, and of the tone and frame of mind which could see no difficulty in the adequacy of the above-cited circumstances of 'external conditions, of habit, of violation,' to change a bear into a whale, to those exemplified in the 'Philosophie Zoologique,' point strongly to the writings of Lamarck as the true suggestor of Mr. Darwin's views of animated nature. We look, however, in vain for any instance of hypothetical transmutation in Lamarck so gross as the one above cited: we must descend to the older illustrators of the favourite idea, to find an equivalent case of the bear in pursuit of water-insects, and we find one in the following*:—

> Flying fish may have developed a greater facility for flying through their desire for prey while hunting or from their fear of death while being hunted, or by being forced ashore by heavy waves and finding themselves trapped in the reeds or grass from which they could not escape back to the sea which, as we all know, can and frequently does happen. Their fins, when no longer bathed in salt water, dried out, cracked, and lost their shape. As long as they found enough food in the reeds and grass where they were trapped, they survived. The fin rays separated from each other, lengthened and became covered with

38. Darwin, p. 184 (1st edition).
* [The following excerpts were translated from the French by D. L. H.]

barbs, or, to be more precise, the membranes which once held the fin
rays together metamorphized. The barbs shaped from this deformed pelli-
cle became extended. Little by little the skin of these animals became
covered with down of the same color as their bodies, and down which
gradually became denser. Their small ventral fins which, like their larger
fins, had once been used for swimming became feet for walking on land.
Still other small changes took place. The nose and neck of some became
enlarged; some became smaller. Similar changes occurred throughout
the rest of the body. Nevertheless, conformity to the original form, which
remains in the overall makeup, is and will always remain easy to
recognize.

Indeed, examine all the species of hen, the large and the small, even
those of India, those which have crests and those which do not, those
which have feathers running the wrong way as one sees in Damiette,
that is, they are inclined from the tail to the head. You will find species
in the sea which are the same, with scales and without. To every species
of parrot, whose plumage is the most diverse, the rarest and the most
singularly marked of all the birds, there corresponds a fish colored like
them, in black, brown, grey, yellow, green, red, violet, gold and blue,
and these markings appear on the same parts of the fish as they do in
such a bizarre manner on the plumage of these very birds.[39]

Demaillet, it must be admitted, enters more fully into the details of
the operation of 'natural selection,' in changing the fish into the bird;
and it is perhaps, from this very 'naivete' in the exposition of his theory
that its weakness has been made so obvious to later zoologists and comparative anatomists. Mr. Darwin rarely shows a fair front to these searching
tests; the facts of the manner of transmutation, as they might have presented themselves to his fancy, are not stated with the 'abandon' of the
old French philosopher. Vague and general as is the illustration based
upon Hearne's remark, it is made still more vague in a later reprint of
the volume 'On the Origin of Species.' It now reads, 'In North America
the black bear was seen by Hearne swimming for hours with widely opened
mouth, thus catching almost like a whale, insects in the water' (Ed. 1860,
p. 184).

'Individuals, it is said, of every species, in a state of nature annually
perish,' and 'the survivors will be for the most part, those of the strongest

39. *Telliamed, ou Entretiens d'un Philosophe Indien avec un Missionaire François,
sur la Diminution de la Mer,* &c., 8vo. Amsterdam, 1768. (An edition in two
volumes of this original and suggestive work was printed, with the life of the
author [Demaillet], at the Hague, in 1755. The passages quoted will be found
at p. 166, vol. II of this edition.)

constitutions and the best adapted to provide for themselves and offspring, under the circumstances in which they exist.' Now, let us test the applicability of this postulate to the gradual mutation of a specific form by some instance in Natural History eminently favourable for the assumed results. In many species nature has superadded to general health and strength particular weapons and combative instincts, which, as, e.g., in the deer-tribe, insure to the strongest, to the longest-winded, the largest-antlered, and the sharpest-snagged stags, the choice of the hinds and the chief share in the propagation of the next generation. In such peculiarly gifted species we have the most favourable conditions for testing one of the conclusions drawn by Messrs. Darwin and Wallace from this universally recognised 'struggle for the preservation of life and kind.' If the offspring inheriting the advantages of their parents, did in their turn, however slightly and gradually, increase those advantages and give birth to a still more favoured progeny, with repetition of the result to the degree required by 'natural selection,'–then, according to the rate of modification experimentally proved in pigeons, we ought to find evidence of progressive increase in the combative qualities of antlers in those deer that for centuries have been under observation in our parks, and still more so in those that have fought and bred from the earliest historical times in the mountain wilds of Scotland. The element of 'natural selection' above illustrated, either is, or is not, a law of nature. If it be one, the results should be forthcoming; more especially in those exceptional cases in which nature herself has superadded structures, as it were expressly to illustrate the consequences of such general struggle for the life of the individual and the continuance of the race.'[40] The antlers of deer are expressly given to the male, and permitted to him, in fighting trim; only at the combative sexual season; they fall and are renewed annually; they belong moreover to the most plastic and variable parts or appendages of the quadruped. Is it then a fact that the fallow-deer propagated under these influences in Windsor Forest, since the reign of William Rufus, now manifest in the superior condition of the antlers, as weapons, that amount and kind of change which the successions of generations under the influence of 'natural selection' ought to have produced? Do the crowned antlers of the red deer of the nineteenth century surpass those of the turbaries and submerged forestlands which date back long before the beginning of our English History? Does the variability

40. 'Individual males have had, in successive generations, some slight advantage over other males in their weapons, and have transmitted these advantages to their male offspring.' (*Darwin*, p. 89.)

of the artificially bred pigeon or of the cultivated cabbage outweigh, in a philosophical consideration of the origin of species, those obstinate evidences of persistence of specific types and of inherent limitation of change of character, however closely the seat of such characters may be connected with the 'best chance of taking care of self and of begetting offspring?' If certain bounds to the variability of specific characters be a law in nature, we then can see why the successive progeny of the best antlered deer, proved to be best by wager of battle, should never have exceeded the specific limit assigned to such best possible antlers under that law of limitation. If unlimited variability by 'natural selection' be a law, we ought to see some degree of its operation in the peculiarly favourable test-instance just quoted.

That the variability of an organism to a certain extent is a constant and certain condition of life we admit, otherwise there would be no distinguishable individuals of a species. The forester, by the operation of this law of variability, is able to distinguish his individual oaks, the shepherd his particular sheep, the teacher his particular scholars. This true and proved law of variability is, in fact, the essential condition of individuality itself. We have searched in vain, from Demaillet to Darwin, for the evidence or the proof, that it is only necessary for one individual to vary, be it ever so little, in order to [be led to] the conclusion that the variability is progressive and unlimited, so as, in the course of generations, to change the species,' the genus, the order, or the class. We have no objection to this result of 'natural selection' in the abstract; but we desire to have reason for our faith. What we do object to is, that science should be compromised through the assumption of its true character by mere hypotheses, the logical consequences of which are of such deep importance.

The powers, aspirations, and missions of man are such as to raise the study of his origin and nature, inevitably and by the very necessity of the case, from the mere physiological to the psychological stage of scientific operations. Every step in the progress of this study has tended to obliterate the technical barriers by which logicians have sought to separate the inquiries relating to the several parts of man's nature. The considerations involved in the attempt to disclose the origin of the worm are inadequate to the requirements of the higher problem of the origin of man; and it may be that the conditions of that problem are beyond our present powers of acquiring certain knowledge.

To him, indeed, who may deem himself devoid of soul and as the brute that perisheth, any speculation, pointing, with the smallest feasibility, to an intelligible notion of the way of coming in of a lower organised species, may be sufficient, and he need concern himself no further about his own relations to a Creator. But when the members of the Royal Institution of Great Britain are taught by their evening lecturer that such a limited or inadequate view and treatment of the great problem exemplifies that application of science to which England owes her greatness, we take leave to remind the managers that it more truly parallels the abuse of science to which a neighbouring nation, some seventy years since, owed its temporary degradation. By their fruits may the promoters of true and false philosophy be known. We gazed with amazement at the audacity of the dispenser of the hour's intellectual amusement, who, availing himself of the technical ignorance of the majority of his auditors, sought to blind them as to the frail foundations of 'natural selection' by such illustrations as the subjoined:–

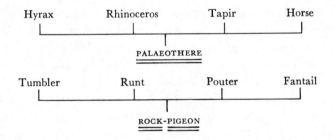

The above diagrams were set before an intelligent audience by a professor, in whom they naturally repose confidence as to facts specially belonging to his science, as parallel instances of departure from type: the one illustrating the extent and directions in which varieties diverge from a type form, in long course of time, by 'natural selection;' the other showing the correlative examples of such divergence, in a short course of time, through human selection. He told them that, in the latter series, the skeleton varied in regard to the number of vertebrae; but did not remark that it was in the variable region of the tail, on which no ornithologist ever depended for a specific character, neither did he state that the alleged difference in the number of dorsal vertebrae[41] was one that is merely simulated by a greater or less extent of the process of anchylosis over a region of the spinal column in which every vertebra was originally distinct. With regard to the parallel diagram, no allusion was made to such differences in the

41. Darwin, p. 22.

relative position of the cranial bones as the following:–viz., that in the palaeotherium, as in the tapir, the maxillary bones intervene between and separate the nasal bones from the intermaxillary bones; whilst in the horse, as in the hyrax, the nasal and intermaxillary bones are united as far as their extremities; that, consequently, the external nostril is bounded by four bones in the horse, but by six in its implied progenitor; that there is as marked a difference in the conformation of the orbit, which is encircled by the union of the malar with the frontal bone in the horse, but is left widely open or incomplete, by the want of such union in the same two cranial bones of the palaeothere. The advocate of the 'natural selection' view exaggerated resemblances and glossed over discrepancies of structure. The resemblance of the palaeothere to its four hypothetical descendants, in respect of their more generalised or more specialised structures, was flippantly affirmed to be as that of a father to his four sons![42] Nothing was said to give his hearers a notion of the important difference between the horse and palaeothere in the structure and implantation of the whole dental system. Yet the horse resembles the elephant in having a long mass of complexly interblended dental substances deeply implanted in a large simple socket; whilst the palaeothere differs from both in having a short mass or crown of differently disposed dental substances implanted by several long fangs in a correspondingly complex socket. To the competent anatomist a score of such anatomical differences would be present to the memory in contrasting the two alleged parallel series of differences from selection natural and human; to which differences in the palaeontological series nothing comparable in essential value has been pointed out in the varieties of *Columba livia*. The competent palaeontologist, moreover, would detect the superficial character of the knowledge that would interpose the tapir in any series leading from palaeotherium: he would point to the eocene lophiodon as the true ancestor of the tapir on the derivative hypothesis.

Neither zoology nor physiology as yet, however, possesses a single fact to support the idea that six incisor and two canine teeth, as in the palaeothere, could be blended or changed, by progressive transmutation, into the pair of large scalpriform teeth, that projects from the fore part of the lower jaw in the hyrax or scriptural coney. The genuine cultivator of science and true representative of the minds on which the glory and greatness of nations depend, would feel bound to illustrate any series of observed varieties of a species by a true parallel. The hoofed mammals

42. Professor Huxley's Lecture 'On Species and Races and their Origin,' Friday, February 10th, 1860, *Journal of the Royal Institution of Great Britain.*

which afford this parallel with the diverging series of pigeons, are the
following:—

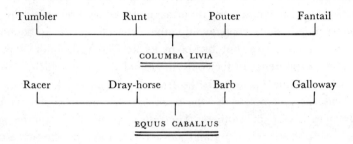

Here the differences in regard to size, colour, development of tegumentary
appendages, number of caudal vertebrae, length or stuntedness of muzzle,
relative length of limb to body, &c., are closely analogous with the diversities
which Mr. Darwin has dwelt upon in the first chapter of his work. And
not only are the subjects of the above diagrams morphologically but physiologically alike; not merely are the differences of form and structure similar
and equivalent, but the powers of procreation are the same. 'The hybrids
or mongrels from between all the domestic breeds of pigeons are perfectly
fertile;'[43] so likewise are the hybrids or mongrels from between all the
domestic breeds of the horse. Now, as this is not the case with the hybrid
between any variety of the horse and of the ass, it may be inferred that
the physiological distinction would be, at least, as great, or more insuperable,
between the horse and the tapir, or the rhinoceros, or the palaeotherium.
The infertility, or very rare fertility, of the solipedous mule, even when
paired with a true horse or ass, and the absolute infertility of such hybrids
inter se, are facts so notorious, that the professorial advocate of 'natural
selection' was compelled to admit that his alleged parallel broke down
at the physiological test,—the most important element of the comparison.

It is assumed by Mr. Darwin that variations, useful in some way to
each being, occur naturally in the course of thousands of generations (p.
80), that such variations are reproduced in the offspring, and, if in harmony
with external circumstances, may be heightened in still further modified
descendants of the species. The transmission and exaggeration of a variety,
step by step, in the generative series, essential to the theory of 'natural
selection,' implies the fertility of the individuals constituting the several
steps of the series of transmutation. But numerous instances, familiar to

43. Darwin, p. 26.

every zoologist, suggest an objection which seems fatal to the theory, since they show extreme peculiarities of structure and instinct in individuals that cannot transmit them, because they are doomed to perpetual sterility.

The most numerous and important members of the hive, which collect the pollen on their peculiarly expanded thighs, and the honey in their peculiarly valvular crop or 'honey-bag,' and which, in the construction of cells of a shape adapted to contain the greatest possible quantity of honey with least possible consumption of wax, have practically solved a recondite mathematical problem, are the neuters, or females with abortive sexual organs,—'non-breeding females' of our great physiologist Hunter. From the hypothetical protoplastic progenitor of all animal species, what an enormous series of 'slight modifications of structure and instinct' must have rolled, snow-ball like, along the articulate line of departure, to have accumulated, according to 'natural selection,' in the *Apis millifica,* which in the days of Moses exercised as now their structures and instincts in the 'land flowing with milk and honey!'

So also in the family of ants, the neuters or sterile females form, in certain species, two, or even three castes,—soldiers, workers, nurses. In *Cryptocerus* the workers of one caste 'carry a wonderful sort of shield on their heads;' in the Mexican genus, *Myrmecocystus,* the workers of one caste are fed by the workers of another caste, and have an enormously expanded abdomen, where a sort of honey is secreted and stored, which, like domestic cattle, they supply to the rest of the community. Mr. Darwin, with one of his usual happy illustrations, compares the workers of the 'driver ant' (*Anomma*), to a 'set of workmen building a house, of whom many were five feet four inches high, and many sixteen feet high; but we must suppose that the larger workmen had heads four instead of three times as big as those of the smaller men, and jaws nearly five times as big;' in short, the most grotesque and extravagant scene in a pantomime is realised in the industrial community of a West African ant.

Yet all these instances of exaggerated peculiarities of structure and instinct are manifested in individuals which never could have transmitted them.

No zoologist, perhaps, is better acquainted with these fatal exceptions to his principle of the organisation of species by hereditary transmission of variation-characters, than Mr. Darwin. He could not, with any pretension to free and candid discussion, pass over the chief instances which have checked the natural disposition of all zoologists to obtain inductively an intelligible idea of the most mysterious phenomena of their science. But

the barrier at which Cuvier hesitated, Mr. Darwin rushes through, and
thus he disposes of the difficulty:–

> We have even slight differences in the horns of different breeds of
> cattle in relation to an artificially imperfect state of the male sex; for
> oxen of certain breeds have longer horns than in other breeds, in compari-
> son with the horns of the bulls or cows of the same breeds. Hence I
> can see no real difficulty in any character having become correlated
> with the sterile condition of certain members of insect-communities: the
> difficulty lies in understanding how such correlated modifications of struc-
> ture could have been slowly accumulated by natural selection.
>
> I have such faith in the powers of selection, that I do not doubt
> that a breed of cattle, always yielding oxen with extraordinary long horns,
> could be slowly formed by carefully watching which individual bulls
> and cows, when matched, produce oxen with the longest horns; and
> yet no one ox could ever have propagated its kind. Thus I believe it
> has been with social insects: a slight modification of structure, or instinct,
> correlated with the sterile condition of certain members of the community,
> has been advantageous to the community: consequently the fertile males
> and females of the same community flourished, and transmitted to their
> fertile offspring a tendency to produce sterile members having the same
> modification (p. 238).

It is a notorious and constant fact, that the castrate bovine has longer
horns than either the perfect male or female. The progressively elongating
result in the case of the oxen, about which our theorist does not doubt,
has not been proved experimentally. It is capable of proof or disproof.
In scientific questions of far less import than the origin of animal species,
involving our own, small value, if any, is attached to supposititious cases.

It is, doubtless, by no means necessary that we should sow a seed of
the very cauliflower we eat in order to get more cauliflowers; seeds of
other individuals of the same stock will suffice. So the bee-keeper feels
satisfied that the progeny of the impregnated young queen will exercise
all the wonderful instincts which result in the production of wax and honey,
as effectively as the virgin-sisters of the queen-mother, who were destroyed
in the preceding winter. And our readers may well wonder what all this
has to do with the explanation of the acquisition of the adaptive structures
and instincts of neuter bees, by homoeopathic doses of Lamarckian transmutation, accumulating through a long series of hereditary transmissions?
We cannot reply; we can only quote, with no less amazement, our author:–

> This difficulty, although appearing insuperable, is lessened, or, as I
> believe, disappears, when it is remembered that selection may be applied

to the family, as well as to the individual, and may thus gain the desired end. Thus a well-flavored vegetable is cooked, and the individual is destroyed; but the horticulturist sows seed of the same stock, and confidently expects to get nearly the same variety; breeders of cattle wish the flesh and fat to be well marbled together; the animal has been slaughtered, but the breeder goes with confidence to the same family (p. 237).

Now every step in the production of the breed or family of cattle may have been observed and recorded; and many of the incidents of the transmutative journey of the edible variety of cabbage from the wild stock may be similarly known; but this is just the knowledge that we disiderate in regard to the creation of the honey-bee by the way of 'natural selection;' and, instead of satisfying our craving with the mature fruit of inductive research, Mr. Darwin offers us the intellectual husks above quoted, endorsed by his firm belief in their nutritive sufficiency!

To more intelligible propositions in support of his hypothesis, we marginally noted, as we perused them, the difficulties or exceptions which rose in our mind. We have still room for a few of these illustrations of the groundwork of 'natural selection.'

> From looking at species as only strongly-marked and well-defined varieties, I was led to anticipate that the species of the larger genera in each country would oftener present varieties, than the species of the smaller genera. To test the truth of this anticipation I have arranged the plants of twelve countries, and the coleopterous insects of two districts, into two nearly equal masses, the species of the larger genera on one side, and those of the smaller genera on the other side, and it has invariably proved to be the case that a larger proportion of the species on the side of the larger genera present varieties, than on the side of the smaller genera (p. 55).

The elephant is, however, a small genus, indeed one of the smallest in the sense of the number of species composing it, which are indeed but two, the *Elephas Indicus* and *Elephas Africanus* of Cuvier. But the range of variety in both African and Indian kinds is by no means inconsiderable. Livingstone adds instances in the elephants of the Zambesi, and the terms 'Dauntelah,' 'Mooknah,' &c., applied by the Indian and Singhalese entrappers of the wild proboscidians to the different varieties that are captured, still more exemplify this tendency to vary in individuals of a 'small genus.' Another exception to Mr. Darwin's rule as strongly and quickly suggests itself in the genus *Pithecus*. Naturalists seem unwilling to admit more species than the Bornean Pongo (*Pithecus Wurmbii*, seu *Satyrus* of Wurmb), and the smaller orang (P. *morio*), since established by Owen.

But the varieties in regard to the cranial crests, to colour, to relative length of arms, appear by a memoir from the pen of the latter naturalist,[44] to be both numerous and well marked. On the other hand, the species of the antelope genus have not hitherto presented any notable varieties to the observation of naturalists; and yet the genus, in respect to the number of these species, is one of the largest in the mammalian class. There may be, of course, a difference in different classes of organisms in this respect. Plants and invertebrates may better exemplify Mr. Darwin's proposition than fishes, reptiles, or quadrupeds. But an hypothesis applied to all living things can only be sustained by laws and rules of a like generality of application.

Mr. Darwin's argument for a common origin of all the varieties of dovecot pigeon, leads him to affirm, 'all recent experience shows that it is most difficult to get any wild animal to breed freely under domestication' (p. 24). But the recent experience at the Zoological Gardens of London tells a different story. Three young individuals, two males and one female, of those most strange exotic quadrupeds, the giraffe, were transported from their African wilderness to the menagerie in Regent's Park in 1836. No sooner had they attained the proper age in 1838 than they bred; and there has been no other interval in the repetition of the act than that which the phenomena of a fifteen months' gestation and seven months' suckling necessarily interpose. 'Nine fawns have been produced without any casualty.'[45] A pair of the largest and wildest of antelopes (the Eland, *A. Oreas*) is transported from the boundless sunny plains of South Africa to the confinement of a park in cloudy and rainy Lancashire; they breed freely there, and become the parents of elands now widely distributed over Great Britain, and promising in another century to be as common in our parks as fallow deer.[46] What conditions might seem more adverse to health and procreative power than such as are exemplified by the contrast of the den and the pond appropriated to the hippopotamus in the Jardin des Plantes, with that noble river where these most uncouth of African wild beasts disported themselves prior to their capture? Before two years elapse after the arrival of the young male and female, they produce a fine offspring.

Such are the signs of defective information which contribute, almost

44. 'Characters of the skull of the male *Pithecus morio,* with remarks on the varieties of *Pithecus satyrus,*' by Professor Owen. *Zoological Transactions,* vol. IV. p. 163 (1856).
45. *Edinburgh Review,* January 1860, p. 179.
46. Ibid., pp. 167–169.

at each chapter to check our confidence in the teachings and advocacy of the hypothesis of 'Natural Selection.' But, as we have before been led to remark, most of Mr. Darwin's statements elude, by their vagueness and incompleteness, the test of Natural History facts. Thus he says:—

'I think it highly probable that our domestic dogs have descended from several wild species.' It may be so; but what are the species here referred to? Are they known, or named, or can they be defined? If so, why are they not indicated, so that the naturalist might have some means of judging of the degree of probability, or value of the surmise, and of its bearing on the hypothesis?

'Isolation also,' says Mr. Darwin, 'is an important element in the process of natural selection.' But how can one select if a thing be 'isolated'? Even using the word in the sense of a confined area, Mr. Darwin admits that the conditions of life 'throughout such area, will tend to modify all the individuals of a species in the same manner in relation to the same conditions' (p. 104). No evidence, however, is given of a species having ever been created in that way; but granting the hypothetical influence and transmutation, there is no selection here. The author adds, 'Although I do not doubt that isolation is of considerable importance in the production of new species, on the whole, I am inclined to believe, that largeness of area is of more importance in the production of species capable of spreading widely' (p. 105).

Now, on such a question as the origin of species, and in an express, formal, scientific treatise on the subject, the expression of a belief, where one looks for a demonstration, is simply provoking. We are not concerned in the author's beliefs or inclinations to believe. Belief is a state of mind short of actual knowledge. It is a state which may govern action, when based upon a tacit admission of the mind's incompetency to prove a proposition, coupled with submissive acceptance of an authoritative dogma, or worship of a favourite idol of the mind. We readily concede, and it needs, indeed, no ghost to reveal the fact, that the wider the area in which a species may be produced, the more widely it will spread. But we fail to discern its import in respect of the great question at issue.

We have read and studied with care most of the monographs conveying the results of close investigations of particular groups of animals, but have not found, what Darwin asserts to be the fact, at least as regards all those investigators of particular groups of animals and plants whose treatises he has read, viz., that their authors 'are one and all firmly convinced that each of the well-marked forms or species was at the first independently created.' Our experience has been that the monographers referred to have rarely committed themselves to any conjectural hypothesis whatever, upon the origin of the species which they have closely studied.

Darwin appeals from the 'experienced naturalists whose minds are stocked with a multitude of facts' which he assumes to have been 'viewed from a point of view opposite to his own,' to the 'few naturalists endowed with much flexibility of mind,' for a favourable reception of his hypothesis. We must confess that the minds to whose conclusions we incline to bow belong to that truth-loving, truth-seeking, truth-imparting class, which Robert Brown,[47] Bojanus,[48] Rudolphi, Cuvier,[49] Ehrenberg,[50] Herold,[51] Kölliker,[52] and Siebold, worthily exemplify. The rightly and sagaciously generalising intellect is associated with the power of endurance of continuous and laborious research, exemplarily manifested in such monographs as we have quoted below. Their authors are the men who trouble the intellectual world little with their beliefs, but enrich it greatly with their proofs. If close and long-continued research, sustained by the determination to get accurate results blunted, as Mr. Darwin seems to imply, the far-seeing discovering faculty, then are we driven to this paradox, viz., that the elucidation of the higher problems, nay the highest, in Biology, is to be sought for or expected in the lucubrations of those naturalists whose minds are not weighed or troubled with more than a discursive and superficial knowledge of nature.

Lasting and fruitful conclusions have indeed, hitherto been based only on the possession of knowledge; now we are called upon to accept an hypothesis on the plea of want of knowledge. The geological record, it is averred, is so imperfect! But what human record is not? Especially must the record of past organisms be much less perfect than of present ones. We freely admit it. But when Mr. Darwin, in reference to the absence of the intermediate fossil forms required by his hypothesis—and only the zootomical zoologist can approximately appreciate their immense numbers—the countless hosts of transitional links which on 'natural selection,' must certainly have existed at one period or another of the world's history—when Mr. Darwin exclaims what may be, or what may not be, the forms yet forthcoming out of the graveyards of strata, we would reply, that our only ground for prophesying of what may come, is by the analogy

47. *Prodromus Florae Novae Hollandiae.*
48. *Anatome Testudinis Europae.*
49. *Mémoires pour servir à l'Anatomie des Mollusques.*
50. *Die Infusionsthierchen, als vollkommene Organismen.*
51. *Disquisitiones de Animalium vertebris carentium,* &c.
52. *Entwickelungsgeschichte des Cephalopoden.*

of what has come to light. We may expect, e.g., a chambered-shell from a secondary rock; but not the evidence of a creature linking on the cuttle-fish to the lump-fish.

Mr. Darwin asks, 'How is it that varieties, which I have called incipient species, become ultimately good and distinct species?' To which we rejoin with the question:–Do they become good and distinct species? Is there any one instance proved by observed facts of such transmutation? We have searched the volume in vain for such. When we see the intervals that divide most species from their nearest congeners, in the recent and especially the fossil series, we either doubt the fact of progressive conversion, or, as Mr. Darwin remarks in his letter to Dr. Asa Gray,[53] one's 'imagination must fill up very wide blanks.'

The last ichthyosaurus, by which the genus disappears in the chalk, is hardly distinguishable specifically from the first ichthyosaurus, which abruptly introduces that strange form of sea-lizard in the Lias. The oldest Pterodactyle is as thorough and complete a one as the latest. No contrast can be more remarkable, nor, we believe, more instructive than the abun-dance of evidence of the various species of ichthyosaurus throughout the marine strata of the oolitic and cretaceous periods, and the utter blank in reference to any form calculated to enlighten us as to whence the ich-thyosaurus came, or what it graduated into, before or after those periods. The Enaliosauria of the secondary seas were superseded by the Cetacea of the tertiary ones.

Professor Agassiz affirms:–

Between two successive geological periods, changes have taken place among plants and animals. But none of those primordial forms of life which naturalists call species, are known to have changed during any of these periods. It cannot be denied that the species of different succes-sive periods are supposed by some naturalists to derive their distinguishing features from changes which have taken place in those of preceding ages, but this is a mere supposition, supported neither by physiological nor by geological evidence; and the assumption that animals and plants may change in a similar manner during one and the same period is equally gratuitous.[54]

Cuvier adduced the evidence of the birds and beasts which had been preserved in the tombs of Egypt, to prove that no change in their specific characters had taken place during the thousands of years—two, three, or

53. *Proceedings of the Linnaean Society*, 1858, p. 61.
54. *Contributions to Natural History*, "Essay on Classification," p. 51.

five—which had elapsed, according to the monumental evidence, since the individuals of those species were the subjects of the mummifier's skill.

Professor Agassiz adduces evidence to show that there are animals of species now living which have been for a much longer period inhabitants of our globe.

'It has been possible,' he writes, 'to trace the formation and growth of our coral reefs, especially in Florida, with sufficient precision to ascertain that it must take about eight thousand years for one of those coral walls to rise from its foundation to the level of the surface of the ocean. There are around the southernmost extremity of Florida alone, four such reefs, concentric with one another, which can be shown to have grown up one after the other. This gives for the beginning of the first of these reefs an age of over thirty thousand years; and yet the corals by which they were all built up are the same identical species in all of them. These facts, then, furnish as direct evidence as we can obtain in any branch of physical inquiry, that some, at least, of the species of animals now existing, have been in existence over thirty thousand years, and have not undergone the slightest change during the whole of that period.[55]

To this, of course, the transmutationists reply that a still longer period of time might do what thirty thousand years have not done.

Professor Baden Powell, for example, affirms:–'Though each species may have possessed its peculiarities unchanged for a lapse of time, the fact that when the long periods are considered, all those of our earlier period are replaced by new ones at a later period, proves that species change in the end, provided a sufficiently long time is granted.' But here lies the fallacy: it merely proves that species are changed, it gives no evidence as to the mode of change; transmutation, gradual or abrupt, is in this case, mere assumption. We have no objection on any score to the change; we have the greatest desire to know how it is brought about. Owen has long stated his belief that some pre-ordained law or secondary cause is operative in bringing about the change; but our knowledge of such law, if such exists, can only be acquired on the prescribed terms. We, therefore, regard the painstaking and minute comparisons by Cuvier of the osteological and every other character that could be tested in the mummified ibis, cat, or crocodile, with those of the species living in his time; and the equally philosophical investigations of the polypes operating at an interval of 30,000 years in the building up of coral reefs, by the profound palaeontologist of Neuchatel, as of far higher value in reference to the inductive

55. Ibid., p. 53.

determination of the question of the origin of species than the speculations of Demaillet, Buffon, Lamarck, 'Vestiges,' Baden Powell, or Darwin.

The essential element in the complex idea of species, as it has been variously framed and defined by naturalists, viz., the blood-relationship between all the individuals of such species, is annihilated on the hypothesis of 'natural selection.' According to this view a genus, a family, an order, a class, a sub-kingdom,–the individuals severally representing these grades of difference or relationship,–now differ from individuals of the same species only in degree: the species, like every other group, is a mere creature of the brain; it is no longer from nature. With the present evidence from form, structure, and procreative phenomena, of the truth of the opposite proposition, that 'classification is the task of science, but species the work of nature,' we believe that this aphorism will endure; we are certain that it has not yet been refuted; and we repeat in the words of Linnaeus, '*Classis et Ordo est sapientiae, Species naturae opus.*

Comments on Owen

Owen was formally educated as a physician, but while practicing in London, he trained himself as a comparative anatomist, eventually joining the staff of the British Museum, becoming the first Superintendent of the Natural History Department, later Director until his retirement in 1884 at the age of eighty. He was the leading comparative anatomist of his time. Though he was frequently termed the British Cuvier, Owen felt an even greater affinity for the speculative philosophies of Lorenz Oken, and Geoffroy Saint-Hilaire. During his long career, Owen published over three hundred and sixty beautifully illustrated monographs on recent and fossil animals, including the Pearly Nautilus, the *Archaeopteryx,* and rare members of the ape family. Owen distinguished between "homology" and "analogy," but in terms of form, function, and ontogenetic development. For example, the dragon-lizard, insects, flying fish, and birds all fly, but with modifications of different organs—the lizards by means of skin stretched between extended ribs, insects with folds in the skin, fish with fins, and birds by using their anterior extremities. These "wings" were analogous, not homologous to each other. Similarly, vertebrate lungs were analogous to the gill slits of fish, homologous to their swim bladders.

Owen the man holds a peculiar place in the scientific community of his day. As a comparative anatomist, he was highly respected, but he seemed incapable of engendering warmth among his fellow scientists. Not infre-

quently he became embroiled in petty controversies. Two of the most un-
palatable character traits, as far as Victorians were concerned, were petti-
ness and dishonesty. Owen seems to have possessed at least the former
in ample measure. For example, his colleagues were dismayed that a man
of Owen's stature should behave as shoddily as he did in a dispute with
Hugh Falconer over priority in naming the fossil elephant *Elephas columbi.*
Owen had legitimate priority in the naming of so many species. Why risk
so high a reputation in a dispute with a considerably less powerful young
naturalist over so trivial a matter? Only Louis Agassiz, the American Cuvier,
seems to have surpassed Owen in his propensity for arousing animosity
among his fellow workers, but Agassiz also had the ability to engender
genuine affection as well. In retrospect, one might suspect that Owen's
personal reputation might have resulted from his bitter, though equivocal,
opposition to evolutionary theory. Such is not the case. His personal reputation had been well established prior to his heated controversy with the
Darwinians.

Darwin's own grievances against Owen stemmed from being treated so
ungenerously by someone whom he had counted a friend. Darwin was
especially pained by Owen's sneers at his use of such phrases as "I believe"
and "I am convinced" in the *Origin,* after Owen himself had discouraged
Darwin from deleting them lest it spoil the "charm" of the book. Owen
may have had "no sympathy whatever with Biblical objectors" and
"sacerdotal revilers" like Adam Sedgwick, but he was not above aiding
Bishop Wilberforce (1860) to attack Darwin on theological grounds. To
make matters worse, Darwin had to tolerate Owen's objections to evolutionary theory, knowing all the while that Owen himself secretly held similar
views. As early as 1838, Owen had proposed an analogy between the production of monsters and the origin of species. In his review of the *Origin
of Species,* Owen is suspiciously gentle with Chambers' *Vestiges of Creation*—and for good reason. His own theory of derivation, published eventually in 1868 (see Bibliography), was very close to that outlined by
Chambers.

If Owen could not grant priority to another man in the naming of
a fossil elephant, one can imagine his dismay at being forestalled in the
matter of evolutionary theory—twice. He himself had adhered to the strict
canons of the inductive method, postponing publication until he could
provide the necessary proof, while Chambers and Darwin rushed into print
without the proper inductive foundation and became famous. At least the
scientific community had had the good sense to denounce Chambers' work,

but it was taking Darwin seriously. The injustice was more than Owen could bear: he found himself in the uncomfortable position of wanting to censure Darwin for enunciating his theory of evolution while simultaneously claiming priority for himself.

Darwin, as usual, did not enter into public combat. Instead, the battle was joined by T. H. Huxley, Darwin's bulldog. Huxley was by no means a disinterested participant. At one point, early in his acquaintanceship with Owen, Huxley became so angered he had to restrain himself from physically assaulting Owen and kicking him into the street. The occasion for Owen's downfall was the hippocampus controversy. In the heat of argument Owen had been led to make some overly inflated remarks about the differences between the brain of man and that of other primates. Huxley, with characteristic audacity, attacked the greatest living anatomist on the details of the comparative anatomy of an organism which Owen himself had studied extensively (Owen, 1862). Seldom had so powerful an opponent been so thoroughly defeated.

In the *Origin* Darwin had claimed that all the leading scientists of his day were publicly committed to the immutability of species. Included on this list of eminent authorities was Owen. Owen took strong exception to this claim. The issue is not clear cut. Darwin was certainly correct in stating that few, if any, scientists believed in the evolution of species prior to the appearance of the *Origin,* but from this fact it did not automatically follow that species arose through the flashing together of elemental atoms. But if scientists did not believe in such a miraculous origin of species, what did they believe? In general, there had been a tacit gentleman's agreement among professional scientists not to discuss the issue publicly until Darwin and Wallace broke the silence in so ungentlemanly a fashion.*

* See also Richard Owen (1849, 1860b) ; Richard S. Owen (1894).

Samuel Haughton (1821–1897)

This speculation of Messrs. Darwin and Wallace would not be worthy of notice were it not for the weight of authority of the names (i.e., Lyell's and yours), under whose auspices it has been brought forward. If it means what it says, it is a truism; if it means anything more, it is contrary to fact. Q.E.D.—Reverend S. Haughton's address to the Geological Society, Dublin, February 9, 1859. (*Life and Letters,* 1:512; C. Darwin, *Autobiography,* 212)

Nevertheless, our joint productions [those of Darwin and Wallace] published in the *Journal of the Proceedings of the Linnaean Society,* 1858 excited very little attention, and the only published notice of them which I can remember was by Professor Haughton of Dublin, whose verdict was that all that was new in them was false, and what was true was old. This shows how necessary it is that any new view should be explained at considerable length in order to arouse public attention.—C. Darwin, *Autobiography,* p. 44.

Have you seen Haughton's abusive article on me? . . . It outdoes even the 'North British' and 'Edinburgh' in misapprehension and misrepresentation. I never knew anything so unfair as in discussing cells of bees, his ignoring the case of Melipona, which builds combs almost exactly intermediate between hive and humble bees. What has Haughton done that he feels so immeasurably superior to all us wretched naturalists, and to all political economists, including that great philosopher Malthus?—C. Darwin to J. D. Hooker, June 5, 1860 (*Life and Letters,* 2:110)

I have lately had many more 'kicks than halfpence.' A review in the last Dublin *Nat. Hist. Review* is the most unfair thing which has appeared,—one mass of misrepresentation. It is evidently by Haughton, the geologist, chemist and mathematician. It shows immeasurable conceit and contempt of all who are not mathematicians . . . The article is a curiosity of unfairness and arrogance; but, as he sneers at Malthus, I am content,

for it is clear he cannot reason. He is a friend of Harvey, with whom
I have had some correspondence. Your article has clearly, as he admits,
influenced him. He admits to a certain extent Natural Selection, yet I
am sure he does not understand me. It is strange that very few do, and
I am become quite convinced that I must be an extremely bad explainer.—
C. Darwin to Asa Gray, Down, June 8, 1860 (*More Letters,* 1:152–153)

Biogenesis*

SAMUEL HAUGHTON

The active and restless mind of man has never been content with the
knowledge of the present, but has always sought to know the future and
the past. The guesses of the Ancients as to the future of man are among
the most interesting, and, at the same time, the most puerile of their philosophical speculations. The reader of the Tusculan Disputations rises from
his task, charmed by the style of the writer, but thankful that a certain
revelation of the future renders him immeasurably superior in knowledge
to the weavers of these pleasant webs of fiction, and though he admire
the skill of the ingenious sophists who live again and dispute in the pages
of Cicero, he would not for an instant exchange his own position for theirs.

The Moderns have resolved, by their speculations on the past, to show
that in ingenuity and oddness of conceit, and probably, also in wideness
from the truth, they are in no respect inferior to the Ancients. The future
being shut out from us, we are resolved to try what we can effect, in
proof of our versatility of imagination, by guessing at the history of the
past.

To establish a character for subtlety and skill, in drawing large conclusions on this subject from slender premises, the first requisite is, ignorance
of what other speculators have attempted before us in the same field;
and the second is, a firm confidence in our own special theory. Neither
of these requisites can be considered wanting in those who are engaged
in the task of reproducing Lamarck's theory of organic life, either as altogether new, or with but a tattered and threadbare cloak, thrown over
its original nakedness.

The sciences of Geology and Political Economy are mainly answerable
for the revival of these exploded and forgotten fancies:—Geology, in supplying the lost history of organic life, which could never be studied profoundly
from the creatures living at any given time; and Political Economy, in

* From the *Natural History Review* (1860), 7:23–32; reprinted with omissions
in the *Annals and Magazine of Natural History* (1863), 11:420–429.

furnishing, from its mean and sordid motives, a Malthusian force, supposed to be sufficient to supply the wants of previous theories.

One of the earliest speculators on the origin of the diversified forms of life we see around us, and class as varieties, species, and genera, was Buffon, who published in 1766[1] his theory of the derivation of all mammal forms by degradation, from fifteen primary and perfect types, and nine special or isolated species.

This theory of βιογένεσις by degradation, although now superseded by the theory of progression, has much to be said in its favour, and derives additional importance from the facts of the history of life made known since Buffon's time, by the science of Geology; the principal of these additional facts are, the degradation of fishes from their first introduction in the Old Red Sandstone period to the present day; the corresponding degradation of the Cephalopods, and, though in a somewhat less degree, of the Reptiles.

The following are the original types of Buffon, from which all mammals are derived by degradation:—

    I. The solid-hoofed type.
       Ex. The Horse.
    II. Large cloven-footed type, with hollow horns.
       Ex. The Ox.
    III. Small cloven-footed type, with hollow horns.
       Ex. The Sheep.
    IV. Cloven-footed type, with solid horns.
       Ex. The Stag.
    V. Doubtful cloven-footed type.
       Ex. The Pig.
    VI. Divided-footed type, carnivorous with retractile claws.
       Ex. The Cat.
    VII. Divided-footed type, carnivorous, with non-retractile claws.
       Ex. The Dog.
    VIII. Divided-footed type, carnivorous, with non-retractile claws, and a pouch under the tail.
       Ex. The Hyaena.
    IX. Divided-footed type, carnivorous, with long bodies: five fingers on each foot, and the thumb separated from the other fingers.
       Ex. The Martin.

1. "Histoire naturelle," vol. XIV.

    X. Divided-footed type, with two large incisor teeth, and no bristles
on the body.
      Ex. The Hare.
  XI. Divided-footed type, covered with bristles.
      Ex. The Porcupine.
 XII. Divided-footed type, covered with scales.
      Ex. The Pangolin.
XIII. Divided-footed type, amphibious.
      Ex. The Seal.
XIV. Quadrumana.
      Ex. The Monkey.
 XV. Divided-footed type, winged.
      Ex. The Bat.

In addition to these, Buffon considered there were nine irreducible species, some of which, according to more recent views, are immediately connected with the preceding, and are even typical examples of them. His nine isolated species are–

    I. The Elephant.
   II. The Rhinoceros.
  III. The Hippopotamus.
  IV. The Giraffe.
   V. The Camel.
  VI. The Lion.
 VII. The Tiger.
VIII. The Bear.
  IX. The Mole.

Some of the classes given by Buffon are as old as the time of Moses, who defines with accuracy the class of Ruminantia, distinguishing it from the Pachydermata and Rodentia, in his classification of "clean" and "unclean" beasts.[2]

Whatever may be thought by the more enlightened moderns of the merits of this classification of mammals, Buffon certainly agrees with them in one respect: he takes the non-reality of species as the starting point of his theory, and, by a continued degradation downwards, develops all the varieties of life we see on the surface of the globe.

To those who love to dwell upon the past, this theory of degradation

2. Leviticus, xi. 2–8.

will afford solace and consolation in the troubles of the present, as they can reflect upon how good and excellent their ancestors were, and congratulate each other upon their superiority to those that will come after them. Every system of philosophy provides its followers with a *"solatium doloris;"* the degradationists find it in the contemplation of the past, and the progressionists in the prospect of the future; to those who are contented with the present, and deny our knowledge of the past or future, both theories appear as the idle dreams of childhood, the awakening from which will disclose a reality totally different from the troubled fancies of the night.

Lamarck is the father of the progressionists, and of the many who quote his name as an authority in support of their systems, or express their disapproval of his doctrine, few have taken the trouble to understand his theory or trace it to its origin. It is apparently founded on the confusion of species, like that of Buffon, but there is in reality an *arrière pensée,* like an unseen presence, which corrupts his reasoning and discloses the motive force of his entire system. This hidden spring of action and theorizing is a profound, and, as many think, a well-founded contempt for humanity, which pervades his writings as thoroughly as it does the "Voyage to the Houyhnhnms." Lamarck was too quick witted and acute an observer, however deficient he may have been as a reasoner, to have believed his own theory, the real mainspring of which is the desire to degrade man into an intelligent baboon, or yahoo; what difference is there in a name! In his desire to do so, he overlooks every fact at variance with his foregone conclusion, and writes of mankind with a virulence which, though devoid of the wit of Swift, springs from the same profound and unalterable conviction of the worthlessness of the creature he describes:—

> Even though Newton, Montesquieu, Voltaire, and so many other human beings have brought honor to the human race by exercising their intelligence and genius, how many senseless and ignorant human beings, who fall prey to the most absurd prejudices and are constantly enslaved by their habits and yet make up the great mass of all nations, have lowered the human race to the level of animals?[3]

Lamarck's contempt for his species is again shown in the strange list of resemblances he selects for his comparison between man and the chimpanzee, a comparison fully as degrading as Swift's mock imitation of a naturalist's description of a yahoo.

3. [Lamarck] *"Recherches sur l' organization des corps vivans,"* p. 127. Paris, 27 Floreal. An. X [Translated by the D. L. H.]

Lamarck's theory consists in the assertion of the following Laws, six in number, which he dignifies with the title of Laws of Nature:–

I. *Law of Specialization of Function,* by which a function at first general, or belonging to the whole body, is determined to a particular organ.

II. *Law of Nutrition producing Death,* by the forced inequality between the materials fixed by assimilation, and removed by excretion. This law is intended to account for death, which is a puzzle to the naturalists.

III. *Law of Movement of Complex Fluids in Canals.*–This law I profess my inability to understand: in the statement of it, Lamarck, who, like most naturalists, was unacquainted with Physics, and untrained in the severe discipline of mathematical reasoning, attributes properties to fluids in motion, which must be considered by lookers-on as little short of miraculous.

IV. *Law of Change of Composition of Fluids in Circulation.*–This law is as obscure, and as miraculous in its results, as the preceding. Natural religion, however, would appear to consider herself entitled to her miracles as well as revealed religion.

V. *Organic Forms acquired under the presiding influence of external circumstances are transmitted by Generation.*–This law involves the famous Law of Natural Selection, attributed within the last few months to Mr. Darwin.

VI. *By the concurrence of the preceding Laws, of a long lapse of time, and an almost inconceivable diversity of surrounding circumstances, all Species have been formed in succession.*–Lamarck's theory is essentially one of Progression, and is totally opposed to that of Buffon, which is one of Degradation; yet it is remarkable that they both rest upon the same foundation—the assumed non-reality of species. Like his successors in the Progression Theory, Lamarck spent his life in the establishment of the reality of species, and it is a humiliating reflection that, at the close of it, he believed himself to have lived under a delusion. Let us hear his confession:

"For a long time I believed that species were constant in nature and that they were made up of the individuals which comprised each of' them. Now I am convinced that I was mistaken in this regard and that only individuals really exist in nature."[4]

What must we think of the principles that guide the speculations of Naturalists, when we find minds like those of Buffon and Lamarck drawing opposite conclusions from the same premises? It matters little in this question whether the premises be true or false, whether species be truly distinct

4. [Translated by D. L. H.]

or not; our surprise at the logic of the Naturalists is natural, and must border on a courteous contempt.

The English revival of Lamarckianism, or "Progress in Organic Life," by Mr. Darwin, involves no idea in advance of those contained in Lamarck's six laws, but gives a greater prominence to the Law of Continuation of Peculiarities by Generation, by the assertion that such peculiarities, and such only, as are useful to the creature, in its struggle for existence, will become hereditary; the reason being, that animals provided with such peculiarity will have the advantage in the battle of life over their fellows in the competition for food, females, and other necessaries for the preservation of the individual and species. This notable argument is borrowed from Malthus' doctrine of Population, and will, no doubt, find acceptance with those Political Economists and Pseudo-Philosophers who reduce all the laws of action and human thought habitually to the lowest and most sordid motives. It is dignified with the title of a Law of Nature, called the *Law of Natural Selection,* and forms the only *bona fide* addition made by Darwin to Lamarck's famous Theory of Progression; in which, however, it is implicitly involved.

I make no account of Mr. Darwin's geological additions to Lamarck, for two reasons. In the first place, the laws of geographical distribution explained by geological change are not *ad rem,* and were previously fully treated of by Buffon and Forbes; and in the second place, Mr. Darwin admits that the facts of Geology are opposed to his (Lamarck's) theory, and they are pleasantly alluded to as the Geological Difficulty! So far as the history of life on the globe indicates a progression, Lamarck is entitled to the benefit of it, as in the case of Mammals and Plants; but certainly not to the exclusion of the facts in favour of degradation,–such as the case of Fishes, Reptiles, and Cephalopods, which must be credited to the account of Buffon and his followers.

Lamarck says distinctly:–

> It is not the organs, that is, the nature and form of the parts of an animal's body, which have occasioned its habits and its particular faculties; but on the contrary its habits, its mode of life, and *the circumstances in which the individuals find themselves* have brought about *with the passing of time* the number and conditions of its organs, and eventually the faculties which it possesses.[5]

This statement implies all that is essential in Mr. Darwin's Law of Natural Selection which, by its prominence, fills in his system the place occupied

5. [Translated by D. L. H.]

by the Law of Imitation in the original theory of Lamarck. This difference arises from the difference of the points of view of the Frenchman and the Englishman,—a difference characteristic of the two races. The Frenchman, with the vivacity and perception of the ridiculous belonging to his nation, seizes upon the quality most likely to elevate a monkey into a man, selects the faculty of Imitation, and, with a bitter satire, endows his monkey with the human desire to better his condition, and lift himself above his brother chatterers. He thus magnifies the monkey power of imitation,—which is truly wonderful, and extends to the most extraordinary actions,—into the position of a Law of Nature, sufficient to create man! The Englishman, on the other hand, firmly believes his theory, and, with a confident faith in the power of food and comfort, equally characteristic of his country, elevates the desire to supply the stomach into a law of sufficient force to convert an eel into an elephant, or an oyster into an orang-outan.

Other theorists, whose name is Legion, have printed their crude fancies, and have met with numerous readers among the young and inexperienced,–the sciolists of science. It is not to be supposed that a public which accepted Mesmerism and Table-turning could judge with accuracy of the pretensions of loose and ill-reasoned speculations on the origin of Life. It has rained, hailed, and poured theories of life,—religious, philosophical, and pseudo-scientific,—with a marvellous rapidity, within the last few years. Some theorists have started from the Nebular Hypothesis of Laplace; others have speculated on the results of superfoetation; and others on the brilliant and seductive theory of the correlation of Physical forces; but they may all be classed as, knowingly or not, the followers of Lamarck. Some have taught that all the planets, being composed of the same mineral constituents as the earth, must produce in succession the same organic phenomena, and weary the reader with the idea of the same pterodactyles and cetacea, the same monads and men, appearing on all the globes that circle round the sun! Others have called to mind the loss of heat of our planet, and, by the correlation of forces, have reproduced it in the increasing intelligence of the successive forms of life that have peopled our globe!! In a word, there is no folly that human fancy can devise, when truth has ceased to be of primary importance, and right reason and sound logic have been discarded, that has not been produced, and preached as a new revelation. Neither have the disciples of Lamarck wanted the martyr spirit supposed to be essential to the apostles of a new faith. They have courted persecution, and reviled their opponents with bitter words, and with such weapons as are permitted by the free civilization under which we live. They argue, with a logic worthy of their system, that because truth has been often in a minority, therefore minorities and theories in a minority must necessarily be true.

It is curious to observe the natural instinct by which Lamarck and his followers appeal from the judgment of their peers to the young, the enthusiastic, and the inexperienced. I shall quote but two instances of this necessary instinct of self-preservation:—

Upon reflection, these considerations can be made to arise in the minds of those who are receptive and are slow to pass judgment! the others would immediately pass judgment in the following manner: they would decide without examination and would choose that which best suits them or best fits in with their conceptions.—*Lamarck,* p. 123.

I by no means expect to convince experienced naturalists, whose minds are stocked with a multitude of facts, all viewed, during a long course of years, from a point of view directly opposite to mine . . . but I look with confidence to the future, to young and rising naturalists, who will be able to view both sides of the question with impartiality.—*Darwin's Origin of Species,* pp. 481–82.

The theories of *Biogenesis,* already described, and many others, are based upon the following three unwarrantable assumptions, the denial of which, until proved, brings to the ground the entire structure, like a child's house of cards—

I. *The indefinite variation of species continuously in the one direction.*

II. *That the causes of variation assigned,* viz., cross-breeding (Buffon); imitation (Lamarck); and natural advantage in the struggle for existence (Darwin), *are sufficient to account for the effects asserted to be produced.*

III. *That succession implies causation.*

On each of these a few words of explanation are necessary.

I. *The indefinite variation of species continuously in the one direction.* This has been expressed by some Lamarckians as a state of unstable equilibrium of nature; but, should we assume the existence of a law, which is contrary to all we know of every other department of nature? If we must have a mechanical analogy to fix our ideas, nature might be better compared to a condition of *dynamic equilibrium,* in which all the parts are in motion, and never return to precisely the same relative positions, but, nevertheless, continually balance round certain definite positions of equilibrium, which never change. What should we think of the astronomer, who from a few years' observation of the precession of the equinoxes, should predict that in due time the north pole of the earth's axis would point to the same position among the stars that the south pole now occupies. Yet this very species of assumption is made by Lamarck and Darwin, in their appeal to the supposed influence of a long lapse of time. Yet, in the writings of the latter progressionist there is this singular inconsistency, that while he shows the utmost effects of human breeding on domestic animals to be capable of production in ten or twenty years; he denies the right of his adversaries to appeal to the unaltered condition of the ass, the ostrich, or the cat, for 3000 years, as a proof that specific forms balance round central types, and have no tendency to depart indefinitely from them.

Is it rational to suppose that Man can alter the head and neck of a pigeon into any desired form in six years, and that Nature, with her greater skill cannot in 3000 years lengthen the ostrich's wings by a single inch, although, according to the theory, it is her evident wish to do so.

II. *The Causes of Variation assigned are not adequate to produce the effects ascribed to them.*–The discussion of the inadequacy of the causes assigned would lead to a treatise longer than that of Buffon, Lamarck, or Darwin—and I must therefore content myself with an example. The humble bee and the hive bee coexist together, and the latter is supposed to be developed from the former by the law of natural selection, breeding, in succession, bees possessed of the talent of economizing more and more of wax in the construction of their cells.

Mr. Darwin appears to be very imperfectly acquainted with the real economy of wax that has occurred in the passage, and I shall endeavour to supply the deficiencies of his theory, and, perhaps, add a great fact to the stock of the progressionists.

1. The humble bee constructs single cells, and uses 100 units of wax.

2. A bee (not known to science, but, doubtless, extinct) was grown, that made cells in the form of *equilateral triangles* placed in double combs, with flat bottoms to the cells. This bee used only 50 units of wax.

3. A bee (also extinct) was grown, that built *square* cells in double combs. This bee used only $41\frac{2}{3}$ units of wax.

4. A bee (also extinct) was grown, forming *hexagonal* cells, with flat bottoms in double combs. This bee used $33\frac{1}{3}$ units of wax.

5. The hive bee (now living side by side with his humble progenitor) was produced by natural selection, dependent on the economy of wax, arising from the contrivance of substituting for the flat bottoms of the hexagonal cells the trihedral angles and planes of the rhombic dodecahedron.

This bee (*our* bee) uses $32\frac{2}{3}$ units of wax.

6. The *Bee of the Future* (not yet produced), which shall have learned how to construct the cells described by the mathematician Lluillier.

This bee will be broader and shorter than the present, the breadth and length admitting of prediction to any degree of approximation.

This Bee of the Future will only require 24½ units of wax!!—*Vivat Geometria!*

Of these six species of bee (the first and the fifth are living), No. 5 using only 32⅔ lbs. of wax in the construction of its cells for every 100 lbs. used by No. 1. According to the Malthusian law, No. 5 has exterminated No. 4, by virtue of the trifling advantage of ⅔rds of a pound of wax in every 100 lbs.; and this slight advantage is gravely alleged as the efficient cause of converting one species of bee into another! This would be all very well, if No. 1, the spendthrift humble bee, were not still living, and holding his ground well against his enemies, to bear witness against this silly theory.

In fact, the whole question of the economy of wax, and other such questions, require a thorough sifting. To my mind, it is evident that economy of wax has nothing whatever to do with the making of the bee's cells, but that this and other properties, such as maximum resistance to fluid pressure, &c., necessarily reside in the bee's cell, because they are the inherent properties of the rhombic dodecahedron, which is the form affected by that cell. The true cause of that shape is the crowding together of the bees at work, jostling and elbowing each other, as was first shown by Buffon. From this crowding together, they cannot help making cells with the dihedral angles of 120° of the rhombic dodecahedron; and the economy of wax has nothing to do with the origin of the cell, but is a geometrical property of the figure named.

III. The most serious logical blunder committed by all who invent a theory of life from the geological succession is, that *Succession implies Causation.* It is argued that the Palaeozoic cephalopoda produced, in some way or other, the Red Sandstone fishes; that these in turn gave birth to the Liassic reptiles; that the non-placental mammals of the upper Oolite *grew* after some fashion, and ultimately produced the Tertiary mammals, some of which, in an unhappy hour, gave birth to man. The only fact at the basis of this astonishing inverted cone of reasoning is that these creatures *did* succeed each other in the manner described, and from this it follows, *post hoc ergo propter hoc,* that they succeeded each other in the way of cause and effect. I propose to test this strange theory by a corresponding theory of the mineralogical succession of igneous rocks, which opens up a fertile field of speculation, hitherto unwrought. The igneous rocks of the Palaeozoic period contain abundance of felspar, whose principal constituent is potash; the Mesozoic igneous rocks abound in soda, replacing potash; and in the tertiary period, soda itself gives way to lime and magnesia. Viewed in the light of the Lamarckian philosophy, here is a distinct indication that soda and lime are only allotropic conditions of potash. We may read the history of their formation in the crust of the globe, if we will only open our eyes and see it written. I may add, by the way, that this theory of the origin of lime is more intelligible than that of many geologists, who would attribute the greater accumulations of calcareous rocks in secondary and tertiary strata to the creation of lime by organic force.

If any chemist or mineralogist were to put forward such a geological theory of the origin of soda and lime as the foregoing, he would be regarded as a lunatic by other chemists and mineralogists.

How does it happen that a theory of the origin of species, which rests on the same basis, is accepted by multitudes of naturalists, as if it were a new Gospel? I believe it is because our naturalists, as a class, are untrained in the use of the logical faculties which they may be charitably supposed to possess in common with other men. No progress in natural science is possible as long as men will take their rude guesses at truth for facts, and substitute the fancies of their imagination for the sober rules of reasoning.

It has been well observed by the greatest of living Palaeontologists, "that past experience of the chance aims of human fancy, unchecked and unguided by observed facts, shows how widely they have ever glanced away from the gold centre of truth!"

Comments on Haughton

Samuel Haughton was a physiologist and a professor of geology at Dublin University. He published numerous papers, primarily in the application of mathematical techniques to problems in physics. For example, he calculated the extent of geological time to be about 200 million years. Haughton had the honor of making the first public reference to evolutionary theory after the Darwin–Wallace papers were read at the Linnaean Society in 1858. In an address to the Geological Society, Dublin, on February 9, 1859, Haughton observed that evolutionary theory was either a truism or false. Two additional negative comments appeared soon thereafter in the

Zoologist, one by Thomas Boyd (1859), the other by Arthur Hussey (1859).

Haughton's opinion of evolutionary theory is so ill-informed that his review would hardly warrant inclusion in this volume if his attitude were not so characteristic of an influential segment of the scientific community—physical scientists trained in mathematics and the classics. In fact, there is almost a direct relation in the reviews of the *Origin* between the number of classical allusions and the amount of misunderstanding of Darwin's ideas. As Darwin himself observed, workers trained exclusively in mathematics and physics seemed peculiarly incapable of understanding him. Similar papers were published by Peter Tait (1869) and Francis Bowen (1860). Tait was a co-worker with Lord Kelvin and published an influential physics text with him (Thomson and Tait, 1867). The second edition of this work (1890) contained appendixes, including two of Kelvin's papers on the age of the earth and a paper on the tides by Darwin's second son, G. H. Darwin. Bowen was a philosopher, encouraged by Louis Agassiz to attack evolutionary theory much as Owen had set Bishop Wilberforce on Darwin. Bowen also had the singular distinction of being the only professor ever to lose his teaching post at Harvard for being politically too conservative. The trustees withdrew his appointment in 1851 for his attacks on the Hungarian revolution (Daniels, 1968).

Haughton, Tait, and Bowen all assumed that the theories of the physical sciences automatically took precedence over those of other sciences and that evolutionary theory could be disproved out of hand by reasoning facilely from simple mathematical principles. Haughton's discussion of the bee cells is typical of this mode of argumentation. Wallace subsequently replied to Haughton, pointing out the extent of his ignorance (Wallace, 1863). The issue was more important than it might at first appear, since it bears on the possibility of the evolution of instincts and the question of teleology.*

* See also Haughton (1846, 1873).

William Hopkins (1793–1866)

Have you seen Hopkins in the new 'Fraser'? the public will, I should think, find it heavy. He will be dead against me, as you prophesied; but he is generously civil to me personally. On his standard of proof, *natural* science would never progress, for without the making of theories I am convinced there would be no observation.—C. Darwin to C. Lyell, Down, June 1, 1860 (*Life and Letters*, 2:108)

In the number just out of *Fraser's Magazine* there is an article or review on Lamarck and me by W. Hopkins, the mathematician, who, like Haughton, despises the reasoning power of all naturalists. Personally he is extremely kind towards me; but he evidently in the following number means to blow me into atoms. He does not in the least appreciate the difference in my views and Lamarck's, as explaining adaptation, the principle of divergence, the increase of dominant groups, and the almost necessary extinction of the less dominant and smaller groups, etc.—C. Darwin to Asa Gray, Down, June 8, 1860 (*More Letters*, 1:153)

Hopkins's review in 'Fraser' is thought the best which has appeared against us. I believe that Hopkins is so much opposed because his course of study has never led him to reflect much on such subjects as geographical distribution, classification, homologies, &c., so that he does not feel it a relief to have some kind of explanation.—C. Darwin to Asa Gray, July 22, 1860 (*Life and Letters*, 2:120)

Physical Theories of the Phenomena of Life*
WILLIAM HOPKINS

Part I

Questions relating to the nature and origin of the phenomena of life, especially when extended to man as an intellectual and spiritual being,

* From *Fraser's Magazine*, June 1860, 61:739–752; July 1860, 62:74–90.

naturally present themselves under two distinct aspects,–first, in the direct and immediate relation which they may bear to the existence and attributes of an Omnipotent Creator and All-wise Governor of the universe; and secondly, in their relation to the secondary causes to which we refer the more ordinary phenomena of inorganic matter. The first may be called the *religious* or *theological* aspect, the second the *scientific* aspect of such questions. It has too frequently happened that men of pious minds and strong religious convictions have regarded this latter view of such subjects as inconsistent with the reverence due to the Deity, and have sometimes, moreover, been ready to brand with hard names those who entertain such views; while the latter, in their turn, have often shown little sympathy with the feelings of reverence and piety in which the opinions of their opponents may have originated. This is deeply to be regretted, and we are anxious to express at once our respect for the sentiments of piety in which we believe opposition to science to have frequently arisen, and our unequivocal belief that all the phenomena which nature presents to us constitute legitimate objects of scientific investigation. At the same time, the subjects to which we are here more especially alluding, require that such investigations be conducted with modesty and reverence, and not with that presumption which would claim even for some of the vaguest conclusions of man's intellect a weight and authority beyond that which may be assigned to higher sources of our knowledge. But at all events, if any one believe that the phenomena of life can be accounted for by the same laws as those which govern the combinations of inorganic matter, let him feel that the subject is freely open to his researches; but while we concede to him entirely the right of conducting his investigations on the same principles as we recognise in all investigations of the laws of inorganic matter, let him also understand that we exact from him in the support of his theories the same logical reasoning and the same kind of general evidence as we demand before we yield our assent to more ordinary physical theories. While we admit the same principles of research, we cannot admit different principles of interpretation, and yield our assent to the naturalist on evidence which we should utterly reject in the physicist. Let the investigator of the causes of vital phenomena state his evidence in support of his theories; it is then for us to test the claim of such theories to our belief by instituting an impartial comparison between the weight of the evidence adduced, and that which we demand in the theories respecting inorganic matter; and moreover, if his theories necessarily involve materialistic or pantheistic conclusions, we are bound to weigh his evidence in the balance against those general considerations which are opposed to materialism and pantheism, and to reject it if found wanting. It is impossible, we repeat, to admit laxity of reasoning to the naturalist, while we insist on rigorous proof in the physicist. He who appeals to Caesar must be judged by Caesar's laws.

Let us, then, consider the logical course of research which will thus be prescribed to the naturalist who would enter with the true spirit of inductive philosophy into what may be termed the *physical* investigations of his subject. The first step will be the generalization of the observed phenomena, which, that we may speak of them the more simply and definitely, we may suppose to be restricted not merely to organic matter, but to animal life. It will embrace the classification of animals, their structural organization, the correlation of their various organs, the adaptation of those organs to the functions which they are required to perform, and, in short, the grouping together of the particular phenomena in such a manner as to indicate any laws by which they may be connected. In every physical investigation the first step is a precisely analogous one. In crystallography, for instance, it consists, of the recognition of the various forms of crystals, and the geometrical laws by which those forms are connected, together with any other charactistic phenomena which may belong to crystalline bodies. In astronomy, the analogous first step (restricting ourselves to the solar system) consists in the observations of the motions of the planets and satellites, with any other phenomena which may present themselves, followed by the grouping together of such observed facts as seem to be connected by general laws. Thus Kepler, after years of observation on the planetary motions, deduced from the observed phenomena the three great laws which characterize those motions—that of elliptic motion, of the uniform description of areas, and that which expresses the relation between the periodic times of the planets and the longest axes of their orbits. Again, in the science of optics, all the phenomena of reflexion, refraction, polarization, &c., were observed, and the corresponding laws of reflexion, refraction, &c., were deduced from them; and so we might enumerate every other branch of physical science. Now, it will be observed that in the three sciences specified above, the laws immediately deduced from the grouping of the observed facts are entirely independent of any theory respecting the *physical* causes to which the phenomena may be referable. They admit for the most part of being enunciated in the language of pure geometry, and are

frequently termed the *geometrical* laws of the phenomena, in contradistinction to the laws by which the physical causes of those phenomena may
be governed.

It is manifest that these geometrical laws in the physical sciences above
specified, are exactly analogous to those which may be inferred from the
generalizations of the phenomena of animal life. Cuvier, as is well known,
made an enormous advance in these generalizations, especially in basing
them not merely on external characters, but on anatomical structure. Hence
he was enabled to deduce his most important law respecting the correlation
of different organs, and their individual and combined adaptation to meet
the peculiar necessities of each class of the animal kingdom. He thus became
the strongest advocate of the doctrine of Final Causes. But before the
close of Cuvier's life, another school of zoologists and anatomists arose,
who were not satisfied with what they regarded as the comparatively narrow
limits of his generalizations. They sought to discover something that might
lead us nearer to the cause and origin of those organs which Cuvier had
regarded chiefly in subservience to the animal wants which they were intended by a beneficent Creator to supply. It was a fundamental idea of
this school of transcendental anatomy, as it was called (of which Geoffroy
St. Hilaire was one of the principal leaders), that the whole animate creation was constructed according to a *uniform plan,* and that consequently
there must be some form of organization which presented a general *type*
of animal structure, every individual form being only a particular modification of this typical one. The attempts to discover this typical form led
to the assertion of resemblances in some cases, which could only be recognized by a wild imagination, and exposed these researches to the charge
of great extravagance. They led, moreover, in many instances to the adoption of pantheistic views, and the utter rejection of the doctrine of final
causes, and of the belief in a Supreme Intelligence and personal Governor
of the universe. Independently, however, of all such collateral views, the
search for a typical form of animal organization, as a higher order of
generalization than had been previously arrived at, was, in a philosophical
point of view, strictly legitimate, and perfectly in accordance with the
rules of inductive philosophy. Restricted as it appears to have been of
late to the vertebrate skeleton, it has led to a generalization of structural
characters which, though it may not in any one precise form command
the universal assent of naturalists, must be recognised as an important
step in the subject.

Let us now consider the next logical step to be taken with the view of establishing a physical theory of these vital phenomena, and in accordance with the process of investigation which is deemed imperative in forming our theories respecting inanimate matter. In the most perfect physical sciences, the observed laws of the phenomena (their geometrical laws) have suggested some physical cause to which the phenomena may be due, or if the mode of expression be preferred, some higher generalization. Then, assuming the truth of our hypothetical cause, we must calculate or investigate by the best means which the subject may supply, the necessary consequences which must result from the action of the cause assigned under such conditions as the problem presents to us. The results thus obtained must then be compared with the observed phenomena, and the proof that our assumed cause is the true one will consist, first, in the accuracy with which we can determine its necessary consequences; and, secondly, on the degree of accordance which we can establish between those consequences and the existing observed phenomena. These two last-mentioned particulars are the important ones of the proof; for whether the hypothesis of the assigned cause be suggested by a careful consideration of the previously observed phenomena, or by any *a priori* considerations, is manifestly of no importance so long as we can establish the accordance between the essential consequences of our hypothetical cause and existing facts.

All this is so well known to scientific men, that these detailed remarks might almost seem to require an apology were it less certain than it is, that however well the abstract philosophical rules of theorizing in physical subjects may be known, many people will still theorize as if Bacon and Newton had never lived. Our special object, too, is to inquire how far those naturalists who have aspired to this higher order of generalization, and have sought to ascertain the physical causes of the origin and nature of vital phenomena, have complied with the essential rules we have indicated, before they claim our acquiescence in their theories. And in order to establish a standard to which the amount of evidence in favour of any particular theory may be referred, let us briefly consider the evidence on which we base our convictions of the truth of the theory of gravitation as applied to the phenomena of the solar system, and that of the Undulatory theory of light, these being the two most perfect physical theories which we possess.

Newton was led (whether by the falling apple or more profound considerations is of little importance) to assign the mutual gravitation of all particles of matter towards each other as the physical cause which governs the motion of the moon, and, by an extension of the hypothesis, the motions

also of all the bodies of the solar system. Moreover, it was assumed that
the intensity of this mutual gravitation or force of attraction varies with
the distance between the mutually attracting particles according to a determinate law—that of the inverse square of the distance between the particles.
In this hypothesis there is no ambiguity. On the contrary, it is so determinate
that mathematicians have been able to deduce with the most refined accuracy innumerable results as the necessary consequences of this assumed
gravitation; while, on the other hand, astronomers have been able to observe
with equally refined accuracy the motions of the various bodies of the
solar system. Hence we are in a position to compare the results of calculation
with those of observation, and their exact accordance in all particulars
affords the most perfect proof we can conceive of the truth of a physical
theory. It is the case to which we must ever recur to understand the degree
of conclusiveness of which physical reasoning is capable, and the amount
of conviction which may be derived from it. Such conviction can scarcely
be said to be less than that derivable from the demonstrations of the properties of numbers and of space.

Again, let us turn to the theory of light, or Physical Optics. We have
an immense number of carefully observed and accurately defined optical
phenomena depending on the reflexion, refraction, polarization, double
refraction, &c., of light, which any theory of light is imperatively called
upon to account for. This is accomplished for an immense majority of
these phenomena by the Undulatory theory. In this theory the fundamental
hypotheses are extremely simple, and constitute a beautiful generalization.
A self-luminous body is assumed to be so by virtue of the extremely rapid
and minute vibrations of its constituent molecules, as a body becomes sonor-
ous in a state of similar though much larger and less rapid vibrations.
All free space and the pores of all transparent bodies are assumed to be
occupied by a highly subtle and refined elastic medium, to which the vibrations of luminous bodies are communicated, and by which they are propagated from one point of space to another, as sound is propagated through
the atmosphere; and finally, the effect of light is assumed to be produced
by the action of the vibrating molecules of this ethereal medium on our
organs of vision, in a manner similar to that in which the aerial vibrations
produce the effect of sound by the action of the contiguous particles of
the air on our organs of hearing. Now from these simple hypotheses we
can deduce by mathematically accurate reasoning a great variety of consequences which are found exactly to agree with the corresponding observed
optical phenomena. This theory, however, is less perfect than that of gravi-

tation, since we cannot calculate all the consequences of our hypotheses with the same certainty as in the latter theory, nor can we assert an equally complete agreement in every case between the results of calculation and those of observation. It is a case of somewhat more imperfect evidence in support of a physical theory; but still the accordance between theory and observation is so great that it might be difficult to draw a line between the conviction we feel of the truth of the Undulatory theory, and that perfect conviction with which we regard the theory of gravitation.

Other physical theories are for the most part far inferior to the two above mentioned in the conclusiveness of the evidence on which they rest. The theories of heat, electricity, and magnetism, for instance, are not yet reduced to the same simple and certain elements as those of gravitation and light; and in all such cases, whatever he may think of the preponderance of probabilities, a sound philosopher will hold his opinions with a certain degree of reticence and reserve, till more perfect evidence shall be adduced. Again, in those branches of science in which he must seek for the causes of the observed phenomena in molecular attractions, we can scarcely be said to have advanced beyond the stage of investigation in which we group and arrange observed phenomena according to their more obvious resemblances, and deduce from them what I have termed in physical astronomy and physical optics, *geometrical laws,* in contradistinction to the *physical causes* to which in those sciences the phenomena are ultimately referred. To take a definite case, the crystallographer has ascertained with great exactness the various crystalline forms and their relations to each other. He has thus done for his science what Kepler did for the astronomy of the solar system. The step analogous to that which Newton made in referring the lunar and planetary motions to gravitation, would consist in the determination of the nature and laws of those molecular attractions from which the phenomena of crystallization would result as necessary consequences. Some attempts have been made to investigate the nature of molecular action, but without any considerable success. In all cases the great difficulty lies in demonstrating that the observed phenomena are the necessary results of the physical causes to which they are attributed. As we have already intimated, the theory of gravitation is the only one in which this has been perfectly effected. But with respect to molecular attractions and their effects the task is infinitely more difficult. Newton and succeeding mathematicians have had to determine the motions of the heavenly bodies as depending on the mutual attractions of the finite and comparatively small number of bodies which compose the solar system.

These are of finite magnitudes, and at finite distances from each other; whereas in determining the effects of molecular actions we have to regard them as the results of an infinite number of infinitely small material particles acting on each other at infinitely small distances. Considering the subject with respect merely to the phenomena of crystallization, as among the best defined and apparently most simple results of molecular forces, the mathematicians who have been able to vanquish all the difficulties of the theory of gravitation have made scarcely any progress in overcoming the far more complex difficulties of the theory of molecular actions. And if such be the difficulties in the simplest branch of the subject, what must they be in the infinitely more complicated branches of the different departments of chemistry. Nearly all that has been effected in that extensive science, must be considered as the generalizations of phenomena equivalent to the geometrical laws above spoken of in astronomy and optics, and still remaining to be referred, in some indefinitely more advanced state of the science, to molecular actions as their physical cause.

It has been our object in these remarks to point out, in the first place, as distinctly as possible, the modes of investigation by which our most perfect physical theories have been established, and by which alone other physical theories possessing a corresponding degree of generalization, can hereafter be established on foundations sufficiently stable to secure general assent to their truth; and, secondly, to indicate the enormous difficulties which present themselves in the application of a similar mode of research to those sciences which involve the action of molecular forces. Without following the investigations of the effects of these forces in some of the simpler cases in which they have been attempted, into the intricate details and mathematical complexities into which they lead us, it is perhaps impossible to realize these difficulties; and if they be so great in reference to the phenomena of mere form and structure, as in crystalline bodies, it seems highly probable that they must be far greater still in reference to all those phenomena which depend on the qualitative properties of bodies, phenomena with which the science of chemistry deals so largely.

In examining the process of reasoning and investigation above described, we observe that the great difficulty in every instance is in the exact determination, whether by mathematical or other accurate modes of reasoning, of the consequences which would necessarily result from the action of the physical causes hypothetically assigned as the true causes of the observed phenomena. In all cases where the phenomena depend on the molecular actions of the constituent particles of matter, this difficulty has been really found invincible; for even in the simplest cases, which have reference only to form and geometrical structure, it has been very partially, if at all, overcome; and in all that infinite variety of cases in which the qualitative properties of bodies are concerned, it has not even been approached. Now if such be the difficulties which beset the completion of our theories of inorganic matter, what must they be with reference to organic matter; and not only organic but animate matter, and further still, with reference to the connexion of mind with matter? When we consider the array of these enormous cumulative difficulties, we cannot but smile at the thought that we should at present hope to arrive at even the most remote approximation to the real solution of the problem of vital phenomena.

But, it may be objected, this is not the solution of the problem, or the mode of effecting it which has ever been contemplated. We believe it. Still it is not less the only complete and demonstrative solution which it admits of; and our especial object in thus clearly and explicitly indicating it, is that we may the more correctly estimate the value of the vague solutions which have been really attempted. We doubt whether many of those who have believed in such solutions have ever formed a distinct conception of what the complete solution is—complete, we mean, in the sense in which Newton's solution of the problem of the planetary motions is complete, for any completeness of solution at which man can arrive must necessarily be relative.[1] We suspect, for instance, that those earlier transcendentalists who sought for this solution in the *unity of type* for the whole animate creation, fell into a confusion of ideas similar to that into which we should fall in confounding Kepler's geometrical laws with Newton's higher generalization, or physical solution of the problem of the planetary motions. At all events, no determination of the nature of vital forces, nor any explanation which they may afford of vital phenomena, can be rendered complete and demonstrative except by the process of reasoning and investigation we have described; and it is only by an appeal to the solution of the problem which would be thus obtained, that we can judge of the incompleteness of other solutions, and of the degree of confidence to which they may be entitled. We shall proceed to examine certain theories which have been offered, and more particularly in the sequel, that put forth by Mr. Darwin in his recent work on *The Origin of Species,* with the view of estimating their value on the principle here stated.

1. A more complete solution than Newton's, for instance, might determine the origin and cause of *gravity* in some more simple and elementary property of matter.

The first theory we have to notice is that of Lamarck, a distinguished French naturalist at the end of the last and beginning of the present century. It is fully developed in his *Zoologie Philosophique,* published in 1809. It embraces the origin of life from inorganic matter, and comprises man, regarded not merely as an animate, but as an intellectual being. We consider his theory of the origin of life altogether worthless as a physical theory, but it serves well to elucidate the point in which all such theories utterly fail; and this, as well as other parts of his general theory, seems to us to afford a curious proof of the union of great acuteness of observation and large views in the grouping and classification of natural objects and phenomena, with the want of that judicial power by which a man is enabled to estimate the true weight of the evidence laid before him, and the degree of confidence with which it ought to inspire us. Is it that the power and habit of minute external observation is so far opposed to the power and habit of abstract reasoning that they are rarely found coexistent in the same mind? Assuredly we should not often expect to find in the abstract philosopher the best observer of external objects; and it might perhaps be equally unreasonable to expect generally in the acute observer the higher powers of abstract reasoning.

According to Lamarck, all animate matter is composed of, or immediately results from, a *cellular tissue,* forming the substance of all vital organs, or surrounding and investing every separate portion of them. It consists of an aggregation of minute cells filled with a more or less compound fluid incapable of permeating the membranous walls of the cells, the whole constituting a soft and flexible mass. But this mass would not be a *living* mass unless acted on by some exciting cause which puts the contained fluids in motion and gives to the mass its properties of animate existence. This exciting cause is supposed to be found in some circumambient ethereal fluid pervading all space in which animate matter can exist. The principal of these *fluids* (as such they are spoken of) are supposed to be heat and electricity, with a certain degree of humidity, these being in fact, as we well know, the convenient agencies by which a bold philosopher in those days overcame the most formidable obstacles which might beset him. These are assumed to produce and maintain a certain *irritability* in the cellular membranes, and also to act upon the contained fluid, thus calling forth the vital actions of the mass.

The above general description of the masses which constitute animal organs would not, we conceive, be materially objected to by physiologists of the present day; and regarding vital phenomena as the result altogether of physical causes, Lamarck committed no violation of the strictest rules of philosophic reasoning in the assumption that the exciting cause—that which gives real vitality to the cellular mass—was some such circumambient medium as above described. He had the same right to assume this as any other philosopher has to assume the existence of that refined ethereal medium by means of which, as already intimated, light is supposed to be propagated from one point of space to another, or to assume that gravity is a universal property of matter and the cause of the planetary motions. But then comes the next step in the proof of any physical theory, always the most difficult, and that in which every false or imperfect theory necessarily fails. It consists in the investigation of the necessary consequences of the cause assigned, acting under probable conditions, and in showing the accordance of the results thus obtained with existing phenomena. Philosophers have not established the theory of gravitation or the Undulatory theory of light by asserting merely their *belief* that the causes assigned according to the fundamental hypotheses of those theories, are capable of producing the phenomena thus referred to them. They are not content to say that it *may* be so, and thus to build up theories based on bare possibilities. They *prove,* on the contrary, by modes of investigation which cannot be wrong, that phenomena exactly such as are observed would *necessarily,* not by some vague possibility, result from the causes hypothetically assigned, thus demonstrating those causes to be the true causes. Lamarck, on the contrary, and those who reason like him, content themselves with the mere *assertion* that phenomena exactly according with existing phenomena will, or rather may, result from the physical causes to which their theories refer them. Thus, Lamarck asserts that from some gelatinous or mucilaginous mass, nature, in some way or other, forms the cellular tissue which becomes a living mass when acted on by the supposed circumambient medium. This constitutes, according to his theory, the *origin of life* or an act of *direct* or *spontaneous generation,* the power of subsequent generation being supposed to be given to the mass in common with the other vital powers thus communicated to it. But here we have nothing beyond bare assertion. There is not an attempt to define the intimate nature of the physical causes assigned or their mode or laws of action, much less to explain even the possibility of their producing the simplest phenomena of life. Where the philosopher who forms his theories of the physical phenomena of inorganic matter feels himself imperatively called upon for *demonstration,* the philosophical naturalist, while he appeals to the same principles of investigation, is here content to offer us his simple *assertion.*

And yet the great naturalist of whose theories we are speaking remarks with the most perfect complacency,

> *One cannot doubt* that when the right kind of inorganic matter is placed in favorable circumstances and is acted upon by natural agents, principally heat and moisture, it can have cellular development impressed upon it and, consequently, to pass from there to the simplest organic state and, finally, to be able to perform the most primitive vital movements.[2]

It is to the total absence of even an attempt at anything resembling demonstration that we would here especially direct the attention of our readers. The same fault characterizes all theories of vital phenomena. It will be our object to point out similar defects in such other theories as we propose to examine.

It has been too much the custom, we fear, among the opponents of views like these above mentioned, to brand their authors indiscriminately with the names of pantheists and atheists. In too many cases such appellations may have been well merited, but certainly not in all, whatever may be the necessary tendency of these doctrines. Lamarck constantly recognises a Supreme Intelligence and great Author and Source of all things; but he maintains that in the act of creation the Creator impressed on matter such properties, and subjected it to such laws as might not only be necessary, but also sufficient for the future maintenance of the universe, and for the production of all subsequent natural phenomena, without any further exertion of His creative or controlling power. At the same time the manner in which he extends his physical theory of life to man as a moral and intellectual being is manifest materialism.

We now come to that part of Lamarck's theory in which he professes to account for the continued organic development of a mass which he supposes to have acquired the properties of vitality in the manner above described. It is this part of his theory which is exactly parallel to, and so much resembles, that recently proposed by Mr. Darwin. The latter author declines entering on any theory, like that above mentioned, of the origin of life, nor does he, like Lamarck, attempt to follow out his views to their legitimate consequences in their application to man as an intellectual and spiritual being. He restricts himself to the explanation of the manner in which he conceives the organization, whether of plants or animals, to have advanced from its lower to its higher stages, thus producing by a persistent advance all their peculiarities of organization. Lamarck also, in the cor-

2. [Translated by D. L. H.; italics are Hopkins'.]

responding part of his theory, endeavours to explain how, according to his views, the same changes have taken place. In examining these theories we shall first point out the common physical hypothesis on which Mr. Darwin's theory and this corresponding part of Lamarck's may be considered to depend, and likewise to indicate the distinction between these and certain other theories which might be erroneously regarded equally as physical theories.

And here it is necessary to have a distinct understanding of the meaning we attach to the word *species,* which has frequently been used by naturalists in two senses, widely different in respect to the points immediately under our discussion. The two meanings may be distinguished by the terms *natural* and *artificial.* By a natural species we mean a group of organic beings which can only have been derived by descent from beings similar to themselves, which possess certain external and anatomical characters distinguishing them from every other group, and whose descendants must necessarily inherit the same distinctive characters. Artificial species consist of groups which are equally distinguished by particular characters, but are such that, so far as the definition is concerned, they may not have been derived from each other or from some common original stock. In the former case the grouping is formed by nature; in the latter it is arbitrary, and only used in subservience to the convenience of classification, though generally made to coincide, as nearly as our knowledge will allow, with a perfectly natural classification. Every natural species must by definition have had a separate and independent origin, so that all theories—like those of Lamarck and Mr. Darwin—which assert the derivation of all classes of animals from one origin, do, in fact, deny the existence of natural species at all. The classifications adopted by those who hold the opposite opinion vary in some measure from each other, and therefore are liable to the charge of error in deviating from the natural lines of demarcation which they believe nature to have drawn. But this is the manifestly necessary consequence of the imperfection of our knowledge, and affords not the slightest argument against the existence of natural species, though some naturalists have appealed to it as such. They might as well contend that the solar spectrum does not consist of various colours, because we are unable to define the exact lines which separate them.

Lamarck's theory, as we have already intimated, if it could be established, would be more complete than Mr. Darwin's, since the latter does not profess to give any account of the origin of vital phenomena. In the former the physical cause assigned for the production of these phenomena must still be considered active, according to the theory, in all the phenomena of the continued existence and progressive advance of animal life. From the utter failure of the author's attempts, however, to establish the most remote relation between the real phenomena of life and his hypothetical cause of circumambient fluids, of heat, electricity, &c., we may altogether dismiss his original physical hypothesis, in the consideration of that part of his theory which embraces the same phenomena as Mr. Darwin's. With this restriction, it may be enunciated for both theories in common, that they refer all the phenomena of the varied organization in the animal kingdom to the continuous operation of the ordinary natural causes to which we ascribe the growth of each individual animal, or of the organs which compose it, and the propagation of its species, but without further consideration of the intimate nature of those causes. They equally maintain that the highest organisms in nature have been derived from the lower ones by some *continuous* and unbroken line of descent. These theories, then, are distinguished by the character of *linear continuity* which they assign to the chain of beings which have connected each existing organism with the simple one from which, according to these theories, it has proceeded. The theory commonly received asserts, on the contrary, the existence of *natural species,* each of which, since by hypothesis they are incapable of being derived from each other, must have had an independent origin. Moreover, it has been usually supposed that each of these species originated in an independent act of the Creator, towards whom the adoption of any other theory has frequently been regarded as an act of irreverence. Thus, as we have already remarked, the religious or theological aspect of the question has been placed in antagonism with its scientific aspect. We have already expressed our regret that such should be the case; and to escape the risk of it at present we shall endeavour to avoid even the phraseology of the theological view, and to state in the language of science that view of the subject which may be regarded as antagonistic to such theories as above described, and which would appear to be our obvious alternative in the case of our rejection of them.

By an act of creation considered with reference to scientific investigation, we mean a result due to some cause beyond and above those secondary causes to which the ordinary phenomena of nature are considered to be due. And we should define the difference between ordinary and extraordinary phenomena to consist in this—that in the former there exists a certain character of *continuity,* and in the latter there exists an equally distinct character of *discontinuity.* To elucidate our meaning, we may remark that the motions of the heavenly bodies are continuous, inasmuch as they are subject only to gradual or continuous variations. If, on the contrary, the velocities of these bodies, or the directions in which they move, or the orbits which they describe, were subject to sudden, instantaneous changes, the phenomena of their motions would have an obvious character of discontinuity. Similar remarks would apply to any other ordinary natural phenomena of inorganic matter. Again, with respect to organic matter, each plant or animal is gradually and continuously developed, and propagates descendants of the same character as itself. Such phenomena have a character of continuity; if the growth, on the other hand, were suddenly arrested or accelerated at particular stages, or if each organic being produced offspring totally unlike itself, the phenomena would be discontinuous. Now, it will be at once admitted that any such discontinuities in the existing phenomena must necessarily indicate a corresponding discontinuity in the action of the physical causes producing them, or in the laws according to which those causes act. But in the phenomena which we all agree to refer to natural or secondary causes, we observe no such discontinuities as those above-mentioned, and we consequently conclude that every cause which we ordinarily recognise as belonging to secondary natural causes, must be supposed to act according to some *continuous law*. This conclusion has probably been tacitly assumed, as it has been by Lamarck, by every one who has speculated on the natural causes to which the phenomena of either the organic or inorganic world are to be attributed. Mr. Darwin thus recognises it in the continuity of the changes which he supposes animals and plants to have undergone in the gradual development of one specific form from another.

Hence, then, the fundamental distinction between the theories we have been discussing, and that which asserts the independent creation and existence of natural species, regarded as a physical theory, and under its merely scientific aspect, is this—that the former recognise only those *continuous* physical causes which produce the ordinary phenomena of nature; whereas the latter, in addition to these causes, recognises a higher order of causation, acting according to some law which, in our ignorance of its nature, we are obliged to describe as *discontinuous*. Moreover, this discontinuous causation may be supposed to act in accordance with the continuous causation before mentioned, and not altogether independent of it. We are of course utterly ignorant of the relation which may exist between these two sets of physical causes, and make no hypothesis respecting it. We are, in fact, equally ignorant of the intimate nature of all the causes, whether continuous or discontinuous, with which we are now concerned. And here it may be remarked, that there is nothing new in the idea of organic matter being under the dominion of two sets of forces, the continuous and the discontinuous. We have in like manner two distinct sets of forces acting on all matter: that of gravitation, producing its effects at finite distances; and molecular forces, which produce their effects only at distances which are infinitely small. These two sets of forces may probably be in some way related to each other; but so far as we are able to contemplate their effects, they must act according to widely different laws. The only difference between this and the above case is that physical action, according to any discontinuous law, is restricted to organic matter.

Let it not be supposed that we are professing to enunciate a new theory. We have only done what might have been done by saying that new species, or any such discontinuous phenomena, are due to separate acts of creative power, but that those acts have a certain reference to previously existing phenomena, according to some self-imposed law in the Divine Mind. We have merely translated some such enunciation as this into the more ordinary language of science.

In order that a theory may have a just claim to be called a *physical* theory, it must assign some determinate physical cause to which the phenomena involved in it are considered to be due. The two theories abovementioned belong to this class, although, as above stated, they do not define the nature and mode of action of the physical causes referred to by them, except by assuming them to be the same as those which produce the ordinary phenomena of nature. Other views, however, which have been offered ought not to be regarded in the same light. The theory of *polarity* suggested by the late Professor Edward Forbes asserts that up to certain epochs in geological time (as, for instance, the termination of the palaeozoic period), the development of *generic* forms gradually decreased from some previous period, and then began to increase; or, as it may be stated, this development increased in going forward in the order of time, or upwards in that of the geological formations from the end of the palaeozoic period, and also increased in proceeding from the same epoch backwards in regard to time, and downwards in regard to geological formations. This he termed *polarity*. But if we regard it merely as a statement of a law which characterizes observed phenomena, it is manifestly not a physical theory in the same sense in which that of gravitation, or those of Lamarck and Mr. Darwin are such, for it asserts, as we are now supposing, no *physical cause* for the phenomena of which it speaks. It is, in this sense, a *generalization* similar to those of which we have spoken as *geometrical* generalizations in the theories of gravitation, physical optics, and crystallography. In these physical theories the proofs of their being true must consist in demonstrating a necessary relation between the observed facts and some physical cause to which those facts are assigned; but in this theory of polarity the proof must consist in determining by simple observation whether the phenomena do or do not follow the law ascribed to them. If, however, the term *polarity* is intended, on the contrary, to convey the idea of some particular physical cause to which the phenomena are due, the theory becomes a physical theory, which we estimate, as such, at the lowest value, since so far from tracing the action of some definite physical cause, it does not even assign such cause in any comprehensible terms.

We may instance another case, which has more the character of a geometrical generalization than of a physical theory. We allude to Mr. Wallace's views respecting the law which has regulated the introduction of new species. The author thus enunciates the law here alluded to—'Every species has come into existence coincident both in time and space with a pre-existing closely allied species.' Mr. Wallace has put forth this view in a clear and striking manner, so far as it is represented as a generalization of observed facts, which show a juxtaposition in time and space of allied species; but to convert it into a physical theory in the proper sense of the expression, some physical cause should be assigned, from the action of which this law of the phenomena would result. But there is no allusion to such cause. And therefore it is that we cannot approve of the assertions that the most singular peculiarities of anatomical structure are explained by it, and that many of the most important facts in nature are almost as necessary deductions from it, as are the elliptic orbits of the planets from the law of gravitation. There appears to us to be some confusion in these modes of expression, between the *law by which phenomena may be characterized,* and *the physical causes to which they may be due,* and which account for the law. We submit that the above assertions are not more correct than the assertion would be that the elliptic motion of the older known planets explains and accounts for that of Neptune or of the tribe of minor planets recently discovered. These latter motions are *in accordance* with the laws of elliptic motions, but are *explained* by the action of mutual gravitation; and the phenomena of organic matter appealed to by Mr. Wallace would, in like manner, be properly explained and accounted for by a theory which should prove the law here stated to be the necessary consequence of some determinate physical cause. We may also add that Mr. Wallace's theory cannot supersede that of polarity, as the author seems to suppose; for no physical cause is assigned which would be necessarily inconsistent with the law of polarity; nor, regarding both theories merely as generalizations of observed phenomena, can Mr. Wallace's law supersede that of Professor Forbes, since it is manifest that the truth of the one does not necessarily involve either the truth or falsehood of the other.

Much of what we have here said may possibly be deemed hypercritical. We think otherwise; for we believe that there is no inconsiderable amount of confusion of language and thought on the higher speculative subjects of which we have been speaking; and we are convinced that we can neither theorize philosophically ourselves on such subjects, nor judge rightly of the theories of others, without a clear and logical conception of those distinctions which we have been desirous of indicating.

Let us now recur to Mr. Darwin's theory and the corresponding part of Lamarck's. Having pointed out the hypothesis on which they may be considered in common to rest—viz., that all vital phenomena are due to the action of ordinary secondary causes—we may proceed to examine how they attempt to deduce from these causes the phenomena in question, but more especially those which relate to the varied forms of animate existence. In the absence of all knowledge of the real nature and laws of vital forces, it is impossible to investigate directly the consequences of those forces under any assigned conditions, as the mathematician calculates all the effects of gravitation; and this constitutes an enormous difference between our best physical theories and the theories of vital phenomena. In the latter, instead of the above direct and perfect mode of investigation, other indirect and far more imperfect modes must be substituted. In the two theories before us, the authors lay down certain principles by which they conceive ordinary natural causes to be regulated in the production of the phenomena referred to them. The principles thus laid down by Lamarck are involved in the three following assumptions:—

1. That any considerable and permanent change in the circumstances in which a race of animals is placed, superinduces in them a real change in their wants and requirements.

2. That this change in their wants necessitates new actions on their part to satisfy those wants, and that finally new habits are thus engendered.

3. That these new actions and habits necessitate a greater and more frequent use of particular organs already existing, which thus becomes strengthened and improved; or the development of new organs when new

wants require them; or the neglect of the use of old organs, which may thus gradually decrease and finally disappear.

The principle on which Mr. Darwin's reasoning rests is that of *Natural selection*. An immensely greater number of animals must be born than can possibly live to what may be regarded as the natural term of their lives. A 'struggle for existence' (to use our author's phrase) must therefore necessarily ensue; and in this struggle the stronger will vanquish the weaker, and live to transmit their species to future generations. Thus nature is supposed to 'select' the best variety[3] of any existing species, however it may have arisen, for the propagation of the race, as a breeder of domestic animals selects the best of those he may be cultivating, to breed from; and as he dooms to more immediate slaughter the inferior portion of his flocks and herds, and thus year by year improves his stock, so nature abandons to early destruction and final extermination those inferior portions of any race of animals which are least able to protect themselves in the 'struggle for existence,' and thus improves the general character of succeeding generations, to which every improvement is supposed to be transmitted by descent. Mr. Darwin does not appear to consider that much influence is due *directly* to the external conditions of climate, food, &c., in the advance of any race of animals. He remarks (p. 85), 'We must not forget that climate, food, &c., probably produce some slight and direct effect. It is, however, far more necessary to bear in mind that there are many unknown laws of correlation of growth, which, when one part of the organization is modified through variation, and the modifications are accumulated by natural selection for the good of the being, will cause other modifications of the most unexpected nature.' He also intimates his opinion that the causes which produce varieties may likewise affect the generative powers of any existing race, so as to enable it to propagate descendants of improved organization.

The great similarity of the notions on which the reasoning of these two theories is based, is obvious; but Lamarck insists more on the direct influence of ordinary external conditions, and Mr. Darwin more especially on that of natural selection. But this difference is perhaps as much in appearance as in reality; for if Lamarck could have been asked whether he included the struggle for existence among the conditions which might influence the wants and habits of animals, and consequently their organization,

---

3. The accurate distinction between *varieties* and *species* consists in this—the former, when crossed, always produce fertile offspring; the latter either do not admit of being crossed at all, or when crossed they produce sterile offspring.

he would necessarily have answered in the affirmative; and on the other hand, we understand Mr. Darwin to affirm that there could be no natural selection without varieties, and it seems difficult to understand how natural selection could operate in perpetuating one variety rather than another, except with reference to those circumstances which, according to Lamarck's view, would modify the wants and constitution of existing species. Again, while Lamarck believed that any modification of the organs of an animal which its altered wants might lead it to desire, would be accorded to it, and thus lead to its indefinitely progressive improvement, Mr. Darwin concludes that any improvement in the form of a variety will be transmitted to future generations, and by a cumulative process, elevate a species in the course of time, from the lowest to the highest order of organization. We confess that we find some difficulty in recognising any very essential difference in the fundamental hypotheses on which these two theories rest as physical theories. In Lamarck's an animal is incited to exertion by some kind of external circumstances, and obtains an improved organization as the reward of his efforts; according to Mr. Darwin's theory the same advantage is gained, but (if we may use the expression) rather by an accident of birth. We are disposed to think the less aristocratic process somewhat the preferable one. However, both are supposed to be equally cumulative, and to lead equally to a continuous slow advance of the race from the lower to the higher organizations. They both, consequently, deny the existence of groups of animals which have descended only from progenitors like themselves. In other words, they equally ignore the existence of natural species, and only recognise specific groups as convenient for the purposes of classification.

Mr. Darwin, in his *Origin of Species,* has entered much more fully than Lamarck into the detail of facts bearing on the subject. In our discussion of it we shall chiefly follow the work just mentioned. And here we would especially remark, that whatever opinion may be formed of the conclusiveness of our author's reasoning, every reader will acknowledge the great value of his observations respecting many of the most interesting facts and speculations which natural history presents to us. It is a work which will doubtless be found in the hands of every professed naturalist, while many others, without necessarily subscribing at all to its general theory, will probably find that its perusal has greatly enlarged their views on the subject, and given them a more adequate conception of the various questions of interest which it presents to the general reader. We are anxious to bear our most cordial testimony to Mr. Darwin's work on these points, as well

as our full participation in the high respect in which the author is universally held, both as a man and a naturalist; and the more so, because
in the remarks which will follow in the second part of this Essay we shall
be found to differ widely from him as regards many of his conclusions
and the reasonings on which he has founded them, and shall claim the
full right to express such differences of opinion with all that freedom which
the interests of scientific truth demand, and which we are sure Mr. Darwin
would be one of the last to refuse to any one prepared to exercise it with
candour and courtesy.

Part II

We proceed to further remarks which we have to offer on theories of
vital phenomena, more especially on that of Mr. Darwin in his *Origin
of Species*.

Mr. Darwin states that he has numerous facts which he still holds in
reserve for future publication, but it is not our intention to follow him
even into the details which he has already given. It would suit neither
our space nor our purpose to do so. Besides, many details are always apt
to perplex the mind and to draw it off from general principles and real
arguments, to which it is here especially our wish to direct the attention
to our readers.

The principal arguments, for both Mr. Darwin's theory and Lamarck's,
are drawn from the influence of domestication on the specific characters
of animals, and that of cultivation on those of plants. All species must
be subject to greater or less variation, for it is all but inconceivable that
any two animals should be exactly alike; and the peculiar circumstances
under which domestic animals are placed, as compared with those in a
state of nature, has doubtless been one great cause of their presenting
so much greater variations than we observe in wild animals. This variability
enables man to modify our domestic breeds still further by his power of
selection, and of thus cultivating any particular variation which he may
wish to promote. 'Like breeds like,' is the great principle of all breeders
who have most improved our domestic animals, though it is in fact by
the variations from this rule that they are enabled to effect their improvements. The principle itself asserts the transmission by descent of any
acquired property, and therefore the general likeness between the parents
and their offspring; but the power of man to improve a race, with reference
to any particular property it may possess, depends on his being able to

select those particular individuals of each successive generation in which this property is more especially developed, in order that they may transmit it in its improved form to their posterity. It has been by this means that breeders have been able within the last three-quarters of a century to effect such marvellous improvements, especially in cattle and sheep. Bakewell, who resided at Dishley, near Loughborough, in Leicestershire, was the first to recognise and apply scientifically this principle to the cultivation of those animals. Breeding became in his hands an art, which has since exercised the most important influence on the science of agriculture in our own country, and is now widely extended in America and France, and, in fact, in all countries where the importance of improving those animals on which man is so largely dependent for the supply of his necessities and enjoyments, is fully recognised. Horses and dogs are also races of animals to which this principle of selection has been largely applied; and pigeons are especially referred to by our author, on account of the curious and extensive modifications which have been effected in them by the continued application of the same art. These and other similar cases are appealed to by Mr. Darwin in proof of the influence which man possesses, by his power of selection, in extending the variations of which any race of animals under his constant control may be susceptible. They are the facts which have suggested our author's fundamental idea of *natural selection,* and form the very cornerstone of his theory. If man has done so much in so short a period of time, what may not nature have effected, he asks, during the illimitable ages in which she has been exercising her powers?

In discussing this question, we must consider whether nature and man act in the same or in different ways. Now it appears to us that man only exercises his influence in modifying organic forms, by counteracting the laws of nature, or rather by altering the conditions under which nature would otherwise act; and that, consequently, we have no right whatever to assume that nature will *necessarily* produce such effects at all when left to her own unobstructed operations, as those which she produces under man's interference. It is not man who is operating, but nature; in the one case under ordinary conditions, in the other, when those conditions are interrupted; and the above assumption is equivalent to the assertion that in either case effects of the same kind will be the necessary result. An elucidation may perhaps occur to the mind of a geologist, of the manner in which man is able to counteract some of the slowly operating though apparently almost irresistible agencies of nature, by the power which he possesses of resisting the ravages of the ocean waves on the shores on which

they are incessantly breaking. Man, by means of works which he is able
to construct, can not only withstand in many cases the efforts of the waves
to encroach on the land, but can, when so minded, reclaim land from
the dominion of the sea; and it might be said that if man can do so
much in a finite time, what may not nature accomplish in rescuing dry
land from the ocean during the enormous period in which her agencies
have been at work? This question, we say, might be asked in this case
as well as in the actual case before us, did we not know that the inevitable
general tendency of the sea is to swallow up the dry land, with little disposition to disgorge it again. It is here, strictly speaking as in the case of
man's influence over the modifications of organic characters, an error of
language to say that *man* recovers dry land from the ocean; it is the
ocean that produces this effect, while acting under the artificial conditions
which man is able to impose upon it. If in this case we should infer that
natural causes, acting under ordinary conditions, would produce effects
of the same kind as those produced under certain artificial conditions,
we should commit a manifest error; and in like manner we may commit
an error in asserting that nature, under ordinary conditions, will produce
modifications in the essential characters of organic beings, such as she may
produce under the artificial conditions to which our domestic animals are
subjected. Certain it is that Leicester sheep and Durham short-horns were
never produced by unaided nature.

But our author asserts that nature is always ready to render any variation
permanent when it is for the good of the individual and its race. We
quite believe that this is one way in which the beneficence of the Creator
may be displayed. But what is *good* for an animal? Is it necessarily good
that a sponge or a polyp should be promoted to the rank of an oyster,
that the ambition of the oyster should aspire to the position of a cephalopod,
or a cephalopod to that of a fish, and so on? We should have thought
it far more likely that the advantage which may thus be accorded to any
organic being was intended for its good in the sphere in which it was
born, but not to raise it above that sphere. At all events, we are sure
that there can be no such probability on the side of our author's view
as to constitute any positive *a priori* argument in its favour. He has a
perfect right, however, to make such an hypothesis; but we maintain, on
the ground of what has been advanced in this and the preceding paragraph,
that, admitting the existence of natural selection, there is no more reason
a priori for believing that it has the power of raising an organic being
from a lower to a higher stage of organic life, than there is for supposing

its power restricted to those limits which the distinctive characters of natural species would impose upon it. The onus of proving the existence of such power must rest with him who asserts it.

It is the opinion of some naturalists that there is a tendency in every variety to return to the original characters of the stock whence it has proceeded, and this tendency Mr. Darwin has designated as the *principle of reversion,* considering it not as depending on crossing with other varieties of the same species, but as an inherent property of every variety, though all intercrossing be prevented. In this sense he does not consider the principle established by any adequate experiment. But this is not the tendency (allowing it really to exist) which we should regard as most effectively antagonistic to that of natural selection. The real antagonistic tendency is that which exists to return to the form of the original stock by the intercrossing of different varieties. It is by the prevention of this process that man imposes artificial conditions so effectively on the operations of nature, and encourages the production of varieties. Remove all these artificial restrictions from any of our domestic races, and there can be no doubt, we conceive, of their returning rapidly through a series of mongrel breeds to a common type, modified only by ordinary local conditions of climate, food, etc. and the struggle for existence with other animals of the same or different species. If it could be proved that there is a predominant tendency in the more perfect and robust of each species to pair with individuals like themselves for the transmission of their kind, then the necessary existence of natural selection as an operative cause must be admitted, and the only question would be as to the extent of its operation. But we have no reason to believe that nature would thus favour the ambition of rising indefinitely in the scale of organization. The despotic will of one of the most despotic monarchs that ever lived could not call forth a race of grenadiers taller than nature chose to furnish him with. The elective affinities were against him, and short women would league with tall men, as if it were on purpose to frustrate the designs of the monarch who presumed to interfere with wills which were stronger in vindicating the law of nature than his own to violate it. Moreover, the experience of breeders of our domestic animals, with the more matured knowledge which later years have given them, indicates, we conceive, a distinct limit to the improvement which they may be able to effect in any direction. We have had considerable opportunities of witnessing the improvements which have been wrought, more especially in our sheep and cattle, and there can be no doubt but that each step in advance, in whatever direction it may take place, becomes more and more difficult, and the greater the perfection obtained, the greater is the attention required to prevent regression towards some inferior type. There are in fact two antagonistic causes at work, producing a tendency on the one hand to ascend, and on the other to descend. The ascent becomes more difficult as we advance, while the tendency to descend increases under the same circumstances; and thus we conceive it to be most probable that these two tendencies come at a certain point to counteract each other, and that therefore no advance can be made beyond that point. We are not asserting this as conclusive against Mr. Darwin's view of the continued advance due to natural selection; but we do assert it as proving that he does not establish a *priori* any necessary probability in its favour. If this cause be really operative to the extent he imagines, the fact must be entirely established by inductive reasoning, and a careful comparison of the consequences which must necessarily flow from the operation of this cause, with the phenomena which nature presents to us. We are not denying our author's right to reason upon his hypothesis as the foundation of his theory. He has an unquestionable right, according to the laws of all philosophical reasoning, to do so. We merely maintain that the soundness of his theory is not to be judged by this foundation, but by the superstructure which can be raised upon it. Let us proceed, then, to consider what must necessarily result from the truth of this hypothesis.

It is manifest that every change which takes place according to Mr. Darwin's theory, and equally so according to Lamarck's, must be strictly *continuous,* so that if we possessed specimens of each organic form which, since the first existence of organic life, has been derived from an immediately preceding form, the series would constitute an unbroken chain. It is not that organic forms of every kind would lie in one and the same unbroken line of descent. There may be an indefinite number of these continuous lines, all springing from the same origin, but branching out in different directions, either from that origin or subsequently from other points, precisely in the same manner as the branches of a tree diverge from its trunk, secondary branches from the primary ones, and so on till we come to the finest twigs and leafstalks. Mr. Darwin has given a diagram fully illustrative of this idea. But in accepting this illustration we must carefully mark the limitation implied by it as to the nature and extent of the law of continuity by which organic forms are connected. This continuity is not general, but what may be termed *linear,* in the sense in which each branch of a tree is continuous in itself without having any immediate connexion with the neighbouring branches. And as the branches of a tree, after their divergence from each other, do not again unite together, so according to this analogy the organic beings which have descended along different lines will be incapable of amalgamating with each other and thus producing intermediate forms. Admitting this kind of linearity, the discontinuity here spoken of would be a necessary result of this law, just as the separation of the ultimate branches of a tree is a necessary consequence of its mode of growth. But here it should be observed that, as the branches of the tree separate entirely at their common point of divergence, so, according to this hypothesis of linearity of descent, organic beings, even in their initial state of divergence from each other, and therefore while nearly identical, must still be incapable of amalgamation, either from their nature or the external conditions to which they may be subjected; for otherwise it is manifest that their union would destroy the separation of the two distinct lines of descent, and thus produce a *general* and not merely a *linear* continuity between all existing organic forms.

It has sometimes been said that, according to any theory like Mr. Darwin's, all different groups of animals must merge into each other, and all distinct lines of demarcation be obliterated. It is manifest, however, that such an objection is not valid if we admit the linearity of development above described; for there might be any amount of difference between two organic forms situated on two distinct lines of descent, though diverging from the same stock. But again, it may be said, that at least along these separate lines of development we ought to find in organic forms now existing, or in those which have existed during the historic period, indications of this gradual transition from one form to a consecutive one. But this argument also, if restricted to the evidence afforded by the historic period, is not necessarily a valid one, for it may be asserted that the variation of organic characters has been so slow as to be altogether inappreciable within the period. Thus we may conceive the horse, the camel, or the crocodile, for instance, to have varied within the last three thousand years in accordance with the law of continuous development, but still so slowly as to render the amount of progression insensible to us. Hence it follows that neither Lamarcks's theory nor Mr. Darwin's, nor any other requiring only a linear and not a general continuity of development, is necessarily inconsistent with those differences which naturalists recognise as constituting generic or specific distinctions. Passing over, then, these objections as untenable, let us proceed first to consider the proof required for the establishment of the fundamental hypothesis—that natural selection does so act as to produce different developments diverging from the same organism.

Varieties, as already stated, are distinguished from natural species by their perfect fertility *inter se*. According to Mr. Darwin's theory, these varieties are supposed to diverge till they become species, *i.e.*, till they can no longer unite with other organic beings which have diverged from the same stock, so as to produce fertile offspring. It might be supposed that this kind of negative property of sterility was acquired directly, according to the theory, by the assumed process of natural selection. This, however, is not our author's view. He supposes other qualities to be acquired by that means, and sterility to result from them as secondary effects, according to the laws of the correlation of growth. In his chapter on Hybridism he remarks (p. 245) :–

> The importance of the fact that hybrids are very generally sterile has, I think, been much under-rated by some late writers. On the theory of natural selection the case is especially important, inasmuch as the sterility of hybrids could not possibly be of any advantage to them, and therefore could not have been acquired by the continued preservation of successive profitable degrees of sterility. I hope, however, to be able to show that sterility is not a specially acquired or endowed quality, but is incidental on other acquired differences.

After these observations the author's object is to break down the line of demarcation between *varieties* and *species,* founded on the perfect fertility of the former when crossed with each other, and the sterility or very imperfect fertility of the latter when similarly crossed. But in doing this he confounds artificial with natural species, to which alone the above distinction from varieties is, of course, asserted. In the chapter just mentioned he says, in speaking of the sterility of species when crossed, or that of their hybrid offspring (p. 246) :–

> It is impossible to study the works of these two conscientious and admirable observers, Kölreuter and Gärtner, who almost devoted their lives to their subject, without being deeply impressed with the high generality of some degree of sterility. Kölreuter makes the rule universal; but then he cuts the knot, for in ten cases in which he found two forms, considered by most authors as distinct species, quite fertile together, he unhesitatingly ranks them as varieties.

Kölreuter is here accused of cutting the knot of a difficulty, but unjustly. The universality of the rule which he asserts is with reference to *natural* and not *artificial* species, and his definition of the term 'species' rendered it imperative upon him to rank the ten exceptional cases as varieties. Again, Mr. Darwin says (p. 267) :–

It may be urged, as a most forcible argument, that there must be some essential distinction between species and varieties, inasmuch as varieties, however much they may differ from each other in external appearance, cross with perfect facility, and yield perfectly fertile offspring. I fully admit that this is almost invariably the case. But if we look to varieties produced under nature, we are immediately involved in hopeless difficulties; for if two hitherto reputed varieties be found in any degree sterile together, they are at once ranked by most naturalists as species. For instance, the blue and red pimpernel, the primrose and cowslip, which are considered by many of our best botanists as varieties, are said by Gärtner not to be quite fertile when crossed, and he consequently ranks them as undoubted species. If we thus argue in a circle, the fertility of all varieties produced under nature will assuredly have to be granted.

It is a grave and very uncomplimentary charge against any one who assumes to himself the possession of a reasoning faculty, to say that he has exercised it only within the narrow limits of a vicious circle. In the above cases, however, Gärtner and the naturalists who have adopted similar views are easily justified. They define species as *natural* species, and propose a classification strictly consistent with their definition. The distinction to be drawn is not, as already remarked, between varities and *artificial* species, or what may, in the imperfection of their knowledge, have been conventionally regarded as species by naturalists, but between varieties and *natural* species. There is here no reasoning in a circle. Kölreuter and Gärtner both assert the absolute generality of this distinction in reference to the vegetable kingdom, adopting their own proper definition of 'species;' and we must deny that the one weakens the force of his conclusion by cutting the knot of a difficulty, or the other by any slip in his reasoning.

In fact we have nothing here to do with artificial lines of distinction between species, but with those which are real or natural; nor does the validity of our reasonings depend on an exact knowledge of the positions of these latter lines, but only on their existence.

But it may be asked, How do we know that such lines actually exist at all? And how do we know, it may be asked in return, that natural selection is actually operative at all in nature? Both these questions are altogether irrelevant. In answer to the latter it would of course be said, and justly, that it is not an asserted truth, but a fundamental hypothesis, which in the building up of the theory founded upon it must be reasoned upon as if it were an essential truth. And so the existence of the distinctions above spoken of is a fundamental hypothesis of any theory which asserts that different groups of organic forms have originated in successive creations; and consequently such distinctions must be reasoned upon as if their existence had been completely proved. Should any such theory be hereafter established, these distinctions will then be recognised as demonstrated truths; but if, on the contrary, any antagonistic theory should be established, though these distinctions might co-exist with lines of continuous descent, they must be given up in the more general sense in which the doctrine of successive creations asserts their existence. To deny *a priori* their existence would be an absurdity in the abstract reasoning of the subject; for it would be to assume at once the truth of some antagonistic theory which it is the object of our reasonings to establish or refute. The ultimate object of the discussion is to prove or disprove the existence of *natural* species, distinguished from varieties as above stated. With *artificial* species we have no concern.

It may be stated as a fundamental proposition of Mr. Darwin's theory, *that varieties distinguished by the property of perfect fertility, when crossed, may so far diverge from each other in the process of time, as to be incapable of producing fertile offspring when similarly crossed.* This is the proposition to be established by experiment and observation. The animals on which we should naturally expect experiments of this kind to be made are those of our domestic breeds, on which man has in fact ever been experimenting, consciously or unconsciously. But of the infinite number of such experiments, we are not aware of one which affords the slightest positive proof of the above proposition. If we continue to breed *in and in,* as Mr. Darwin has somewhere remarked, the variety may in turn dwindle away and disappear; but on being crossed with another variety of the same stock, it seems immediately to regain its pristine vigour and fertility, in exact opposition to the above proposition. If we allow full weight to all our author's arguments in his chapter on Hybridism, we only arrive at the conclusion that natural selection *may* possibly have produced changes of organization, which *may* have superinduced the sterility of species; and that therefore the above proposition *may* be true, though not a single positive fact be adduced in proof of it. And it must be recollected that this is no proposition of secondary importance—a mere turret, as it were, in our author's theoretical fabric—but the chief corner-stone which supports it. We confess that all the respect which we entertain for the author of these views has inspired us with no corresponding feeling towards this *may be* philosophy, which is content to substitute the merely possible for the probable, and which, ignoring the responsibility of any approximation to rigorous demonstration in the establishment of its own theories, complacently assumes them to be right till they are rigorously proved to be wrong. When Newton in former times put forth his theory of gravitation, he did not call on philosophers to believe it, or else to show that it was wrong, but felt it incumbent on himself to prove that it was right. And this must be the rule for the philosopher who advances his theories on the phenomena of life, equally with him who speculates on the properties of inanimate matter. We know that, in the case before us, no investigation can at present be founded on the action of the molecular forces which govern the combinations of animate matter, on account of our total ignorance of such forces; and so far these theories must be comparatively imperfect; but what we require is some proof of fundamental propositions, founded on observation or experiment, as above stated. Till this or something equivalent to it be done, we maintain that no theory like Mr. Darwin's can have a real foundation to rest upon, or can ever secure the general convictions of philosophers in its favour.

But let us pass to another point, one also of the first importance in any theory involving continuity of development. We have already pointed out that no unanswerable argument against any such theory is deducible from the absence of evidence to prove that any appreciable change has taken place in existing organisms during the historic period, simply because it may require a longer period to effect such change. But the question assumes a very different aspect when considered with reference to the geological period, during which such important changes in animal organization have undoubtedly taken place. It is, then, in the records of geology that we must seek for the evidence of that continuity by which according to the theories we are considering, these changes must have been characterized. Now, all are agreed that palaeontology affords no evidence of this continuity; and it is in the absence of this evidence that we meet with one of the gravest objections to any theory of continuous organic variation. Mr. Darwin has stated this difficulty with all that fairness and impartiality which so eminently distinguish his work in the statement of facts opposed to his views. He allows that the number of organic forms which must have existed intermediate to those which are known must have been enormous, and admits that it may naturally be asked (p. 280), 'Why, then, is not every geological formation and every stratum full of such intermediate links? Geology surely does not reveal any such finely graduated organic chain; and this, perhaps, is the most obvious and gravest objection which can be urged against my theory.' He then proceeds in his ninth chapter to develop his reasons for believing that the answer to this difficulty is to be found in the imperfection of the geological record.

In the first place we would remark that we entirely agree with our author as to the immense lapse of time indicated by geological phenomena, more especially by the enormous mass of fossiliferous strata which must have been transported by water from one locality and deposited in another, and by a process which must have been so slow that the whole work done must have required an almost inconceivable length of time for its accomplishment; and Mr. Darwin has well remarked, that no one, perhaps, who is not a practical geologist, can adequately realize this impression of the lengths of geological periods. Independently of this conclusion, there is little in this chapter from which we do not gravely dissent, as arguments for destroying almost entirely our confidence in the records of geology. If the discredit which it is here attempted to throw on these records be admitted, we cannot conceive what value can attach to any of the general conclusions of palaeontology; and that science which has already effected so immense a revolution in our opinions on some of the most interesting speculative subjects which can occupy the thoughts of men, must become comparatively valueless.

Mr. Darwin's purpose is to account for the want of such an immense number of the links of his chain of graduated organisms, by showing that either these intermediate forms have not been preserved at all in the sedimentary strata, or, if they ever were thus entombed, the record has been obliterated by subsequent denudation. But to arrive at satisfactory conclusions on this subject would essentially require a far more ample discussion of the phenomena of elevation, denudation, and deposition than our author has entered upon. Our space will allow us to make only a few general remarks upon it.

Mr. Darwin observes that a small portion only of animals existing at any past period can have been embedded and preserved in sedimentary beds. No perfectly soft organisms could be thus preserved, and many others thus entombed may have been subsequently destroyed by percolation of water or other solvents. This is undoubtedly true, but it is not the question. We are here essentially concerned with those animals which are best calculated to be thus preserved; and the question is, why, in these particular cases, and not in those in which preservation is comparatively rare, so few of the links of a continuous chain of organisms should have been preserved, when an immense number of contiguous links must have been equally capable of preservation. 'But,' our author further remarks, 'the imperfection of the geological record mainly results from another and more important cause than any of the foregoing—namely, from the several formations being separated from each other by wide intervals of time.' There are doubtless many instances of this kind, as indicated either by an extensive unconformability of stratification, or an entire change in the general character of the organic contents; but these are not the places in which we should look for the continuity of the organic chain; and when no such proofs of discontinuity of deposition on the scale here contemplated are observed, we do not admit any probability of its having occurred. Where two successive strata are perfectly conformable, a difference of mineral character is certainly no necessary or probable indication of a long interval between the periods of their deposition. Our author talks, too, of the frequent and entire destruction of sedimentary beds. Now we believe the entire destruction of any sedimentary bed of considerable horizontal extent to have been of rare occurrence. All the more important denudations of which we have any evidence have been preceded by large upheavals, by which the strata have been tilted; and thus while those portions of each stratum which have been most elevated may have been exposed to enormous denudation, those portions which have been least elevated, or perhaps depressed, have been thus kept out of the reach of the denuding agencies. The entire obliteration of a stratum would require in general that it should be upheaved in such a manner as never to deviate sensibly from a horizontal position. In fact, this approximation to horizontality must be closer than it frequently may be during the time of deposition, for the smallest dip in an extensive stratum would place it in a condition as to denudation similar to that above described as due to large upheavals. The higher portions might be denuded while the lower remained uptouched. The Weald affords one of the best elucidations of denudation accompanied only by the partial destruction of strata. We have no reason to suppose that a single stratum has been obliterated by this denudation, which, while it has left scarcely a remnant of the removed beds in the central portion of the district, has left portions of them untouched on its borders, where they dip beneath the existing surface. The same remarks apply to the oolitic beds of this country, as well as to the enormous denudations of Wales, or the northern parts of England, and in fact to all the great denudations of which geology affords us a record. It is for these reasons that we entertain a strong conviction of the permanence of some portions of all sedimentary

beds of considerable horizontal extent. Beds of limited extent may of course
have been so deposited as to admit of their subsequent entire obliteration;
but with respect to those of larger extent, it is certain that if all the areas
of existing land were denuded to the present level of the surface of the
ocean, scarcely a single sedimentary bed would be obliterated, simply because there is probably no such bed some part of which does not lie lower
than that level. This may elucidate the extreme difficulty of getting entirely
rid of any extensive sedimentary bed after it had once been desposited.

During the periods of denudation there could, of course, have been no
preservation of organic remains over the denuded area; but deposition
of the matter transported from such an area must necessarily have taken
place simultaneously with the denudation, and *at no remote distance;* for
it is certain that none but the smallest portion of the transported sediment
could float to any great distance, in either deep or shallow water. Denudation and deposition must not only be connected by contemporaneity in
time, but by proximity in space. Thus deposition of sediment will always
be proceeding within areas which may probably have been small compared
with those over which the organic forms of any particular kind, existing
at that time, may have extended; and in this manner, likewise, the preservation of organic remains may probably have proceeded without intermission,
when considered with reference to the whole of one of these areas, though
it may have been intermittent in any particular part of it. Hence we maintain that within one of these areas there must be a far more continuous
record of its whole geological history than Mr. Darwin admits. Consequently, we cannot but suppose that a far greater number of the intermediate links of his unbroken chain of organisms must have been preserved
than those of which we have at present the smallest evidence. Moreover,
they must in number have exceeded immensely those which have been
discovered, and must have been equally capable of preservation; and since
it is admitted that each must have existed a long period of time, there
would be the greater chance of some individuals being preserved as monuments of the former existence of their race. It should also be observed
that the number of individuals thus preserved may be almost infinitely
small compared with all those of the same kind which may have existed;
and also, that our argument only requires this evidence of continuity (if
Mr. Darwin's theory be true) in those species which best admitted of
preservation.

But again, it has been contended that so small a portion of any stratum
has been examined, that we can know little of what it really contains;

and this sometimes seems to be stated, as if the value of our knowledge of the organic contents of a given formation ought to be estimated by the proportion which the part of it actually examined bears to the whole formation. If such were the case, it would indeed be a cheerless prospect for the palaeontologist, for assuredly the value of his science would be so small as to be scarcely appreciable for centuries to come. But the fact is, that men will have their irresistible convictions in defiance of opinions adduced in support of particular theories. At the last meeting of the British Association, Sir Roderick Murchison asserted his belief, on strong stratigraphical evidence, that a certain bed in the north of Scotland belonged to the old red sandstone of that country. But it was found that this same bed contained reptilian remains; and Professor Huxley showed that nothing similar to these remains had been found except in strata considerably higher in the geological scale, and never near the level here assigned to them; and therefore contended that the stratigraphical evidence must be at fault. Now, if palaeontological evidence were to be taken at Mr. Darwin's valuation, this reasoning would have been little better than an unworthy quibble; and yet it was actually deemed so important by all the geologists and palaeontologists who were present, as to balance the stratigraphical evidence, though the latter, independently of some strong antagonistic view, would have been deemed conclusive. It is true that every question of this kind must be a question of probabilities, and an opinion formed upon it to-day may be upset to-morrow; but that does not lessen the existing probability, on which alone our present convictions (subject, it may be, to a certain reservation) must be founded. We do not, in fact, derive our convictions at all, respecting either the composition or contents of the sedimentary strata, from any attempt to accurate comparison between the portions of them which have and those which have not been examined. How do we know that in every part of the chalk formation calcareous matter enormously preponderates? or that argillaceous matter predominates in the gault, and siliceous in the new red sandstone? On what portion of the whole mass of these formations has the eye of man ever rested, except the almost infinitesimally small portions at their external surfaces? And yet our convictions are almost as strong on these points as those which we feel respecting the demonstrated properties of number and space. The reason is obvious to any one who gives a moment's thought to the subject. It is by a process—sometimes perhaps an unconscious one—of induction, by which we conclude the whole of these masses to have respectively the same characters as those which we invariably see at so many points of their surfaces. And so it is with respect to organic remains. When we find the same distinctive organisms in any formation, in a sufficient number of different localities, we conclude that those organisms are characteristic of the formation in general, and that those organic forms which are absent in every locality examined, are generally absent in other parts of the formation, just as we believe the mineral substances characteristic of any one of the above-named formations to be generally absent from the other two. And these are valid conclusions according to any theory of probabilities by which their validity can be tested.

From these considerations, which we cannot now stay to develop, we regard the probability of the present existence of these intermediate forms in the fossil state, or their former existence as living beings, to be indefinitely small. A great number may yet be found, and still leave the continuity of the series nearly as incomplete as at present; and therefore, without adding in any appreciable degree to the evidence in favour of Mr. Darwin's theory, or of any other equally involving a continuity of change in organic forms.

Mr. Darwin has also attempted to meet the argument against such theories as his own, founded on the sudden appearance of whole groups of allied species. According to his theory, these organic beings must have had their line of ancestors, and the difficulty lies in the total want of evidence of their former existence. He at once affirms that they must have existed in some distant locality, where their remains are now buried far beneath our reach. Thus, in speaking of the first appearance of the group of teleostean fishes in the early period of the chalk, comprising a large majority of existing fishes, he says,

> Assuming that the whole of them did appear, as Agassiz believes, at the commencement of the chalk formation, the fact would certainly be highly remarkable; but I cannot see that it would be an insuperable difficulty on my theory, unless it could likewise be shown that the species of this group appeared suddenly and simultaneously throughout the world at the same period . . . Some few families of fish now have a confined range; the teleostean fish might formerly have had a similarly confined range, and after having been largely developed in some sea, might have spread widely.

It is thus by a vague hypothesis, entirely unsupported by facts, that our author meets a difficulty which appears to us, as it has appeared to many others, to be of the gravest magnitude. He does not himself venture to say more than that he does not think it altogether insuperable. This kind of argument demands our assent to a proposition, because it is not impossible. It is to found a theory, not on our knowledge, but on our ignorance. Nor is this the only instance in which he seems to have adopted similar reasoning. In discussing the question of the sudden introduction of groups of allied species into the lowest known fossiliferous strata, our author remarks (p. 306)—

> Most of the arguments which have convinced me that all the existing species of the same group have descended from one progenitor, apply with nearly equal force to the earliest known species. For instance, I cannot doubt that all the silurian trilobites have descended from some one crustacean, which must have lived long before the silurian age, and which probably differed greatly from any known animal.

And again (p. 307)—

> If my theory be true, it is indisputable that before the lowest silurian stratum was deposited, long periods elapsed, as long as, or probably far longer than, the whole interval from the silurian age to the present day; and that during these vast yet quite unknown periods of time, the world swarmed with living creatures.

We confess ourselves to have been somewhat astonished at this bold manner of disposing of difficulties. It may be that our freedom of thought has been shackled by an habitual adherence to the principles of philosophising in such matters. We had not dreamt that because the objections to a theory could not be proved to be absolutely insuperable, we were called upon to accept it as true. We had fancied that the laws of reasoning in such matters, to which Newton and Laplace, Fresnel and Faraday, in their mathematical or experimental investigations, have bowed in reverential obedience, were still in force; and that every one who proposed a physical theory acknowledged the obligation of supporting it by positive reasons, which should at least put the balance of probabilities in its favour. We had imagined, too, that the facts reasoned upon ought to be real, and not hypothetical. Our author is perplexed with the existence of trilobites, comparatively highly-organized animals, in almost the earliest recognised fossiliferous strata, and to make the fact square with his theory, he at once creates a hypothetical world of indefinite duration for the due elaboration of the ancestral dignity of these intrusive crustaceans. If this supposition were one of the requirements of a theory resting on foundations which could not be moved, we might be obliged to admit it; but we confess that the adoption of such conclusions, unsupported by any positive and independent evidence, merely on the demand of an unproved theory, appears to us little consistent with the sobriety and dignity of philosophical investigation.

Tant pis pour les faits, is a taunt which has frequently been thrown out against philosophers whose boldness has somewhat exceeded their discretion; and we cannot help thinking that Mr. Darwin must be prepared to submit to it largely. It must be recollected, too, that the geological evidence, which he has so entirely set at nought, is precisely that by which alone any theory like his can be adequately tested, since it is that evidence which bears directly on the great, if not the only, distinctive inference which can be drawn from such theories—the linear continuity of specific characters. The two great points which require to be established (as already intimated) are those which we have discussed in the preceding paragraphs of this part of our essay. First, it is essential to show by experiment that varieties characterized by fertility, may diverge into species distinguished by sterility, in the sense already explained; and secondly, to prove by palaeontological evidence that linear continuity, which must necessarily have existed, if any theory of continuous development be true. To us, however, it appears that no approximation has been made to the establishment of these fundamental points; and we can no more believe that the elaborate theory which Mr. Darwin has endeavoured to construct can rest securely on the base on which he has placed it, than we should believe that a massive tower rising above the surrounding mist, really rested on the vapour which surrounded its base. We venture to assert, without fear of contradiction, that any physical theory of inorganic matter which should rest on no better evidence than the theory we are considering, would be instantly and totally rejected by every one qualified to form a judgment upon it.

Among the physical theories of vital phenomena, there is one which we have not yet noticed—that brought forward some years ago in the *Vestiges of Creation.* It is well known that the more highly organized animals, in the course of their embryonic development, pass through a series of states resembling in some measure those through which the lower animals pass, only that the development of the latter is arrested, as it were, at some earlier stage than that of the former. This forms the basis of the theory in question. Each specific organism is supposed to be connected with another immediately preceding it by the fact of foetal development having been carried on during gestation an additional step from one specific form to the next. This theory, like those already spoken of, requires the assumption of linear and diverging development, if, at least, each successive step be supposed to be the least which observed specific differences will admit of; for without this linear divergency it would not account for those larger differences which may be supposed to exist between animals whose lines of descent have been different. Thus, then, we should have chains of organisms as in the other theories; but here they might be *discontinuous,* the cause producing these finite steps in advance being supposed to act discontinuously at long intervals of time, and not necessarily by slow and continuous efforts, like the ordinary causes in nature. It will thus be seen that this theory gives a determinate form to the discontinuous physical action which we have above insisted on, in some form or other (vol. lxi. p. 748), as necessary to account for the discontinuities presented to us by the specific differences in nature. So far as regards the effects resulting from it, the theory will be identical with Lamarck's or Mr. Darwin's, if we suppose the successive advancing steps in organization to be *indefinitely small,* in which case they would constitute continuous chains of organic forms.

The great defect of this theory is the want of all positive proof, a defect, however, which it only shares in common with the other theories we have been discussing. The facts of embryonic development asserted as the foundation of the theories are exaggerations of those recognised by physiologists; nor do we possess the smallest indication of any new species ever having been produced in the manner supposed. But authors of theories, it would seem, are far less sensitive to the want of proof in their own speculations than in those of others. The absence of proof in his own theory did not prevent the author of the *Vestiges* from speaking contemptuously of Lamarck's as one of the follies of the wise; and at present his own theory is generally, we suppose, regarded as the folly of one who was *not* wise—not wise, we mean, as tested by the only means which, in the preservation of his *incognito,* he has afforded the world of judging of his wisdom. There is a strange mutability in the fate of vague and doubtful theories. That of the *Vestiges* attracted a large share of public attention, and is now almost as much forgotten as Lamarck's. Mr. Darwin's is unquestionably of the same kind. The history of those which have preceded it appears to us to be ominous of its future destiny.

In our preceding remarks we have designedly restricted ourselves as much as possible to those parts of the subject which have a more or less immediate reference to the theories discussed, as *physical* theories. Lamarck's work and the *Vestiges* are a good deal restricted in the same way; but Mr. Darwin's work contains much that will be read, as we have already in-

timated, with great interest, independently of the theory which it is its
great purpose to establish, both by the general reader, and those who may
be following the subject as a special object of study. It indicates a large
acquaintance with the facts of natural history, and still larger stores of
knowledge in reserve. In the statement of facts, the author is uniformly
impartial. It is difficult to conceive a fairer advocate. But when, in his
judicial capacity, he comes to the discussion of facts in their theoretical
bearings, we recognise a want of strict adherence to philosophical and
logical modes of thought and reasoning. There is one great and plausible
error of this kind which pervades nearly his whole work. He constantly
speaks of his theory as explaining certain phenomena, which he represents
as inexplicable on any other theory. We altogether demur to this statement.
A theory resting on unproved hypotheses can explain little except in that
vague hypothetical sense in which a theory that really explains little may
often be made apparently to explain much. There is a remarkable ductility
about theories unshackled by the rigours of demonstrative proofs. A phenomenon is properly said to be *explained,* more or less perfectly, when
it can be proved to be the necessary consequent of preceding phenomena,
or more especially, when it can be clearly referred to some recognised
cause; and any theory which enables us to do this may be said, in a precise
and logical sense, to explain the phenomenon in question. But Mr. Darwin's
theory can explain nothing in this sense, because it cannot possibly assign
any necessary relation between phenomena and the causes to which it refers
them. A great number of facts are mentioned as being only explicable
on this theory, and might thus appear to an inattentive reader to constitute
a large amount of inductive evidence. But all that is attempted to be done
is to *assert,* not to *prove,* that the facts are consistent with the theory;
and so far from being explicable, in some imperfect sense, by Mr. Darwin's
theory alone, they are certainly equally so, in general, by the other theories
we have mentioned. In fact, those of Lamarck and of the *Vestiges* immediately resolve themselves into Mr. Darwin's, so far as regards the effects
produced by the causes severally assigned by those theories, except that
the author of the *Vestiges* would account for the discontinuities between
species which Mr. Darwin's does not admit. Hence the phenomena in question are in no way distinctive in proof of the theory of the latter writer
more than of the others. The more express object, however, in Mr. Darwin's
book, is here to show that his theory will account for phenomena of which
the doctrine of successive creations renders no account at all—*i.e.,* no scientific account; since it is supposed to refer the phenomena to the arbitrary

will of the Creator. But we do not admit that this is the correct representation of the doctrine, as appears at once when enunciated and explained,
as in a preceding part of this essay (vol. lxi. p. 749), where it is pointed
out that the discontinuous physical cause there assumed must be supposed
to act, not arbitrarily, but in a certain accordance with the ordinary causes
with which it is conjoined; or, adopting the more usual phraseology, the
creative power is not to be supposed to act arbitrarily, in the strict sense
of the term, but in a certain accordance with ordinary causes, and in
a certain relation to pre-existing organisms. But, in fact, the doctrine of
successive creations, any more than that of final causes, does not pretend
to be a *physical theory,* since it pretends not to define the nature of any
physical cause to which certain extraordinary phenomena of nature may
be referred. It may be regarded as a mode of expressing our disbelief in
the assertion that such phenomena are, like ordinary phenomena, due solely
to the action of ordinary natural causes, together with our belief that there
is some higher order of causation in nature than that which is usually
recognised in secondary causes, whether that higher causation is to be sought
in properties originally impressed on matter by the Creator, or is rather
to be regarded as a more immediate emanation from the Divine mind.
It professes to be a negation of other theories rather than a theory of
itself, and therefore cannot be called upon to account for phenomena at
all in the physical sense in which we necessarily call upon a definite physical
theory to account for them. The attempt, therefore, to establish a comparison between the modes in which a phenomenon may be accounted for
according to the doctrine of successive creations, and by any definite physical
theory, is equivalent to the attempt to compare two things of which the
properties and qualities are totally different. We do not object to the assertion that all natural phenomena must be referred to *natural causes,* provided
always we give a proper interpretation to this last expression; but we maintain that neither Mr. Darwin nor any other writer, has yet given the slightest
proof that the recognised discontinuities in specific forms can be accounted
for by the action of ordinary secondary causes. Indeed, our author makes
at any time but little use of the verb 'to prove,' in any of its inflections.
His formula is 'I am convinced,' 'I believe,' and not 'I have proved.' We
are not finding fault with these more modest forms of expression; but
we may be allowed, perhaps, to remark, that they are the formulae of
a creed, and not of a scientific theory. Views like his rest, in fact, on
no demonstrative foundation, but, as we conceive, on *apriori* considerations,
and on what appears to us a restricted, instead of an enlarged, view of

the physical causes, operations, and phenomena which constitute what we term Nature.

A large portion of the phenomena which our authors declares to be inexplicable on the theory of successive creations, are alluded to in the latter part of his work, under the heads of 'Geographical Distribution,' 'Classification,' 'Morphology,' 'Embryology,' &c. We have explained the incorrectness of this view, and beyond it the facts spoken of have no direct bearing on the real establishment of this particular theory. We decline, therefore, entering into any of their details, though they present many interesting subjects of discussion, which, independently of the author's particular theory, we cannot too strongly recommend for perusal.

We have not yet alluded to the most difficult and most important point connected with any theory which asserts the continuous variation of all organic forms from the lowest to the highest, including man as the last link in the chain of being. It is the transition in passing up to man from the animals next beneath him, not to man considered merely as a physical organism, but to man as an intellectual and moral being. Lamarck and the author of the *Vestiges* have not hesitated to expose themselves to a charge of gross materialism in deriving mind from matter, and in making all its properties and operations depend on our physical organization. Mr. Darwin, on the contrary, seems carefully to have shunned all allusion to the immediate point in question. He has said much on the subject of animal instinct and the successive derivation of its higher from its lower forms, but as far as we recollect, has not even alluded to that infinitely more difficult transition from the instinct of the brute to the noble mind of man. This, however, is the great difficulty to be grappled with, especially in theories which involve the necessity of continuous change in both the material and immaterial portions of animal life, and include man in common with the lower animals. We know not precisely how far our author would extend his theory with respect to man, for he has been reserved on this point. But it appears to us inevitable that it should be so extended by most of those who adopt it; and we would gladly therefore know something more of this mysterious process of metamorphosis from the Quadrumana to the Bimana, according to any theory of continuous development. It may be said perhaps that man may have existed long before the term indicated by any known records of his existence, and that he began his earthly career under a character much more approximate to that of the ape than that which he has now acquired by the progressive advancement of his faculties. But then, where are the missing links in the chain of intellectual and moral being? What has become of the aspirants to the dignity of manhood whose development was unhappily arrested at intermediate points between the man and the monkey? It will not be doubted, we presume, that there exists at present an enormous gap between the intellectual capabilities of the lowest race of men and those of the highest race of apes; and if so, we ask again, why should the creatures intermediate to them—exalted apes or degraded men—have been totally exterminated, while their less worthy ancestors have successfully struggled through the battle of life? But there are other questions which seem to us far more difficult to answer than these. We believe that man has an immortal soul, and that the beasts of the field have not. If any one deny this, we can have no common ground of argument with him. Now we would ask, at what point of his progressive improvement did man acquire this spiritual part of his being, endowed with the awful attribute of immortality? Was it an 'accidental variety' seized upon by the power of 'natural selection' and made permanent? Is the step from the finite to the infinite to be regarded as one of the indefinitely small steps in man's continuous progress of development, and effected by the operation of ordinary natural causes? We can scarcely suggest these questions without an apprehension of their being deemed irreverent. But they force themselves irresistibly upon us in considering these theories in regard to their legitimate and almost necessary extension to man. The difficulty of passing, according to any theory of development, from the finite to the infinite—from the mortal to the immortal—cannot be avoided by any advocate of such theories, except by denying the immortality of man or admitting that of a sponge or a polyp. We have stated emphatically our opinion that all natural phenomena are subjects for scientific investigation, so far as such investigation can be applicable to them; but what answer can science possibly give to such questions as those above suggested? It is at this stage of his scientific researches, where the reasonings and methods of science are no longer applicable, that the candid and earnest investigator of truth will turn to any other source from which he conceives that further knowledge is to be derived. He is thus driven as it were, not merely as a man of certain religious convictions, but as a philosopher, to regard the subject under its religious aspect, and to compare the conclusions at which he may thus arrive with those to which his physical theories may have conducted him. We are not pretending here to say what ought to be the result of this comparison—that forms no part of the task we have undertaken in this essay; but we do assert that it must be made. We must look upon this picture and upon that, and decide on their respective claims. We know not how the difficulties arising from the spiritual nature of man are answered by the advocates of any unrestricted theory of continuous development, nor is it our desire to question too curiously any one's abstract religious views in relation to the physical theories he may advocate on vital phenomena. Considerations may present themselves to other minds which have not occurred to our own, tending to lessen difficulties which to us appear so insurmountable; still, we cannot but feel a strong conviction that many minds are so constituted that it would be difficult or impossible for them to hold such theories as we have been discussing, without being led to doubt some of the highest hopes and aspirations which it hath entered into the heart of man to conceive respecting his future destiny. Surely, then, we are bound to scrutinize such theories with all possible care, and in doing so, not merely the religious dogmatist, but every true philosopher, will give full weight to his faith in spiritual matters, and feel that the convictions which he derives from it are true elements in his decision for the acceptance or rejection of these unrestricted theories of development.

Mr. Darwin, it appears, expects converts to his views not among experienced naturalists, but among the young and rising ones. 'It is so easy,' he remarks, 'to hide our ignorance under such expressions as "plan of creation," "unity of design," &c., and to think that we give an explanation when we only re-state a fact,' (p. 482). We believe that this has frequently been true. But if the sentence just quoted be intended as a warning to younger naturalists, and not merely as an explanation of the pertinacious adherence of older ones to preconceived opinions, the warning might perhaps be strengthened by the insertion of another phrase as one which may not be without its self-deceptive influences—'natural selection.' Assuredly the theory founded on this notion does not give explanations, in any rigorous sense of the term, of many things which it professes to explain, but really only asserts. Our author also remarks that all who believe in the mutability of species will do good service by the conscientious expression of their convictions. We heartily agree with him respecting the value of a conscientious expression of opinion; but we would remind the young naturalist that no convictions on scientific subjects can be conscientious unless arrived at by accurate inquiry, and careful and continued thought; and that of all such subjects the one we have been discussing is the most mysterious and difficult. We would also further remind him that the philosophical naturalist must not only train the eye to observe accurately, but the mind to think logically; and the latter will often be found the harder task of the two. With respect to all but the exact sciences, it may be said that the highest mental faculty which they call upon us to exert is that by which we separate and appreciate justly the *possible,* the *probable,* and the *demonstrable.* It is the exercise of this faculty which, in our opinion, renders the natural sciences, when pursued in a proper spirit, an excellent mental training, and a corrective to that which is derived from the too exclusive study in the exact sciences of the *demonstrable* alone. But let not the young student suppose that this faculty comes by intuition. It can only be acquired by the habit of patient, continuous, and impartial thought on the more difficult questions on which he may have to form his opinions, and will often be found more especially difficult for those whose minds have had none of the rigorous training which the study of the exacter sciences affords. Of all things, too, we would warn the young man of science against prematurely pledging himself to the adoption of doubtful theories. Let him be careful lest, while he imagines that he is only liberating himself from former prejudice, and exercising a proper freedom of thought, he may be imposing shackles on himself for some future time, when the desire to maintain his own consistency may stand opposed to that higher freedom of thought which may become necessary when enlarged knowledge and more matured judgment may possibly suggest large modifications of preconceived opinions. It may be difficult to avoid the influence exerted over us by the prejudices of others, but it is infinitely more difficult to rid ourselves of those which may attach to our own early and immature conclusions. A philosophical reservation of opinion on doubtful points will not diminish the zeal of an earnest lover of nature's truths. Let him pursue his researches, recollecting that Biological science requires at present its Keplers rather than its Newtons—the discovery of the more obvious laws according to which its phenomena may be arranged, rather than attempts at that higher generalization which may account for such laws by the operation of physical causes. The naturalist who carries on his researches on this principle, not necessarily without reference to, but without the professed adoption of, vague and uncertain theories, will preserve himself in the best temper and frame of mind for the impartial investigation of truth, and the final adoption of those theories which may hereafter appear to be best established.

Comments on Hopkins

William Hopkins (1793–1866) took his degree from Cambridge in 1827 and then settled down for a long and successful career as a mathematical

"coach." From a list of his distinguished students it is easy to see just
how successful he was. Among the most prominent were Sir William
Thomson, later Lord Kelvin, Peter Tait (see comments on Haughton),
Henry Fawcett, James Clerk-Maxwell, and Isaac Todhunter, Whewell's
biographer. Like Haughton, Hopkins wrote numerous papers applying
mathematics to physics, especially geology. He acquired a taste for geology
from none other than Adam Sedgwick. With such a background it is not
surprising that he would come out against Darwin's theory. There seemed
to be something about a classical education in mathematics that predisposed
a person to find evolutionary theory decidedly unsatisfactory.

Hopkins' review was especially damaging because it was impartial. He
did not dismiss evolutionary theory simply because it conflicted with accepted views in biology, nor did he rant against it because of any theological
objections he might have had. Rather, he argued that all phenomena which
nature presents are legitimate objects of scientific investigation, including
the origin of man. He proposed to judge evolutionary theory according
to the same canons of reasoning and evidence applied to physical theories.
He found it impossible "to admit laxity of reasoning to the naturalist,
while we insist on rigorous proof in the physicist." If naturalists wanted
to be considered scientists, then they had to meet the standards of science,
and these standards were exemplified by the theory of gravitation and
the undulatory theory of light. His was no casual reference to "Baconian
induction," but a detailed criticism of evolutionary theory on the basis
of the best views then current on the nature of science. Except for a few
peculiar paragraphs devoted to man's "counteracting" natural law, Hopkins' paper is a set piece in nineteenth-century philosophy of science.

Biologists tend to become indignant when their own efforts are judged
by the standards of excellence set in physics. Regardless of whether the
practice of judging biology on standards derived from physics is inherently
reprehensible, it should be undertaken only by someone with sufficient
knowledge of both biology and physics, and such men are hard to come
by. More often than not, the critic finds fault with biology either because
he is ignorant of the biological theory he is criticizing or because he lacks
sufficient understanding of the physical paradigm. As Darwin observed,
Hopkins did not understand evolutionary theory very well. But it is of
greater significance that Hopkins' canons of reasoning and evidence were
drawn from only two physical theories, and he held an extremely distorted
(though prevalent) view of these theories.

As noted in the introduction to this anthology, the philosophy of science
extracted from a cursory knowledge of physics by such men as Whewell

and Mill was untrue not only for science in general but also for the physical theories which served as paradigms. Hopkins demanded that evolutionary theory be proved in the same manner that gravitational theory and the undulatory theory of light had been proved. In the first place, neither had been proved because both were false. In the second place, these two theories were hardly typical of physics. There were other physical theories— for example, the theory of matter erected on Maxwell's equations—that fitted the paradigm derived from these two theories little better than evolutionary theory. As Peirce observed, when statistical theories, like the theory of gases, were taken as paradigms, evolutionary theory did not come off so badly. Given Hopkins' parochial selection of physical theories and his unrealistic interpretation of them, evolutionary theory was bound to be judged inadequate, no matter how fair he might be.

Instructive in the diagnosis of the faults of nineteenth-century philosophy of science is the distinction which Hopkins draws between geometrical laws (like those of Kepler) and physical laws (like those of Newton). According to Hopkins, Kepler's laws were deduced from the phenomena after years of observation and were purely descriptive, being independent of any physical theory. In this respect, Kepler's laws were like Cuvier's law of the correlation of parts and Forbes's principle of polarity. Newton's laws, like those of Lamarck and Darwin, provided physical explanations of the phenomena described in these geometrical laws.

In retrospect, it is easy to appreciate the difficulty Hopkins was having in his struggle to make the appropriate distinctions among scientific statements—between empirical generalizations and theoretical laws, between an axiomatic system as an uninterpreted calculus and as interpreted to be a physical theory. It is also easy to point out Hopkin's error in formulation. Kepler's and Newton's laws were equally geometrical as raw formulae, equally physical when interpreted to refer to material bodies. Kepler did not deduce his laws from the phenomena after years of patient observation. He used a handful of observations on a single planet, and these observations were made, not by him but by Tycho Brahe. The years which Kepler spent were not devoted to fact-gathering but in working out what we would now call theological and aesthetic considerations and in uncovering an error in calculation which he had made in his early work. Kepler assigned a cause for his laws just as Newton did, though it was not a cause which was to gain much currency. Similarly, Cuvier's laws were not theory-free but were part and parcel of his general physiological views. Only Forbes's principle of polarity was properly descriptive.

Hopkins recognized Darwin's theory as suggesting a possible causal explanation for the existence and diversity of organic species, and stated repeatedly that his use of the hypothesis of evolution by natural selection was perfectly legitimate. It was not to be judged simply by direct evidence as to its truth but on the total superstructure built upon it. It fell short in the matter of proof. Evolutionary theory was not a pyramid resting on its apex but a tower resting on the vapor which surrounds its base. "We venture to assert, without fear of contradiction, that any physical theory of inorganic matter which should rest on no better evidence than the theory we are considering, would be instantly and totally rejected by everyone qualified to form a judgment upon it."

Hopkins remarks that Newton freely used the verb "to prove" whereas Darwin tended to use verbs like "believe" (see also Owen). Physical scientists like Newton "are not content to say that it *may* be so, and then build up theories based on bare possibilities. They *prove*, on the contrary, by modes of investigation which cannot be wrong." Newton's laws were expressed as universal statements and were sufficiently accurate to warrant the retention of this form for some time. Hence, they could give the (erroneous) impression of having been deductively proved. Darwin's laws were trend and tendency statements stating what usually happened *ceteris paribus*. No deductions to singular events were possible. Hence, Darwin could not even pretend to deductive proof. Hopkins objects to Darwin's repeated reference to his theory "explaining" phenomena. What, he asks, is this "imperfect sense" in which phenomena are explicable by Mr. Darwin's theory? The question is legitimate. Hopkins also observes that there "is a remarkable ductility about theories unshackled by the rigours of demonstrative proofs." Hopkins could not have known that even physicists would soon be forced to accustom themselves to such ductility of inference. Hopkins' review represents the best evaluation that could be made of evolutionary theory, given the naturalistic, empiricist philosophy of science of the day (see review by Fawcett for replies to Hopkins' criticisms).*

* See also Hopkins (1839–40).

Henry Fawcett (1833–1884)

I feel that I ought not to have so long delayed writing to thank you for your very kind letter to me about my article on your book in *Macmillan's Magazine*.

I was particularly anxious to point out that the method of investigation pursued was in every respect philosophically correct. I was spending an evening last week with my friend Mr. John Stuart Mill, and I am sure you will be pleased to hear from such an authority that he considers that your reasoning throughout is in the most exact accordance with the strict principles of logic. He also says the method of investigation you have followed is the only one proper to such a subject.

It is easy for an antagonistic reviewer, when he finds it difficult to answer your arguments, to attempt to dispose of the whole matter by uttering some commonplace as 'This is not a Baconian induction.'

I expect shortly to be spending a few days in your neighborhood, and if I should not be intruding upon you, I should esteem it a great favour if you will allow me to call on you and have half an hour's conversation with you.

As far as I am personally concerned, I am sure I ought to be grateful to you, for since my accident nothing has given me so much pleasure as the perusal of your book. Such studies are now a great resource to me.—Henry Fawcett to C. Darwin, Bodenham, Salisbury, July 16, 1861 (*More Letters,* 1:189–190)

You could not possibly have told me anything which would have given me more satisfaction than what you say about Mr. Mill's opinion. Until your review appeared I began to think that perhaps I did not understand at all how to reason scientifically.—C. Darwin to Henry Fawcett, Down, July 1861 (*More Letters,* 1:189)

I wondered who had so kindly sent me the newspapers, which I was very glad to see; and now I have to thank you sincerely for allowing me to see your MS. It seems to me very good and sound; though I am

certainly not an impartial judge. You will have done good service in calling
the attention of scientific men to means and laws of philosophizing. As
far as I could judge by the papers, your opponents were unworthy of
you. How miserably A. [A. W. Williamson] talked of my reputation, as
if that had anything to do with it . . . How profoundly ignorant B. [Edwin
Lankester] must be of the very soul of observation! About thirty years
ago there was much talk that geologists ought only to observe and not
theorise; and I well remember some one saying that at this rate a man
might as well go into a gravel pit and count the pebbles and describe
the colours. How odd it is that anyone should not see that all observation
must be for or against some view if it is to be of any service!—C. Darwin
to Henry Fawcett, September 1861 (*More Letters,* 1:194–195); *Life of
Henry Fawcett* (1886), pp. 100–101.

A Popular Exposition of Mr. Darwin on the Origin of Species*
HENRY FAWCETT

No scientific work that has been published within this century has excited
so much general curiosity as the treatise of Mr. Darwin. It has for a time
divided the scientific world into two great contending sections. A Darwinite
and an anti-Darwinite are now the badges of opposed scientific parties.
Each side is ably represented. In the foremost ranks of the opposition against
Darwin have already appeared Professor Owen, Mr. Hopkins, Sir B. Brodie,
and Professor Sedgwick; whilst Professor Huxley, Professor Henslowe, Dr.
Hooker, and Sir Charles Lyell, have given the new theory a support more
or less decided. We shall endeavour most carefully to avoid the partiality
of partisanship; and, as our object is neither to attack nor to defend,
but simply to expound, we shall have no necessity to assume the tone
of ungenerous hostility exhibited in the *Edinburgh Review* [by Richard
Owen], or to summon from theology the asperities contained in the *Quarterly* [by Bishop Wilberforce]. Such may be appropriate to controversy,
but can give those who are unacquainted with Mr. Darwin's work no
idea of his theory; which, all must agree, has been stated with the most
perfect impartiality, and is the result of a life of most careful scientific
study.

It will be, in the first place, advisable to enunciate, as clearly as possible,
the problem which the treatise on the Origin of Species suggests for solution.
And this cannot be done unless we possess a distinct conception of the
words we employ. Let us therefore inquire, what is the meaning of the

* From *Macmillan's Magazine* (December 1860), 3:81–92.

word species? The necessity of classifying the various objects in the animal
and vegetable kingdoms was fully recognised by Socrates when he applied
his dialectical mode of investigation to test the meaning of general terms.
The object of classification was to carve out the organic world into distinct
groups, each of which possessed some common property. "Family" was
the most comprehensive, then "Genus," then "Species," and then "Variety."
A Family would thus include many Genera, a Genus many Species, and
a Species many Varieties. These divisions are to some extent arbitrary and
artificial; for in nature many of the distinctions, which in certain cases
seem most marked and decided, are not universally preserved, but fade
gradually away. Thus no distinction might appear to be more easily recognisable than that which exists between animals and vegetables; but, as
we descend to the less highly-organised forms of creation, the most distinctive characteristics of animals and vegetables become fainter, and at length
we meet with organisms with regard to which even the highest scientific
acumen finds it difficult to decide whether they are vegetables or animals.
And, similarly, it is impossible accurately to define the exact amount or the
exact kind of difference which is sufficient to place two organisms in two
distinct Families or two distinct Genera. This difficulty and doubt increases
as we descend to the lower classificatory divisions, and it is admitted that it
is impossible to frame any exact definition of a specific difference. This
difficulty is strikingly exhibited in botany. A most eminent botanist, Mr.
Bentham, maintains that there are only one thousand two hundred species
of English plants. An equally high botanical authority, Mr. Babington,
affirms that there are two thousand species. A similar difference of opinion
exists amongst naturalists. For instance, it was long a disputed point whether
or not the dog and wolf were varieties of one species.

The following definition of a species is sometimes given: that two animals
or vegetables belong to different species when they are infertile with each
other. This hardly deserves the name of a definition. It is enunciated in
deference to pre-conceived notions, and assumes the incorrectness of the
theory which it is afterwards used to disprove. This definition can manifestly
have little influence in diminishing that difficulty, which has been above
alluded to, of deciding what is a specific difference; for it requires a test
which can rarely be applied to the existing organic world, and is entirely
inapplicable to those numerous species which have passed away. Thus it
would be almost impossible to ascertain whether different molluscs, or insects, or testacea, are fertile with each other; and, manifestly, such an
imperfect experiment in breeding cannot be made upon those animals and

plants of which we have solely a geological record. Therefore it would seem that the classification of species must remain so arbitrary, that equally high scientific authorities may continue to dispute whether the plants of a limited area like England should be held to constitute two thousand or one thousand two hundred species. The question of species may thus, at the first sight, appear to be a dispute about an arbitrary classification, and it may naturally be asked, Why, therefore, does the problem of the Origin of Species assume an aspect of supreme scientific interest?

The common assumption that species are infertile with each other, and that the descendants of any particular species always belong to that species, at once suggests this difficulty:–How can a new species be introduced into the world. There is abundant evidence that new species have been introduced. If we go back to a comparatively modern geological epoch, it will be found that all the fossil animals belong to undoubtedly distinct species from any which exist at the present time. This is admitted by every naturalist and geologist, whatever may be his opinion on the origin of species. The geological record shows us that past species have died out, and that existing species have been gradually introduced. What is the cause which has produced this extinction of species? What is the agency by which the new species have been placed upon the earth? These are the questions to which Mr. Darwin has sought to give an answer. The same question has been asked again and again, and it admits only two kinds of answers. If the ordinary assumption is admitted that no two members of different species can be the progenitors of a mixed race, and if it is also supposed that the descendants of any varieties of a particular species must always be considered as belonging to this species, and that, however much in succeeding ages such descendants may differ from the parent stock, this difference can never entitle them to be ranked as distinct species—if these propositions are admitted, it then becomes quite manifest that the statement that a new species has appeared is tantamount to the assertion that a living form has been introduced upon the earth which cannot have been generated from anything previously living. It therefore becomes necessary to suppose that the same effort of Creative Will, which originally placed life upon this planet, is repeated at the introduction of every new species; and thus a new species has to be regarded as the offspring of a miraculous birth. We are as powerless to explain by physical causes this miracle as we are any other. To hope for an explanation would be as vain as for the human mind to expect to discover by philosophy the agency by which Joshua made the sun and moon stand still. Our ignorance, therefore, of the origin of species is absolute and complete, if every new species is supposed to require a distinct and independent act of the will of an Omnipotent Creator. When it was supposed that every heavenly body had its path guided by a direct omnipotent control, all are now ready to admit that the cause of the motion of these bodies was unknown, and that this want of knowledge was not the less complete because it was disguised under such expressions as the harmonies of the universe. Those, therefore, who attempt to render unnecessary the belief in these continuously-repeated creative fiats, seek to explain hitherto unexplained phenomena of the highest order of interest and importance in natural history. Whenever this explanation shall have been given, a similar service will have been done to this science, as was performed by Newton for astronomy, when he enunciated his law of gravitation. Newton's discovery is now found in numerous religious works as a favourite illustration of the wisdom of the Creator; and it is now considered that a hymn of praise is sung to God when we expound the simplicity of the Newtonian laws. The day will doubtless come when he who shall unfold, in all their full simplicity, the laws which regulate the organic world, will be held, as Newton is now, in grateful remembrance for the service he has done not only to science, but also to religion.

Aptly, indeed, has the origin of species been described as the mystery of mysteries; for, as long as a phenomenon is accounted for by creative fiats, it is enshrouded in a mystery which the human mind is powerless to penetrate. Mr. Darwin has endeavoured to bring this subject within the cognisance of man's investigations, by supposing that every species has been produced by ordinary generation from the species which previously existed. Such a supposition is the only alternative for those who reject the doctrine of creative fiats.

We shall now proceed to expound the agency by which Mr. Darwin conceives that this development of new species from previous ones has taken place. We think our exposition will indicate the great difference between the speculations of Mr. Darwin and those of other theorisers upon the transmutation of species, such as Lamarck and the author of the "Vestiges of Creation." But it may perhaps conduce to clearness to remark beforehand upon a very unfair and very erroneous test which has been applied to Mr. Darwin's work. Every hostile criticism repudiates the theory, because, as it is asserted, it is not based upon a rigorous induction. There is much philosophic cant about this rigorous induction. An individual who is supremely ignorant of science finds no difficulty in uttering some such salvo as, "This is not the true Baconian method." Such expressions, which too frequently are mere meaningless phrases, were repeated *ad nauseam* at the British Association. They are revived in an article on Mr. Darwin in the *Quarterly Review*. There we find it reiterated, "This is not a true Baconian induction." In reply to all this, it should at once be distinctly stated that Mr. Darwin does not pretend that his work contains a proved theory, but merely an extremely probable hypothesis. The history of science abundantly illustrates that through such a stage of hypothesis all those theories have passed which are now considered most securely to rest on strict inductive principle. Dr. Whewell has remarked, "that a tentative process has been the first step towards the establishment of scientific truths." Some association perchance, as the falling apple, first aroused in Newton's mind a suspicion of the existence of universal gravitation. He then had no proof of the particular law of this gravitating force; he made several guesses. The inverse square was the only one which caused calculation to agree with observation; the inverse square was therefore assumed to be the true law. The most complicated calculations were based upon this assumption; they have been carefully corroborated by observation, and in this manner the law of gravitation has been proved true beyond all dispute. Those who attack the philosophic method of Mr. Darwin ought explicitly to state how they would proceed to establish a theory on the origin of species by what they term a rigorous induction. Is such an example to be found in the doctrine of creative fiats? The greatest of logicians has remarked,[1] "The mode of investigation which, from the proved inapplicability of direct methods of observation and experiment, remains to us as the main source of the knowledge we possess, or can acquire, respecting the conditions and laws of recurrence of the more complex phenomena, is called in its most general expression the deductive method, and consists of three operations—the first, one of direct induction; the second, of ratiocination; and the third, of verification." The method here indicated Mr. Darwin has most rigorously observed. A life devoted to the most careful scientific observations and experiments, and to the accumulation of a most comprehensive knowledge of the details of natural history, has suggested to Mr. Darwin's mind a certain hypothesis with regard to the origin of species. The results which have been deduced from this hypothesis he has endeavoured to verify by a comparison with observed phenomena. Mr. Darwin has been himself most careful to point out that this verification is not yet complete. Until it becomes so, Mr. Darwin's theory must be

1. *Mill's Logic,* vol. i. book iii, chap. xi. p. 491.

ranked as an hypothesis. The eminently high authorities who have already welcomed Mr. Darwin's theory as a probable hypothesis, should induce the general public to welcome it as a legitimate step towards a great scientific discovery; and those who cannot take any special part in the controversy will render science a great service if they resent bigoted prejudice, and earnestly seek to give both parties in the dispute a fair hearing . . . .*

Few can deny the reality of this struggle for existence, and few can dispute the method of its action and the tendency of its results. The main ground of controversy is, Will this constant accumulation of inherited variations ever constitute a specific difference? The most hostile critics of Mr. Darwin acknowledge the value of his theory so far as it accounts for the origin of varieties, but maintain that he has failed to prove that the accumulated inheritance of these small variations, singled out by a process of natural selection, will ever constitute a specific difference, or, in other words, would produce in any organism a variation so great as to cause it to be infertile with the stock from which it originated. As we have before remarked, Mr. Darwin's theory cannot pretend to be completely proved; we are therefore bound not to apply to it those tests supplied by logic to which every proof ought to be submitted. It is a question of probability; and, as long as it remains so, everything which can be either said in support or contradiction ought to be fairly stated and maturely weighed. We will therefore endeavour to give a correct statement of the leading arguments on both sides.

Many common animals are sculptured upon ancient Egyptian tombs; many have also been preserved as mummies; and, when it is found that, during the three thousand years which have elapsed, the ibis, for instance, has remained unchanged, it is maintained that the process of development supposed by Mr. Darwin's theory cannot have occurred.

An individual would excite a smile of ridicule who, having discovered that Mont Blanc three thousand years ago was of the same altitude as it is at the present time, should consider that he had refuted those theories of modern geology which suppose that the stupendous peaks of Switzerland were lifted from off their ocean bed, and that every physical change in this earth's appearance has been produced by the indefinitely prolonged operation of the same physical causes which on every side around us continue in ceaseless activity. The extinction of species and the introduction of new ones are associated with periods which can only be described as geological epochs; and the time which has elapsed since the occurrence

* [Fawcett's description of evolutionary theory has been deleted.]

of the most remote recorded historical event is but an instant compared with the period which is indicated by the deposition of one of the strata which tell the history of this planet's structure.

The three thousand years which have elapsed since the animals were sculptured upon the Egyptian tombs have not sufficed to produce any change in the physical geography of the valley of the Nile. We have no reason to suppose but that the soil was then of the same fertility as it is now, the temperature of the same warmth, the air of the same moisture, and that the mighty river itself rolled down the same volume of water and sedimentary matter to the sea. The struggle for existence was carried on then under precisely the same conditions as it is now; particular animals and plants possessed then as now the same relative advantages and disadvantages in their structure; and the causes which determine success in this struggle for life have remained absolutely unaltered. But let us look forward to the geological epoch. Egypt may not be then what she is now; the land may be upheaved; the Nile may have changed its course; and many animals and plants which flourish there now will be unsuited to these changed physical circumstances; they will then probably not prevail in the struggle for existence, but will pass away as extinct species, and their place will be occupied by other organisms which are adapted to the changed conditions of life. The fossils in every stratum unmistakeably indicate such successive revolutions in the animal and vegetable world. No one will dispute that old forms of life are thus succeeded by new ones. The question to be determined is, Must we continue to confess complete ignorance of the laws which regulate the introduction of these new forms of life? A confession of this ignorance is made whenever we resort to the doctrine of creative fiats.

Those who, like Mr. Darwin, endeavour to explain the laws which regulate the succession of life, do not seek to detract one iota from the attributes of a Supreme Intelligence. Religious veneration will not be diminished, if, after life has been once placed upon this planet by the will of the Creator, finite man is able to discover laws so simple that we can understand the agency by which all that lives around us has been generated from those forms in which life first dawned upon this globe.

The distinctness of the groups of the fossil animals which compose the geological records supplies the most formidable argument against Mr. Darwin's theory; and, unless the difficulty thus suggested can be explained away, the main support of the theory is gone. We cordially rejoice that this theory is ultimately to be refuted or established by the principles of geology. We were therefore not a little astonished, that in the discussions upon Mr. Darwin at the British Association at Oxford geology was not even alluded to. It was sad, indeed, to think that the opponents of the theory sought to supply this omission by summoning to their aid a species of oratory which could deem it an argument to ask a professor if he should object to discover that he had been developed out of an ape. The professor aptly replied to his assailant by remarking, that man's remote descent from an ape was not so degrading to his dignity as the employment of oratorical powers to misguide the multitude by throwing ridicule upon a scientific discussion. The retort was so justly deserved, and so inimitable in its manner, that no one who was present can ever forget the impression it made. Happy are we to be able to escape from such recriminations, for there is some chance of a satisfactory solution when we can appeal to physical principles.

The argument to which allusion has just been made shall be stated in Mr. Darwin's own words; for, so singular is his impartiality, and so sincere his love of truth, that he has himself advanced, in their utmost force, all the most important arguments which can be opposed to his theory. "The number of intermediate varieties which have formerly existed on the earth must be truly enormous. Why, then, is not every geological formation, and every stratum, full of such intermediate links? Geology assuredly does not reveal any such finely graduated organic chain; and this, perhaps, is the most obvious and gravest objection to my theory. The explanation lies, as I believe, in the extreme imperfection of the geological record."

The mode therefore is plainly indicated by which the incorrectness of Mr. Darwin's speculations can be completely established. If the physical philosopher can demonstrate that the geological record has not this character of extreme imperfection, Mr. Darwin will doubtless be amongst the first to admit that his theory can then be no longer maintained. Mr. Hopkins[2] of Cambridge, than whom no one can be better qualified, has commenced this mode of attack; and such is the spirit with which Mr. Darwin receives a fair and generous antagonist, that we believe he was amongst the first to welcome and acknowledge the hostile criticism of Mr. Hopkins.

Mr. Darwin attributes imperfection to the geological record upon two different grounds:–

First.–An extremely incomplete examination has yet been made of any existing strata, and the animals and plants which are preserved in any strata can only form a very small portion of those which were living when the strata were deposited.

2. See *Fraser's* [*Magazine*] of June and July 1860.

Secondly.–The strata which now exist form but a small proportion of those which have been deposited. Between the strata now remaining numerous intermediate ones have been completely removed by denudation.

The first of these propositions rests upon the following consideration:– Strata have only been examined with a scientific purpose for the last few years. The geology of many countries is as yet unknown; only those portions which now happen to be dry land are exposed to view. Certain conditions are requisite for the preservation of an animal or plant in a fossil state. For instance, when a dead body sinks to the bottom of the sea it will decay, unless there is a sufficiently rapid deposition of sediment to surround and inclose the animal before decay commences. In every case, also, the soft tegumentary portions of an organism must perish. The late Professor Forbes has remarked, "Numbers of our fossil species are known and named from single and broken specimens, or from a few specimens, collected on some one spot." These considerations are important, and would suffice to account for the non-appearance of *complete* series of transitional forms. But Mr. Darwin frankly admits that the explanation of the almost *entire* absence of transitional forms must be mainly based upon those other causes which, according to his explanation, have made the geological record so extremely imperfect. The efficiency of denudation in completely removing every trace of a stratum, impressed itself upon Mr. Darwin's mind when, in the *Beagle,* he examined the western coast of South America. This coast along many hundred miles of its length is, by subteranneous agency, gradually rising with a uniform velocity of about three feet in a century. Suppose this upheaval has been continuous, the time will then have been comparatively recent since many portions of this shore which are now dry land were undoubtedly covered with the ocean. Along this coast the rocks are being ceaselessly worn by the action of the waves, and rivers are constantly bearing to the sea sedimentary matter. It will therefore inevitably follow that upon the bed of the ocean strata are being continuously accumulated. Similar strata must have been formed upon the adjacent dry land before it was upheaved from the ocean; therefore it might be expected that here would be found a considerable accumulation of tertiary strata. Such, however, is not the case; the tertiary strata are so poorly developed, that they will be inevitably removed by rain and other atmospheric agencies before the expiration of a comparatively brief geological epoch. What, then, has become of those considerable strata which were undoubtedly accumulated upon this land when it formed the bed of the ocean? There is one way, and only one, by which we can account for this removal of strata. As the bed of the ocean became gradually upheaved, different portions of the land, before emerging from the ocean, became subject to the action of coast-waves. Denudation consequently occurred; and the power and extent of this denudation is recorded by the fact, that of these stratified deposits the remains are too small to enable any permanent record of their former existence to be long preserved. Such considerations, Mr. Darwin maintains, may be extended to the whole world; for modern geology requires us to suppose that in every portion of this globe there have been alternate periods of depression and upheaval. There is reason to suppose that life can rarely be maintained beneath water of a certain depth. It has, for instance, been clearly demonstrated that those minute animals which build up our coral reefs require a certain amount of light, and that therefore they must work at a fixed distance beneath the surface of the ocean. Coral reefs exist on the coast of Australia many hundreds of feet in perpendicular height. The bed of the ocean, therefore, must have subsided with exactly the same rapidity as the walls of these coral reefs have risen in perpendicular height. In a similar manner, a deposit of great thickness filled with the same kind of fossil shells probably indicates that during the formation of this deposit the ocean remained of a uniform depth; or, in other words, the subsidence kept pace with the deposition of sedimentary matter. When this subsidence ceased, and an upheaval commenced, the rate of this upward movement may perhaps have been uniform with the former downward motion. The strata would then, as they approach the surface of the ocean, be subjected to a denudation by the coast-waves during a period equal to that which sufficed for their deposition. By this denudation Mr. Darwin maintains that we have every reason to suppose that a series of strata might be completely removed in a similar manner to those tertiary strata which, as we have before remarked, have been washed from off the rising coast of South America. Against this theory of denudation, Mr. Hopkins has advanced an argument which is well worthy of serious consideration. In order that it may be stated in its full force, Mr. Hopkins's own words shall be used:–

> We believe the entire destruction of any sedimentary bed of considerable horizontal extent to have been of rare occurrence. All the more important denudations of which we have any evidence have been preceded by large upheavals, by which the strata have been tilted; and thus, while those portions of each stratum which have been most elevated may have been exposed to enormous denudation, those portions which have been least elevated or perhaps depressed, have been thus kept out

of the reach of the denuding agencies. The entire obliteration of a stratum would require in general that it should be upheaved in such a manner as never to deviate sensibly from a horizontal position. In fact, this approximation to horizontality must be closer than it frequently may be during the time of depositon, for the smallest dip in an extensive stratum would place it in a condition as to denudation similar to that above described as due to large upheavals. The higher portions might be denuded, while the lower remain untouched. The Weald affords one of the best elucidations of denudation accompanied only by the partial destruction of strata. We have no reason to suppose that a single stratum has been obliterated by this denudation, which, while it has left scarcely a remnant of the removed beds in the central portion of the district, has left portions of them untouched on its borders, where they dip beneath the existing surface.

We cannot here enter more fully into this deeply interesting question. Our object has been to indicate as much of the character of the argument on both sides as would convince the reader that the solution of the problem which is here suggested involves many of the most complicated and profound principles of physical geology. And yet Mr. Darwin admits that with the solution of this question his whole theory must either stand or fall. Why, then, have his speculations in some quarters been received in so unscientific a spirit, when he maintains that they are based upon scientific principles, and boldly challenges these same principles to prove their incorrectness?

It might appear according to the geological record that whole groups of allied species have suddenly come upon the earth. This suggests another difficulty, which deserves careful consideration. Low down in the chalk, groups of teleostean fishes are found in great numbers, and it has been supposed that before them no traces are preserved of any species allied to these teleostean fishes. Yet such allied forms must undoubtedly have existed, if this new species has been introduced by a process of gradual development. The whole question, therefore, turns upon the degree of perfection which is to be attributed to the geological record. Mr. Darwin's position upon this subject has been most powerfully strengthened by a recent discovery in paleontology. Sessile cirripedes are found largely distributed over all tertiary strata, and they are of the most ubiquitous families of testacea existing at the present time. Until within a very few years not the slightest trace of a sessile cirripede had been found in any secondary strata, and the sessile cirripedes might have been quoted against Mr. Darwin with even more effect than the teleostean fishes. It was, in fact, repeatedly said, "Here are a group of animals so easily preserved that they are fossilized in great numbers in all tertiary strata. They cannot be found in any secondary strata. Is it not therefore evident that they could not have been gradually developed, but that they were suddenly created at the commencement of the tertiary period?"

Mr Darwin would then, as he does now, have in vain besought his opponents not to place too implicit confidence in the perfection of the geological record. But within the last few months a skilful paleontologist, M. Bosquet, has sent Mr. Darwin a drawing of a perfect specimen of an unmistakeable sessile cirripede, which he had himself extracted from the chalk of Belgium. And, as if to make the case as striking as possible, this sessile cirripede was a Chthamalus, a very large and ubiquitous genus, of which not one specimen has as yet been found, even in any tertiary stratum. Hence we now positively know that sessile cirripedes existed during the secondary period; and these cirripedes might have been the progenitors of our many tertiary and existing species.

Since, before this discovery, nothing appeared more improbable than that sessile cirripedes were to be found in the secondary period, ought we to regard the difficulty suggested by the telostean fishes as insuperable? For who can say that in a similar manner advancing knowledge may not some day remove it? Already M. Pictet has carried the existence of the teleostean fishes one stage beyond the period when it has been supposed they were suddenly created. Other eminent paleontologists incline to the belief "that some much older fishes, whose affinities are as yet imperfectly known, are really teleostean."

Mr. Darwin appears desirous to maintain, as a probable inference from his theory, that every past and present organism has descended from four original forms. Such an inference is at once met by a very obvious objection; for it requires us to suppose that life existed upon this planet long previous to the deposition of those Silurian rocks which afford us the first traces of fossil remains. Mr. Darwin is consequently compelled to assert that fossil-bearing rocks of a date long anterior to the Silurian period were once deposited, but have been either removed or transformed.

Many of those who may be inclined to agree with Mr. Darwin that all organisms have descended from a few original forms will, perhaps, think that it is unfortunate to lay stress upon such a supposition. It is not involved in the theory, nor is it a necessary inference from it; it cannot, therefore, be advisable to allow speculative difficulties to add to the obstacles against which the theory has to contend. There is a great problem to be solved, and its *enunciation* may involve nothing which can even be disputed;

for it is a demonstrated truth, that those organisms of which we have the first record were succeeded by new and distinct species, and that the same process has been again and again repeated. What has been the agency to affect this succession of life? All must admit that such a problem really presents itself for solution. Why, then, attempt to make the solution likewise indicate the form in which life was first introduced upon this earth? Transmutationists and non-transmutationists must agree that life was originally introduced by an act of Creative Will, and a transmutationist need not necessarily concern himself with the number of forms which were thus first spontaneously created.

In an earlier part of this paper we have endeavoured to point out the analogy between the process of natural selection and the method which is pursued by the horticulturist and the breeder of animals. The question will very probably arise, What has the horticulturist and the breeder of animals effected towards the creation of a new species? It is important to consider this question, because it will lead to the perception of that imperfection in the common definition of species which we have already alluded to. Breeds of pigeons which have undoubtedly descended from the same original stock have been made by the pigeon-fancier, so different in their structure that, if they were found as fossils, they would be undoubtedly classed as distinct species. But they are not regarded as distinct species, because they are fertile with each other. Man has never yet made two varieties from the same stock to differ so much from each other as to possess the great characteristic of distinct species, namely, infertility. If this is ever effected, it will be the greatest experimental corroboration of Mr. Darwin's theory. Man, therefore, has already produced what may be termed a morphological species, but he has not produced a physiological species. This is a distinction which ought to be kept carefully in view; for there is not any amount of structural difference which would enable us a priori to predict whether two animals were infertile with each other— this being the only reliable test of physiological species. Thus the horse differs from the ass in only two particulars. The horse has a bushy tail; the ass a tufted tail. The horse has callosities on the inner side of both the fore and hind legs; the ass has callosities only on the inner side of the fore legs.[3] No trace of these characteristic differences could be found in fossil specimens of the horse and ass. If, therefore, found fossil, they could but be classed as belonging to the same species; but, when the physio-

3. Further information upon this subject will be found in a most able Essay on Mr. Darwin, in the *Westminster Review* for April, 1860.

logical test is applied, they are at once ranked as distinct species, because
the offspring of the horse and the ass are infertile. And, therefore, with
reference to the classification of species, we often argue in a vicious circle.
Thus, formerly, every botanist considered the cowslip and the primrose
as belonging to two distinct species; but within the last few years horticulturists have unmistakeably produced the cowslip from the primrose; and,
therefore, it would appear that they might claim to have produced a species
by the accumulation of the differences presented by varieties. Thus one
species would have been generated from another species; and the great
species question could be regarded as solved. But then it is at once rejoined,
"This has not, by an ocular demonstration of the efficiency of development,
been done. Our original classification was wrong. It is true we should
not have discovered its error unless you had made your experiment because,
without such an experiment we must have continued to believe that the
cowslip and the primrose were specifically distinct."

We have now exhausted all the space we can claim; we trust we have
devoted it to a candid exposition of the leading points of Mr. Darwin's
theory, and that we have fairly stated the most important arguments on
the other side. Our object will be fully attained if we induce those who
do not know the work itself to peruse it with an unprejudiced mind. It
is not for us to hazard a prediction as to the ultimate fate of the theory
itself. Dr. Hooker, a man of the highest scientific reputation, when closing
a most remarkable discussion at the late meeting of the British Association,
used emphatic words to the following effect:–"I knew of this theory fifteen
years ago; I was then entirely opposed to it; I argued against it again
and again; but since then I have devoted myself unremittingly to natural
history; in its pursuit I have travelled round the world. Facts in this science
which before were inexplicable to me became one by one explained by
this theory, and conviction has been thus gradually forced upon an unwilling
convert." Other minds may perhaps pass through similar stages of primary
doubt and ultimate belief; but, be that as it may, if Mr. Darwin's theory
were disproved to-morrow, the volume in which it has been expounded
would still remain one of our most interesting, most valuable, and most
accurate treatises on natural history.

Comments on Fawcett

In spite of being blinded as a young man by his father in a shooting
accident (one pellet lodged in each eye), Henry Fawcett became a professor

of political philosophy at Cambridge and from 1880 to 1889 was Postmaster General of England. It was a letter from William Hopkins that roused him from the depression that followed the accident. His defense of Darwin stemmed from the indignation he felt over the Baconian cant he had endured during the 1860 meetings of the British Association. In response, he published the preceding paper and then repeated his defense of Darwin's methodology at the 1861 meetings at Manchester. In his paper Fawcett attempts to show that Darwin's methodology was in accordance with the best canons of the philosophy of science—those of William Whewell and John Stuart Mill. Fawcett's understanding of these philosophies of science was not deep, but the case he presents was sufficiently well-informed and forceful to counter the outright nonsense being expounded by many of the opponents of evolutionary theory. As Hopkins' review shows, it was possible to understand the philosophies of Herschel, Whewell, and Mill and still find Darwin's methodology inadequate.

Like many scientists of the day, Fawcett saw no conflict between evolutionary theory and religion. In time it would be assimilated as Newton's theory had been and like it would be used to show the power and glory of the Almighty. God's creative power was reserved for the initial creation and need not be invoked for each species. But as Bishop Wilberforce and a host of other theologically minded authors realized, the gradual evolution of man from other animals confronted the absolutist ethics then popular with a considerable obstacle. How can morals be eternal and immutable if man as a moral agent was a child of time and gradually evolved? If man has a soul and his earliest ancestors did not, when did he get it? Did a soulless ape give birth to a morally responsible man-child? The conflict was not so easily resolved as Fawcett might think.

Like Carpenter, Fawcett objects to the interfertility definition of "species." It seemed to assume the falsity of evolutionary theory at the outset. He even goes so far as to list a whole series of circumstances which make it impossible to decide whether or not two groups are interfertile. To use the modern term, he argues that the interfertility definition is not "operational." This will come to be one of the most popular criticisms of the biological definition of Dobzhansky and Mayr (Hull, 1968). In spite of all the difficulties which Fawcett finds with the interfertility definition of "species," he nevertheless agrees with Huxley that the production from the same stock of two physiological species would provide the greatest experimental corroboration of Darwin's theory.*

* For details of Fawcett's life see Stephen (1886).

Frederick Wollaston Hutton (1836–1905)

I quite agree with what you say on Lieutenant Hutton's Review (who he is I know not); it struck me as very original. He is one of the very few who see that the change of species cannot be directly proved, and that the doctrine must sink or swim according as it groups and explains phenomena. It is really curious how few judge it in this way, which is clearly the right way.—C. Darwin to J. D. Hooker, Down, April, 1861 (*Life and Letters*, 2:155)

I hope you will permit me to thank you for sending me a copy of your paper in *The Geologist,* and at the same time to express my opinion that you have done the subject a real service by the highly original, striking, and condensed manner with which you have put the case. I am actually weary of telling people that I do not pretend to adduce direct evidence of one species changing into another, but that I believe that this view in the main is correct, because so many phenomena can be thus grouped together and explained.

But it is generally of no use; I cannot make persons see this. I generally throw in their teeth the universally admitted theory of the undulation of light,—neither the undulation nor the very existence of ether being proved, yet admitted because the view explains so much. You are one of the very few who have seen this, and have now put it most forcibly and clearly. I am much pleased to see how carefully you have read my book, and, what is far more important, reflected on so many points with an independent spirit. As I am deeply interested in the subject (and I hope not exclusively under a personal point of view) I could not resist venturing to thank you for the right good service which you have done.—C. Darwin to F. W. Hutton, Down, April 20, 1861 (*Autobiography,* p. 265)

Review of the Origin of Species*

[F. W. HUTTON]

We could scarcely let this year pass away without some notice of a book which at least will make 1860 remarkable in the annals of natural history

* From *The Geologist* (1860), 3:464–472. This paper may not have been written by Hutton; the following clearly were: *The Geologist* (1861), 4:132–136, 183–188, 286–288.

science. Whatever opinion may be entertained of the speculations on the origin of species sketched out by Mr. Darwin in his introductory work to the fuller and more explicit one he announces for some future day there is no doubt that in its entirety his theory is one which for many years to come must receive the earnest attention of the scientific world; for whether the law of the necessity of organic variation and development as dependent on external circumstances attendant on the general "struggle for life" be universal in application or not, Mr. Darwin has at any rate opened out a new vein of reflection and investigation which must be followed out until the new theory be either disproved or proved from its first causes to its final results.

Nor must we be prevented from the true examination of its value and merit by any previous prejudices, nor deterred by the objections and abuse of those who are ever ready to attack new opinions on the old and ridiculous grounds of a real or pretended dread of an antagonism to Holy Scripture, as if the Word was not based on the sure foundations of truth. "I must express by detestation of the theory," says one opponent,*

> because of its unflinching materialism; because it denotes the demoralized understandings of its advocates. Look, too, at their credulity. Why Darwin actually believes that a white bear, by being confined to the slops of the Polar Sea might be turned into a whale; that a lemur might be turned into a bat; that a three-toed tapir might be the great-grandfather of a horse; or the progeny of an ass may have gone back to a buffalo.

Such, however, are mere verbosities, baseless assertions, unwarranted attributions of irreligion and gross ironical misrepresentations of an author's writings, too transparent not to be seen through by the well-versed student of Nature. There is, however, a speciousness of appearance in the positiveness of diction of this style of attack which misleads the unreflecting as the flame allures unwary moths, and which often causes such inflated pomposities to be mistaken for acknowledged facts. Time was, and not so long since, either, when fossils were enigmas even to the learned; when thoughtful and sapient men discussed with heat of temper and with angry tones whether such organic remains of past creations embedded in the soil were really shells, and bones, and plants, or whether they were plastic forms modelled in the dark recesses of the ground. Even now a-days some literary adventurers and crack-brained sages—and we are sorry to say, some men too, of better note but mistaken views—now and then attempt to palm

---

* Each of the quotations that follow are slightly modified from Adam Sedgwick's *Spectator* review—D. L. H.

off this long ago exploded whim under a specious guise upon an intelligent world. The danger from such productions is small, and few indeed of those worth caring for would think a fossil bone or shell aught else than the treasured fragment of some ancient living being.

More dangerous, however, are the wilful pervertors who argue with a specious show of knowledge; and such detractors Darwin's theory, like every other, is sure to bring forward against itself. "Species have been constant," says one, "ever since they first existed; change the conditions, and the old species would disappear. New species would come in and flourish. But how? by what causation? By *creation*. What is meant by creation? The operation of a power quite beyond the power of a pigeon fancier, a cross-breeder, or a hybridizer, in which one can believe by the legitimate conclusion of sound reason drawn from the laws and harmonies of nature, and, believing, can have no difficulty in the repetition of new species."

Dickens, in one of his novels very shrewdly remarks that the advice given to street-boys about to fight "to go in and win" is very excellent if they only knew how to follow it; and when one naturally asks how new species which geology shows us appearing from time to time *first* began, the answer, by *creation* is as easy to give and about as useless as the advice offered to the street-boys. It is, after all, a mere assertion, an evasion of the question, a cloak for ignorance. We see different races from time to time leaving their relics entombed in the solid rocks of the earth, we see the remains, however, only of the *perished* race; we have no proof, no trace, no evidence whatever in those great entombments of the origin or *first* appearance of the *progenitors* of those races. Those races might have sprung from single pairs, or the primitive individuals might have been *created* in hundreds. Few, we think, would incline to the opinion of the direct creation of hundreds of the like animals or plants at one time; but if, on the other hand, we incline to the *direct creation* of a single pair, we must admit that that pair must have been created ages before its race could be useful or necessary on the face of the earth; must have been created in fact in advance of those changes of physical conditions of our planet, which all admit to have been brought about in the lapse of time by natural operations, in order to provide for the necessary propagation of their descendants in sufficient numbers at the period when they should *usefully* abound. We should incline to think that a theory which proposed to view the development of the required races or species as concurrent with the physical changes rendering necessary their presence,—and as consequently necessarily developed by natural laws, like we see everywhere else around us so wisely and immutably preordained, apparently from the beginning of all things, by the Almighty Designer,—would be preferable to the idea of direct creations, and affording a more reasonable reply than the mere assertions of the miraculous agency with which our query is so commonly met.

But "the assumption of the *direct creation* of species is an hypothesis," says another, "which does not suspend or interrupt any established law of nature. It does not suppose the introduction of new phenomena unaccounted for by the operation of any known law; and it appears to be a power *above* established laws, and yet acting in conformity with them." It may be due to the astuteness of our intellect, but we cannot see how a power can be *above* and not be necessarily *antagonistic* to established laws, and consequently how it can be possible for such a power to be in conformity with such established laws.

"The pretended physic and philosophy of modern days," says a third, "strips man of all his moral attributes, and holds as of no account his origin and place in the created world. A cold atheistical materialism pervades the sentiments of modern philosophy. The new doctrine is untrue and mischievous. It is opposed to the obvious course of nature, and the very opposite of inductive truth."

Why should it be considered atheistical to believe the laws of the Great Perfection to be *perfect*. The inscrutable Eternal cannot err; why then should His laws be so defective and imperfect as to require repeated efforts of creative energy? Is this world like an old watch so much out of order as to require continual oilings and repeated repairs? Why, too, should it be objectionable to consider the laws He has given to nature as worthily and incessantly subservient to His will? Or why should it be thought irreligious to believe the Maker of all things in His *first* designs should have foreseen the necessity of future modifications to future altered conditions, and have provided accordingly in His *first type-plans* for their future illimitable adaptations to the ever-changing scenes presented in the progress of our earth's ever-altering conditions? Why, indeed, may we not look around us and believe in the universal bowing of all nature hourly, daily, unceasingly to the unerring laws and sustaining power of God? Why should we not see in every change His presence and His will? Why should the high position of man be brought in on all occasions in our natural history researches when we do not at present know of any link which binds him to the brute creation?

If these remarks on our part seem strong, let it however be known that we are not professedly defending Mr. Darwin's doctrines, but attempting to pourtray as forcibly as we can the unjustness and uncharitablesness of

such attacks upon a new and well-studied theory. Let a new doctrine be
always well and impartially examined, and justly accepted or rejected according to our honest opinions of its merits or failings. Mr. Darwin's theory
briefly resolves itself into this.*

The evident modifications of primitive type-plans, which indisputedly
we see in past and present forms of life, are undoubtedly the strongest
arguments in favour of Darwin's theory of progressive development by
natural selection. But as geology alone must be the sole source of knowledge
for testing or learning the effects of great periods of time in the gradual
transmutation of species, so will our efforts be resultant of efficient proof
only in proportion to the perfection or imperfection of the geological record.
This record Darwin justly says *is* defective. No doubt, it is; no doubt
there are great gaps in the earth's past history of which no trace remains—
and many, and far more numerous gaps which scientific investigations have
not yet filled up. Still, we may hope to find, and by patience, and research
no doubt we ultimately shall mark out, the great points in the picture
around which the details may reliably be filled in by correctly drawn inferences. If we tabulate the number of known species of any particular class
of animals or plants, we find the number invariably in the aggregate ranging
higher until we attain a maximum in the present creation, notwithstanding
there are occasional deficiencies of individual genera between certain geological formations which shows that we have not yet a perfect knowledge
of all the forms living during those eras in which such deficiencies occur.
Such results, however, are important in their bearing on the doctrine of
the natural development of species. Taking for example the totals of known
mollusca, we commence in the Silurian period with 317, and close in the
recent with 16,000. It follows, then, that if in the pre-Silurian age life
began on our planet with the same number of definite type-plans, such
as the globular, the radiate, soft-bodies, the vertebrate, &c., which we see
so prominently defined in the existing races; and taking the mollusca, for
example, of that pre-Silurian period at unity, or as the first commencement
of their special or direct creation; and regarding such special or direct
creations as the miraculous interference of the Deity, then we have as
a result an ever increasing ratio of miraculous interferences, and we must
also regard creative energy as sixteen thousand times more active in our
time than in the pre-Silurian period. A condition of things few of us would
be inclined to admit. On the other hand, this natural radiation of numbers—let us put it down by diverging figure (*see woodcut*, p. 297) each

* [Hutton's description of evolutionary theory has been deleted.]

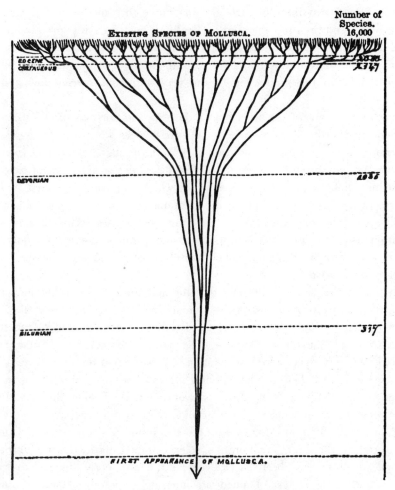

First Appearance of Life

Diagrammatic view of the actual increase of species of Mollusca, illustrating the increasing intensities of creative action at the periods stated, if the theory of the direct creations of species be admitted. (Each line of the Tree of Numerical Increase of Specific Forms represents 100 species.)

297

line of which is representative of hundreds—is so representative of the natural radiation of life-forms by the splitting up of species by natural variations into new species,—one species first naturally divided into two species, these two into four, these four into eight, and so on, as would naturally result by the operation of natural laws carrying on gradually and incessantly the transmutation and subdivision of old into new species,— as to incline the mind at first sight to faith in the past operations of such natural laws in the production of the very numerous species now living around us.

The proportions of the level lines indicating the horizons of the various periods which the tree of increase of species does not cover, represents, of course, also the successively available spaces for the geographical spread of the species existent at those dates, for the earth's surface cannot exceed a definite limited area the maximum of which may be considered to be represented by the border lines of the diagram.

The struggle for existence by the multiplied species seems thus to be continuously increased by a continuous and rapid decrease of available terrestrial space by the ever increasing sub-division and restriction of the geographical area.

It should also be borne in mind that the diagram shows only the increase of specific forms of one class of the animal kingdom—the mollusca. Taking these as having increased from unity to sixteen thousand; and taking the increase of all the other classes—the Radiata, Crustacea and Insects, the Vertebrates terrestrial, aerial, or aquatic, &c.,—as equal only to this sum in the aggregate of their similar specific increase, we have for the animal kingdom an assumed total of thirty-two thousand. Taking the like sum as representative of a similar increase of species in the vegetable kingdom; and we obtain then, as a final result, the conclusion that the creative action as exerted in the direct creation of species, commencing in the pre-Silurian period at unity, has been successively and continuously intensified ever since, until now it has obtained an intensity sixty-four thousand times greater than when it commenced; or upwards of twenty-one thousand times greater than it was at that point in past time—the Silurian era—to which we can trace back the records of its action.

The radiata, the vertebrata, or any other class, exhibit the like results with the mollusca, confirming the impression of the prolificness of existing species as due to the natural sub-division by natural transmutating operations during past ages of a fewer number of original forms. We cannot follow Mr. Darwin through all his arguments in support of his theory, nor do we always agree with his teachings, still so many important problems can be feasibly solved by the application of his doctrine, as go far to convince one that it has a really good foundation for three great natural truths—the undoubted influence of the struggle for life; the necessitous interference of external physical circumstances upon the varieties and conditions of life and vegetation at all periods of the earth's history; and the existence, at least, of a principle of natural selection. We should on some points be inclined to go further than Mr. Darwin, especially in regard to the matter of time. Granting his position that the changes produced by natural selection *usually* require great ages of time, we are still disposed to consider that such changes might, under favourable or active circumstances be rapidly accomplished, and that in some cases they might even be brought about in the range of two or three or even of a single generation.

The greatest objection, it seems to us, which can be brought against the theory is its reliance on *natural* causes and *chance* in effecting the changes.

We should be more inclined to refer the modifications which species of animals or plants have undergone to the direct will of God, for it seems difficult to conceive how a being totally ignorant of its own structure or conditions of living should so commence modifying its structure, form, or habits, as to adapt not itself, but successively its progeny to new forms and conditions of life. Take ourselves—some few who have undergone severe anatomical studies excepted—and how much do we know of our bodies? What do we know of the organs in their interiors? Do we know how often in a day our heart beats, or our lungs palpitate? How many ounces of blood run in our veins? If we are ill can we tell what organ is affected? or diseased internally can we say where or why? Do we of ourselves, untaught even, know either the existence or use of one of our unseen and not external organs? Even of those which are visible what do we know? can we tell why the will causes the hand to write, or the feet to walk? Or what is the means of communication between our will and our limbs? Did our progenitors, however remote, conceive the idea of nails to our fingers, eyelids to our eyes, or lashes for our eyelids? Do we conceive any improvement in our offspring? Could we suggest any possible improvement of our present structure? Could we add one beautiful line to the face? or one more efficiently constructed limb? Could we suggest any more convenient arrangement, or disposition of our parts? And if *we,* standing at the highest pinnacle of knowledge, cannot suggest a sportive variety, even, of ourselves, how much less can we consider that mere brutes, or insensate plants, should have any innate power of themselves to cause the slightest improvement of their organization? If we could not suggest one improvement of our condition, how much less can we believe that the alpine partridge effected his own power of changing the colour of its feathers, or the insect assume the colour of the leaf it feeds upon; and still less can we conceive how the peach could assume of itself its downy surface, or the plum its purple bloom. Such results if naturally produced can only emanate from divine laws. The beautiful perfection of our bodies—the wonderful adaptations in the forms of animals to render them efficient for their purposes of life seem so skilfully planned, that it is impossible to regard them as effects of chance, and not as inapproachably perfect designs. If we could accept the transmutation doctrines, we must concede the transmutatory laws as of pre-eminently divine origin and maintenance, purposely conceived to be ever forcibly acting in direct antagonism to the necessity of destruction and change, to which all nature seems subject. In this light we might accept it, and trace back the natural divergence of life-forms to the first vital force thrown off from the hand of the Creator, who threw off with an eternal and ever enduring force the vast clouds of vapours that have in the roll of ages collapsed into the myriads of worlds and suns that swarm in the heavens above and around us—of which we can neither see the limits nor conceive the expanse—but which may yet be the smallest and least wonderful of all the myriads of world-clusters with which the same great Creator has stardusted His course through the realms of boundless and interminable space.

Comments on Hutton

F. W. Hutton (1836–1905) had a varied and bizarre career which took him all over the world. His occupations ranged from a military officer to a professor of biology and geology at Canterbury College, New Zealand. To the contemporary reader, the most significant feature of Hutton's paper is his denunciation of sacerdotal revilers. It is interesting to note, however, that all of his quotations are from Sedgwick's *Spectator* paper. Darwin, however, was impressed by Hutton's list of objections to evolutionary theory, the replies to these objections, and the long list of phenomena explicable on the assumption of descent from a common prototype. (This section has been omitted from this anthology.)

Although Hutton agreed with Darwin's views on the manner in which a scientific theory was to be accepted or rejected, he held the same views on nature and mind that led so many others to reject evolutionary theory. The strongest objection which Hutton could bring against Darwin's theory was its total reliance on natural causes and chance. He felt inclined to refer species modification to the direct will of God, expressed in divine laws. The difference between Darwin's laws and those instituted by God is that divine laws would show foresight. Earlier reviews in this anthology tended to ignore in their criticisms the role that "chance" played in evolutionary theory. For example, Owen never mentions it. In later years it was to become one of the main objections to Darwin's theory.

Hopkins also thought that life was immaterial and independent of matter. Thus, though species might evolve from one another, the first origin of life would have to be miraculous. In spite of the thoroughness and good will with which Hutton had studied the *Origin*, he still did not understand it very well. He was able to ask, for example, how a progenitor could have conceived in advance the structure that would be useful to its descendants so that it might commence to modify it. Seldom has a theory which seemed so straightforward proved so easy to misunderstand as evolutionary theory.*

* See also Hutton (1861 and 1899).

Fleeming Jenkin (1833–1885)

It is only about two years since the last edition of the *Origin,* and I am fairly disgusted to find how much I have to modify, and how much I ought to add; but I have determined not to add much. Fleeming Jenkin has given me much trouble, but has been of more real use to me than any other essay or review.—C. Darwin to J. D. Hooker, Down, January 16, 1869 (*More Letters,* 2:379)

I have been interrupted in my regular work in preparing a new edition of the 'Origin,' which has cost me much labour, and which I hope I have considerably improved in two or three important points. I always thought individual differences more important than single variations, but now I have come to the conclusion that they are of paramount importance, and in this I believe I agree with you. Fleeming Jenkin's arguments have convinced me.—C. Darwin to A. R. Wallace, Down, January 22, 1869 (*Life and Letters,* 2:288)

Will you tell me *where* are Fleeming Jenkin's arguments on the importance of single variations? Because I at present hold most strongly the contrary opinion that it is the individual differences or *general variability* of species that enables them to become modified and adapted in new conditions.

Variations or 'sports' may be important in modifying an animal in one direction, as his *colour* for instance, but how it can possibly work in changes requiring coordination of many parts as in Orchids for example, I cannot conceive.—A. R. Wallace to C. Darwin, January 30, 1869, in Vorzimmer (1963).

I must have expressed myself atrociously; I meant to say exactly the reverse of what you have understood. F. Jenkin argued in the 'North British Review' against single variations ever being perpetuated, and has convinced me, though not in quite so broad a manner as here put. I always thought

individual differences more important; but I was blind and thought that
single variations might be preserved much oftener than I now see is possible
or probable. I mentioned this in my former note merely because I believed
that you had come to a similar conclusion, and I like much to be in
accord with you. I believe I was mainly deceived by single variations offering such simple illustrations, as when man selects.—C. Darwin to A. R.
Wallace, Down, February 2, 1869 (*Life and Letters,* 2:288–289)

The Origin of Species*
[FLEEMING JENKIN]

The theory proposed by Mr. Darwin as sufficient to account for the
origin of species has been received as probable, and even as certainly true,
by many who from their knowledge of physiology, natural history, and
geology, are competent to form an intelligent opinion. The facts, they think,
are consistent with the theory. Small differences are observed between animals and their offspring. Greater differences are observed between varieties
known to be sprung from a common stock. The differences between what
have been termed species are sometimes hardly greater in appearance than
those between varieties owning a common origin. Even when species differ
more widely, the difference they say, is one of degree only, not of kind.
They can see no clear, definite distinction by which to decide in all cases,
whether two animals have sprung from a common ancestor or not. They
feel warranted in concluding, that for aught the structure of animals shows
to the contrary, they may be descended from a few ancestors only,—nay,
even from a single pair.

The most marked differences between varieties known to have sprung
from one source have been obtained by artificial breeding. Men have selected, during many generations, those individuals possessing the desired
attributes in the highest degree. They have thus been able to add, as it
were, small successive differences, till they have at last produced marked
varieties. Darwin shows that by a process, which he calls natural selection,
animals more favourably constituted than their fellows will survive in the
struggle for life, will produce descendants resembling themselves, of which
the strong will live, the weak die; and so, generation after generation,
nature, by a metaphor, may be said to choose certain animals, even as
man does when he desires to raise a special breed. The device of nature
is based on the attributes most useful to the animal; the device of man
on the attributes useful to man, or admired by him. All must agree that

* From *The North British Review,* June 1867, 46:277–318.

the process termed natural selection is in universal operation. The followers of Darwin believe that by that process differences might be added even as they are added by man's selection, though more slowly, and that this addition might in time be carried to so great an extent as to produce every known species of animal from one or two pairs, perhaps from organisms of the lowest known type.

A very long time would be required to produce in this way the great differences observed between existing beings. Geologists say their science shows no ground for doubting that the habitable world has existed for countless ages. Drift and inundation, proceeding at the rate we now observe, would require cycles of ages to distribute the materials of the surface of the globe in their present form and order; and they add, for aught we know, countless ages of rest may at many places have intervened between the ages of action.

But if all beings are thus descended from a common ancestry, a complete historical record would show an unbroken chain of creatures, reaching from each one now known back to the first type, with each link differing from its neighbour by no more than the several offspring of a single pair of animals now differ. We have no such record; but geology can produce vestiges which may be looked upon as a few out of the innumerable links of the whole conceivable chain, and what, say the followers of Darwin, is more certain than that the record of geology must necessarily be imperfect? The records we have show a certain family likeness between the beings living at each epoch, and this is at least consistent with our views.

There are minor arguments in favour of the Darwinian hypothesis, but the main course of the argument has, we hope, been fairly stated. It bases large conclusions as to what has happened upon the observation of comparatively small facts now to be seen. The cardinal facts are the production of varieties by man, and the similarity of all existing animals. About the truth and extent of those facts none but men possessing a special knowledge of physiology and natural history have any right to an opinion; but the superstructure based on those facts enters the region of pure reason, and may be discussed from all doubt as to the fundamental facts.

Can natural selection choose special qualities, and so breed special varieties as man does? Does it appear that man has the power indefinitely to magnify the peculiarities which distinguish his breeds from the original stock? Is there no other evidence than that of geology as to the age of the habitable earth? and what is the value of the geological evidence? How far, in the absence of other knowledge, does the mere difficulty in classifying organized beings justify us in expecting that they have had a common ancestor? And finally, what value is to be attached to certain minor facts supposed to corroborate the new theory? These are the main questions to be debated in the present essay, written with a belief that some of them have been unduly overlooked. The opponents of Darwin have been chiefly men having special knowledge similar to his own, and they have therefore naturally directed their attention to the cardinal facts of his theory. They have asserted that animals are not so similar but that specific differences can be detected, and that man can produce no varieties differing from the parent stock, as one species differs from another. They naturally neglect the deductions drawn from facts which they deny. If your facts were true, they say, perhaps nature would select varieties, and in endless time, all you claim might happen; but we deny the facts. You produce no direct evidence that your selection took place, claiming only that your hypothesis is not inconsistent with the teaching of geology. Perhaps not, but you only claim a 'may be,' and we attack the direct evidence you think you possess.

To an impartial looker-on the Darwinians seem rather to have had the best of the argument on this ground, and it is at any rate worth while to consider the question from the other point of view; admit the facts, and examine the reasoning. This we now propose to do, and for clearness will divide the subject into heads corresponding to the questions asked above, as to the extent of variability, the efficiency of natural selection, the lapse of time, the difficulty of classification, and the value of minor facts adduced in support of Darwin.

Some persons seem to have thought his theory dangerous to religion, morality, and what not. Others have tried to laugh it out of court. We can share neither the fears of the former nor the merriment of the latter; and, on the contrary, own to feeling the greatest admiration both for the ingenuity of the doctrine and for the temper in which it was broached, although, from a consideration of the following arguments, our opinion is adverse to its truth.

Variability.—Darwin's theory requires that there shall be no limit to the possible differences between descendants and their progenitors, or, at least, that if there be limits, they shall be at so great a distance as to comprehend the utmost differences between any known forms of life. The variability required, if not infinite, is indefinite. Experience with domestic animals and cultivated plants shows that great variability exists. Darwin calls special attention to the differences between the various fancy pigeons, which, he says, are descended from one stock; between various breeds of

cattle and horses, and some other domestic animals. He states that these
differences are greater than those which induce some naturalists to class
many specimens as distinct species. These differences are infinitely small
as compared with the range required by his theory, but he assumes that
by accumulation of successive difference any degree of variation may be
produced; he says little in proof of the possibility of such an accumulation,
seeming rather to take for granted that if Sir John Sebright could with
pigeons produce in six years a certain head and beak of say half the bulk
possessed by the original stock, then in twelve years this bulk could be
reduced to a quarter, in twenty-four to an eighth, and so farther. Darwin
probably never believed or intended to teach so extravagant a proposition,
yet by substituting a few myriads of years for that poor period of six years,
we obtain a proposition fundamental in his theory. That theory rests on
the assumption that natural selection can do slowly what man's selection
does quickly; it is by showing how much man can do, that Darwin hopes
to prove how much can be done without him. But if man's selection cannot
double, treble, quadruple, centuple, any special divergence from a parent
stock, why should we imagine that natural selection should have that power?
When we have granted that the 'struggle for life' might produce the pouter
or the fantail, or any divergence man can produce, we need not feel one
whit the more disposed to grant that it can produce divergences beyond
man's power. The difference between six years and six myriads, blinding
by a confused sense of immensity, leads men to say hastily that if six
or sixty years can make a pouter out of a common pigeon, six myriads,
may change a pigeon to something like a thrush; but this seems no more
accurate than to conclude that because we observe that a cannon-ball has
traversed a mile in a minute, therefore in an hour it will be sixty miles
off, and in the course of ages that it will reach the fixed stars. This really
might be the conclusion drawn by a savage seeing a cannon-ball shot off
by a power the nature of which was wholly unknown to him, and traversing
a vast distance with a velocity confusing his brain, and removing the case
from the category of stones and arrows, which he well knows will not
go far, though they start fast. Even so do the myriads of years confuse
our speculations, and seem to remove natural selection from man's selection;
yet, Darwin would be the first to allow, that the same laws probably or
possibly govern the variation, whether the selection be slow or rapid. If
the intelligent savage were told, that though the cannon-ball started very
fast, it went slower and slower every instant, he would probably conclude
that it would not reach the stars, but presently come to rest like his stone

and arrow. Let us examine whether there be not a true analogy between this case and the variation of domestic animals.

We all believe that a breeder, starting business with a considerable stock of average horses, could, by selection, in a very few generations, obtain horses able to run much faster than any of their sires or dams; in time perhaps he would obtain descendants running twice as fast as their ancestors and possibly equal to our race-horses. But would not the difference in speed between each successive generation be less and less? Hundreds of skilful men are yearly breeding thousands of racers. Wealth and honour await the man who can breed one horse to run one part in five thousand faster than his fellows. As a matter of experience, have our racers improved in speed by one part in a thousand during the last twenty generations? Could we not double the speed of a cart-horse in twenty generations? Here is the analogy with our cannon-ball; the rate of variation in a given direction is not constant, is not erratic; it is a constantly diminishing rate, tending therefore to a limit.

It may be urged that the limit in the above case is not fixed by the laws of variation but by the laws of matter; that bone and sinew cannot make a beast of the racer size and build go faster. This would be an objection rather to the form than to the essence of the argument. The existence of a limit, as proved by the gradual cessation of improvement, is the point which we aim at establishing. Possibly in every case the limit depends on some physical difficulty, sometimes apparent, more often concealed; moreover, no one can a priori calculate what bone and sinew may be capable of doing, or how far they can be improved; but it is unnecessary further to combat this objection, for whatever be the peculiarity aimed at by fancy-breeders, the same fact recurs. Small terriers are valuable, and the limit below which a terrier of good shape would be worth its weight in silver, perhaps in gold, is nearly as well fixed as the possible speed of a race-horse. The points of all prize cattle, of all prize flowers, indicate limits. A rose called 'Senateur Vaisse' weighs 300 grains, a wild rose weighs 30 grains. A gardener, with a good stock of wild roses, would soon raise seedlings with flowers of double, treble, the weight of his first briar flowers. He or his grandson would very slowly approach the 'Cloth of Gold' or 'Senateur Vaisse,' and if the gradual rate of increase in weight were systematically noted, it would point with mathematical accuracy to the weight which could not be surpassed.

We are thus led to believe that whatever new point in the variable beast, bird, or flower, be chosen as desirable by a fancier, this point can be rapidly approached at first, but that the rate of approach quickly diminishes, tending to a limit never to be attained. Darwin says that our oldest cultivated plants still yield new varieties. Granted; but the new variations are not successive variations in one direction. Horses could be produced with very long or with very short ears, very long or short hair, with large or small hooves, with peculiar colour, eyes, teeth, perhaps. In short whatever variation we perceive of ordinary occurrence might by selection be carried to an extravagant excess. If a large annual prize were offered for any of these novel peculiarities, probably the variation in the first few years would be remarkable, but in twenty years' time the judges would be much puzzled to which breeder the prize should fall, and the maximum excellence would be known and expressed in figures, so that an eighth of an inch more or less would determine success or failure.

A given animal or plant appears to be contained, as it were, within a sphere of variation; one individual lies near one portion of the surface, another individual, of the same species, near another part of the surface; the average animal at the centre. Any individual may produce descendants varying in any direction, but is more likely to produce descendants varying towards the centre of the sphere, and the variations in that direction will be greater in amount than the variations towards the surface. Thus, a set of racers of equal merit indiscriminately breeding will produce more colts and foals of inferior than of superior speed, and the falling off of the degenerate will be greater than the improvement of the select. A set of Clydesdale prize horses would produce more colts and foals of inferior than superior strength. More seedlings of 'Senateur Vaisse' will be inferior to him in size and colour than superior. The tendency to revert, admitted by Darwin, is generalized in the simile of the sphere here suggested. On the other hand, Darwin insists very sufficiently on the rapidity with which new peculiarities are produced; and this rapidity is quite as essential to the argument now urged as subsequent slowness.

We hope this argument is now plain. However slow the rate of variation might be, even though it were only one part in a thousand per twenty or two thousand generations, yet if it were constant or erratic we might believe that, in untold time, it would lead to untold distance; but if in every case we find that deviation from an average individual can be rapidly effected at first, and that the rate of deviation steadily diminishes till it reaches an almost imperceptible amount, then we are as much entitled to assume a limit to the possible deviation as we are to the progress of a cannon-ball from a knowledge of the law of diminution in its speed.

This limit to the variation of species seems to be established for all cases of man's selection. What argument does Darwin offer showing that the law of variation will be different when the variation occurs slowly, not rapidly? The law may be different, but is there any experimental ground for believing that it *is* different? Darwin says (p. 153), 'The struggle between natural selection, on the one hand, and the tendency to reversion and variability on the other hand, will in the course of time cease, and that the most abnormally developed organs may be made constant, I can see no reason to doubt.' But what reason have we to believe this? Darwin says the variablity will disappear by the continued rejection of the individuals tending to revert to a former condition; but is there any experimental ground for believing that the variability *will* disappear; and, secondly, if the variety can become fixed, that it will in time become ready to vary still more in the original direction, passing that limit which we think has just been shown to exist in the case of man's selection? It is peculiarly difficult to see how natural selection could reject individuals having a tendency to produce offspring reverting to an original stock. The tendency to produce offspring more like their superior parents than their inferior grandfathers can surely be of no advantage to any individual in the struggle for life. On the contrary, most individuals would be benefited by producing imperfect offspring, competing with them at a disadvantage; thus it would appear that natural selection, if it select anything, must select the most perfect individuals, having a tendency to produce the fewest and least perfect competitors; but may be urged that though the tendency to produce good offspring is injurious to the parents, the improved offspring would live and receive by inheritance the fatal tendency of producing in their turn parricidal descendants. Yet this is contending that in the struggle for life natural selection can gradually endow a race with a quality injurious to every individual which possesses it. It really seems certain that natural selection cannot tend to obliterate the tendency to revert; but the theory advanced appears rather to be that, if owing to some other qualities a race is maintained for a very long time different from the average or original race (near the surface of our sphere), then it will in time spontaneously lose the tendency to relapse, and acquire a tendency to vary outside the sphere. What is to produce this change? Time simply, apparently. The race is to be kept constant to all appearance, for a very long while, but some subtle change due to time is to take place; so that, of two individuals just alike in every feature, but one born a few thousand years after the other, the first shall tend to produce relapsing offspring, the second shall not. This seems rather like the idea that keeping a bar of iron hot or cold for a very long time would leave it permanently hot or cold at the end of the period when the heating or cooling agent was withdrawn. This strikes us as absurd now, but Bacon believed it possibly true. So many things may happen in a very long time, that time comes to be looked on as an agent capable of doing great and unknown things. Natural selection, as we contend, could hardly select an individual because it bred true. Man does. He chooses for sires those horses which he sees not only run fast themselves, but produce fine foals. He never gets rid of the tendency to revert. Darwin says species of pigeons have bred true for centuries. Does he believe that it would not be easier by selection to diminish the peculiarities of the pouter pigeon than to increase them? and what does this mean, but that the tendency to revert exists? It is possible that by man's selection this tendency may be diminished as any other quality may be somewhat increased or diminished, but, like all other qualities, this seems rapidly to approach a limit which there is no obvious reason to suppose 'time' will alter.

But not only do we require for Darwin's theory that time shall first permanently fix the variety near the outside of the assumed sphere of variation, we require that it shall give the power of varying beyond that sphere. It may be urged that man's rapid selection does away with this power; that if each little improvement were allowed to take root during a few hundred generations, there would be no symptom of a decrease of the rate of variation, no symptom that a limit was approached. If this be so, breeders of race-horses and prize flowers had better change their tactics; instead of selecting the fastest colts and finest flowers to start with, they ought to begin with very ordinary beasts and species. They should select the descendents which might be rather better in the first geseration, and then should carefully abstain from all attempts at improvement for twenty, thirty, or one hundred generations. Then they might take a little step forward, and in this way, in time, they or their children's children would obtain breeds far surpassing those produced by their overhasty competitors, who would be brought to a stand by limits which would never be felt or perceived by the followers of the maxim, *Festina lente*. If we are told that the time during which a breeder or his descendants could afford to wait bears no proportion to the time used by natural selection, we may answer that we do not expect the enormous variability supposed to be given by natural selection, but that we do expect to observe some step in that direction, to find that by carefully approaching our limit by slow degrees, that limit would be removed a little further off. Does any one think this would be the case?

There is indeed one view upon which it would seem natural to believe that the tendency to revert may diminish. If the peculiarities of an animal's structure are simply determined by inheritance, and not by any law of growth, and if the child is more likely to resemble its father than its grandfather, its grandfather than its great-grandfather, etc., then the chances that an animal will revert to the likeness of an ancestor a thousand generations back will be slender. This is perhaps Darwin's view. It depends on the assumption that there is no typical or average animal, no sphere of variation, with centre and limits, and cannot be made use of to prove the assumption. The opposing view is that of a race maintained by a continual force in an abnormal condition, and returning to that condition so soon as the force is removed; returning not suddenly, but by similar steps with those by which it first left the average state, restrained by the tendency to resemble its immediate progenitors. *A priori* perhaps, one view is as probable as the other; or in other words, as we are ignorant of the reasons why atoms fashion themselves into bears and squirrels, one fancy is as likely to meet with approval as another. Experiments conducted in a limited time point as already said to a limit, with a tendency to revert. And while admitting that the tendency to revert may be diminished though not extinguished, we are unaware of any reason for supposing that pouters, after a thousand generations of true breeding, have acquired a fresh power of doubling their crops, or that the oldest breed of Arabs are likely to produce 'sports' vastly surpassing their ancestors in speed. Experiments conducted during the longest time at our disposal show no probability of surpassing the limits of the sphere of variation, and why should we concede that a simple extension of time will reverse the rule?

The argument may be thus resumed.

Although many domestic animals and plants are highly variable, there appears to be a limit to their variation in any one direction. This limit is shown by the fact that new points are at first rapidly gained, but afterwards more slowly, while finally no further perceptible change can be effected. Great, therefore, as the variability is, we are not free to assume that successive variations of the same kind can be accumulated. There is no experimental reason for believing that the limit would be removed to a greater distance, or passed, simply because it was approached by very slow degrees, instead of by more rapid steps. There is no reason to believe that a fresh variability is acquired by long selection of one form; on the contrary, we know that with the oldest breeds it is easier to bring about a diminution than an increase in the points of excellence. The sphere of variation is a simile embodying this view;—each point of the sphere corresponding to a different individual of the same race, the centre to the average animal, the surface to the limit in various directions. The individual near the centre may have offspring varying in all directions with nearly equal rapidity. A variety near the surface may be made to approach it still nearer, but has a greater tendency to vary in every other direction. The sphere may be conceived as large for some species and small for others.

Efficiency of Natural Selection.-Those individuals of any species which are most adapted to the life they lead, live on an average longer than those which are less adapted to the circumstances in which the species is placed. The individuals which live the longest will have the most numerous offspring, and as the offspring on the whole resemble their parents, the descendants from any given generation will on the whole resemble the more favoured rather than the less favoured individuals of the species. So much of the theory of natural selection will hardly be denied; but it will be worth while to consider how far this process can tend to cause a variation in some one direction. It is clear that it will frequently, and indeed generally, tend to prevent any deviation from the common type. The mere existence of a species is a proof that it is tolerably well adapted to the life it must lead; many of the variations which may occur will be variations for the worse, and natural selection will assuredly stamp these out. A white grouse in the heather, or a white hare on a fallow would be sooner detected by its enemies than one of the usual plumage or colour. Even so, any favourable deviation must, according to the very terms of the statement, give its fortunate possesor a better chance of life; but this conclusion differs widely from the supposed consequence that a whole species may or will gradually acquire some one new quality, or wholly change in one direction and in the same manner. In arguing this point, two distinct kinds of possible variation must be separately considered: *first,* that kind of common variation which must be conceived as not only possible, but inevitable, in each individual of the species, such as longer and shorter legs, better or worse hearing, etc.; and, *secondly,* that kind of variation which only occurs rarely, and may be called a sport of nature, or more briefly a 'sport,' as when a child is born with six fingers on each hand. The common variation is not limited to one part of any animal, but occurs in all; and when we say that on the whole the stronger live longer than the weaker, we mean that in some cases long life will have been due to good lungs, in others to good ears, in others to good legs. There are few cases in which one faculty is pre-eminently useful to an animal beyond all other faculties, and where that is not so, the effect of natural selection will simply be to kill the weakly, and insure a sound, healthy, well-developed breed. If we could admit the principle of a gradual accumulation of improvements, natural selection would gradually improve the breed of everything, making the hare of the present generation run faster, hear better, digest better, than his ancestors; his enemies, the weasels, greyhounds, etc., would have improved likewise, so that perhaps the hare would not be really better off; but at any rate the direction of the change would be from a war of pigmies to a war of Titans. Opinions may differ as to the evidence of this gradual perfectibility of all things, but it is beside the question to argue this point, as the origin of species requires not the gradual improvement of animals retaining the same habits and structure, but such modification of those habits and structure as will actually lead to the appearance of new organs. We freely admit, that if an accumulation of slight improvements be possible, natural selection might improve hares as hares, and weasels as weasels, that is to say, it might produce animals having every useful faculty and every useful organ of their ancestors developed to a higher degree; more than this, it may obliterate some once useful organs when circumstances have so changed that they are no longer useful, for since that organ will weigh for nothing in the struggle of life, the average animal must be calculated as though it did not exist.

We will even go further: if, owing to a change of circumstances some organ becomes pre-eminently useful, natural selection will undoubtedly produce a gradual improvement in that organ, precisely as man's selection can improve a special organ. In all cases the animals above the average live longer, those below the average die sooner, but in estimating the chance of life of a particular animal, one special organ may count much higher or lower according to circumstances, and will accordingly be improved or degraded. Thus it must apparently be conceded that natural selection is a true cause or agency whereby in some cases variations of special organs may be perpetuated and accumulated, but the importance of this admission is much limited by a consideration of the cases to which it applies: first of all we have required that it should apply to variations which must occur in every individual, so that enormous numbers of individuals will exist, all having a little improvement in the same direction; as, for instance, each generation of hares will include an enormous number which have longer legs than the average of their parents although there may be an equally enormous number who have shorter legs; secondly, we require that the variation shall occur in an organ already useful owing to the habits of the animal. Such a process of improvement as is described could certainly never give organs of sight, smell or hearing to organisms which had never possessed them. It could not add a few legs to a hare, or produce a new organ, or even cultivate any rudimentary organ which was not immediately useful to any enormous majority of hares. No doubt half the hares which are born have longer tails than the average of their ancestors; but as no large number of hares hang by their tails, it is inconceivable than any change of circumstances should breed hares with prehensile tails; or, to take an instance less shocking in its absurdity, half the hares which are born may be presumed to be more like their cousins the rabbits in their burrowing organs than the average hare ancestor was; but this peculiarity cannot be improved by natural selection as described above, until a considerable number of hares begin to burrow, which we have as yet seen no likelihood of their doing. Admitting, therefore, that natural selection may improve organs already useful to great numbers of a species, does not imply an admission that it can create or develop new organs, and so originate species.

But it may be urged, although many hares do not burrow, one may, or at least may hide in a hole, and a little scratching may just turn the balance in his favour in the struggle for life. So it may, and this brings us straight to the consideration of 'sports,' the second kind of variation above alluded to. A hare which saved its life by burrowing would come under this head; let us here consider whether a few hares in a century saving themselves by this process could, in some indefinite time, make a burrowing species of hare. It is very difficult to see how this can be accomplished, even when the sport is very eminently favourable indeed; and still more difficult when the advantage gained is very slight, as must generally be the case. The advantage, whatever it may be, is utterly outbalanced by numerical inferiority. A million creatures are born; ten thousand survive to produce offspring. One of the million has twice as good a chance as any other of surviving; but the chances are fifty to one against the gifted individuals being one of the hundred survivors. No doubt, the chances are twice as great against any one other individual, but this does not prevent their being enormously in favour of *some* average individual. However slight the advantage may be, if it is shared by half the individuals produced, it will probably be present in at least fifty-one of the survivors, and in a larger proportion of their offspring; but the chances are against the

preservation of any one 'sport' in a numerous tribe. The vague use of
an imperfectly understood doctrine of chance has led Darwinian supporters,
first, to confuse the two cases above distinguished; and, secondly to imagine
that a very slight balance in favour of some individual sport must lead
to its perpetuation. All that can be said, is that in the above example the
favoured sport would be preserved once in fifty times. Let us consider
what will be its influence on the main stock when preserved. It will breed
and have a progeny of say 100; now this progeny will, on the whole,
be intermediate between the average individual and the sport. The odds
in favour of one of this generation of the new breed will be, say $1\frac{1}{2}$ to
1, as compared with the average individual; the odds in their favour will
therefore be less than that of their parent; but owing to their greater
number, the chances are that about $1\frac{1}{2}$ of them would survive. Unless
these breed together, a most improbable event, their progeny would again
approach the average individual; there would be 150 of them, and their
superiority would be say in the ratio of $1\frac{1}{4}$ to 1; the probability would
now be that nearly two of them would survive, and have 200 children,
with an eighth superiority. Rather more than two of these would survive;
but the superiority would again dwindle, until after a few generations it
would no longer be observed and would count for no more in the struggle
for life, than any of the hundred trifling advantages which occur in the
ordinary organs. An illustration will bring this conception home. Suppose
a white man to have been wrecked on an island inhabited by negroes,
and to have established himself in friendly relations with a powerful tribe,
whose customs he has learnt. Suppose him to possess the physical strength,
energy, and ability of a dominant white race, and let the food and climate
of the island suit his constitution; grant him every advantage which we
can conceive a white to possess over the native; concede that in the struggle
for existence his chance of a long life will be much superior to that of
the native chiefs; yet from all these admissions, there does not follow the
conclusion that, after a limited or unlimited number of generations, the
inhabitants of the island will be white. Our shipwrecked hero would probably become king; he would kill a great many blacks in the struggle for
existence; he would have a great many wives and children, while many
of his subjects would live and die as bachelors; an insurance company
would accept his life at perhaps one-tenth of the premium which they
would exact from the most favoured of the negroes. Our white's qualities
would certainly tend very much to preserve him to a good old age, and
yet he would not suffice in any number of generations to turn his subjects'

descendants white. It may be said that the white colour is not the cause of the superiority. True, but it may be used simply to bring before the senses the way in which qualities belonging to one individual in a large number must be gradually obliterated. In the first generation there will be some dozens of intelligent young mulattoes, much superior in average intelligence to the negroes. We might expect the throne for some generations to be occupied by a more or less yellow king; but can any one believe that the whole island will gradually acquire a white, or even a yellow population, or that the islanders would acquire the energy, courage, ingenuity, patience, self-control, endurance, in virtue of which qualities our hero killed so many of their ancestors, and begot so many children; those qualities, in fact, which the struggle for existence would select, if it could select anything?

Here is a case in which a variety was introduced, with far greater advantages than any sport ever heard of, advantages tending to its preservation, and yet powerless to perpetuate the new variety.

Darwin says that in the struggle for life a grain may turn the balance in favour of a given structure, which will then be preserved. But one of the weights in the scale of nature is due to the number of a given tribe. Let there be 7000 A's and 7000 B's, representing two varieties of a given animal, and let all the B's, in virtue of a slight difference of structure, have the better chance of life by $\frac{1}{7000}$th part. We must allow that there is a slight probability that the descendants of B will supplant the descendants of A; but let there be only 7001 A's against 7000 B's at first, and the chances are once more equal, while if there be 7002 A's to start, the odds would be laid on the A's. True, they stand a greater chance of being killed; but then they can better afford to be killed. The grain will only turn the scales when these are very nicely balanced, and an advantage in numbers counts for weight, even as an advantage in structure. As the numbers of the favoured variety diminish, so must its relative advantage increase, if the chance of its existence is to surpass the chance of its extinction, until hardly any conceivable advantage would enable the descendants of a single pair to exterminate the descendants of many thousands if they and their descendants are supposed to breed freely with the inferior variety, and so gradually lose their ascendency. If it is impossible that any sport or accidental variation in a single individual, however favourable to life, should be preserved and transmitted by natural selection, still less can slight and imperceptible variations, occurring in single individuals be garnered up and transmitted to continually increasing numbers; for if a very highly-favoured white cannot blanch a nation of negroes, it will hardly be contended that a comparatively very dull mulatto has a good chance of producing a tawny tribe; the idea, which seems almost absurd when presented in connexion with a practical case, rests on a fallacy of exceedingly common occurrence in mechanics and physics generally. When a man shows that a tendency to produce a given effect exists he often thinks he has proved that the effect must follow. He does not take into account the opposing tendencies, much less does he measure the various forces, with a view to calculate the result. For instance, there is a tendency on the part of a submarine cable to assume a catenary curve, and very high authorities once said it would; but, in fact, forces neglected by them utterly alter the curve from the catenary. There is a tendency on the part of the same cables, as usually made, to untwist entirely; luckily there are opposing forces, and they untwist very little. These cases will hardly seem obvious; but what should we say to a man who asserted that the centrifugal tendency of the earth must send it off in a tangent? One tendency is balanced or outbalanced by others; the advantage of structure possessed by an isolated specimen is enormously outbalanced by the advantage of numbers possessed by the others.

A Darwinian may grant all that has been said, but contend that the offspring of 'sports' is not intermediate between the new sport and the old species; he may say that a great number of offspring will retain in full vigour the peculiarity constituting the favourable sport. Darwin seems with hesitation to make some such claim as this, and though it seems contrary to ordinary experience, it will be only fair to consider this hypothesis. Let an animal be born with some useful peculiarity, and let all his descendants retain his peculiarity in an eminent degree, however, little of the first ancestor's blood be in them, then it follows, from mere mathematics, that the descendants of our gifted beast will probably exterminate the descendants of his inferior brethen. If the animals breed rapidly the work of substitution would proceed with wonderful rapidity, although it is a stiff mathematical problem to calculate the number of generations required in any given case. To put this case clearly beside the former, we may say that if in a tribe of a given number of individuals there appears one super-eminently gifted, and if the advantage accruing to the descendants bears some kind of proportion to the amount of the ancestor's blood in their veins, the chances are considerable that for the first few generations he will have many descendants; but by degrees this advantage wanes, and after many generations the chances are so far from being favourable to his breed covering the ground exclusively, that they are actually much against his having any descendants at all alive, for though he has a rather better chance of this than any of his neighbours, yet the chances are greatly against any one of them. It is infinitely improbable that the descendants of any one should wholly supplant the others. If, on the contrary, the advantage given by the sport is retained by all descendants, independently of what in common speech might be called the proportion of blood in their veins directly derived from the first sport, then these descendants will shortly supplant the old species entirely, after the manner required by Darwin.

But this theory of the origin of species is surely not the Darwinian theory; it simply amounts to the hypothesis that, from time to time, an animal is born differing appreciably from its progenitors, and possessing the power of transmitting the difference to its descendants. What is this but stating that, from time to time, a new species is created? It does not, indeed, imply that the new specimen suddenly appears in full vigour, made out of nothing; but if offers no explanation of the cause of the divergence from the progenitors, and still less of the mysterious faculty by which the divergence is transmitted unimpaired to countless descendants. It is clear that every divergence is not thus transmitted, for otherwise one and the same animal might have to be big to suit its father and little to suit its mother, might require a long nose in virtue of its grandfather and a short one in virtue of its grandmother, in a word, would have to resume in itself the countless contradictory peculiarities of its ancestors, all in full bloom, and unmodified one by the other, which seems as impossible as at one time to be and not to be. The appearance of a new specimen capable of perpetuating its peculiarity is precisely what might be termed a creation, the word being used to express our ignorance of how the thing happened. The substitution of the new specimens, descendants from the old species, would then be simply an example of a strong race supplanting a weak one, by a process known long before the term 'natural selection' was invented. Perhaps this is the way in which new species are introduced, but it does not express the Darwinian theory of the gradual accumulation of infinitely minute differences of every-day occurrence, and apparently fortuitous in their character.

Another argument against the efficiency of natural selection is, that animals possess many peculiarities the special advantage of which it is almost impossible to conceive; such, for instance, as the colour of plumage never displayed; and the argument may be extended by pointing out how impossible it is to conceive that the wonderful minutiae of, say a peacock's tail,

with every little frond of every feather differently barred, could have been
elaborated by the minute and careful inspection of rival gallants or admiring
wives; but although arguments of this kind are probably correct, they admit
of less absolute demonstration than the points already put. A true believer
can always reply, 'You do not know how closely Mrs. Peahen inspects her
husband's toilet, or you cannot be absolutely certain that under some unknown circumstances that insignificant feather was really unimportant;'
or finally, he may take refuge in the word correlation, and say, other
parts were useful, which by the law of correlation could not exist without
these parts; and although he may have not one single reason to allege
in favour of any of these statements, he may safely defy us to prove the
negative, that they are not true. The very same difficulty arises when a
disbeliever tries to point out the difficulty of believing that some odd habit
or complicated organ can have been useful before fully developed. The
believer who is at liberty to invent any imaginary circumstances, will very
generally be able to conceive some series of transmutations answering his
wants.

He can invent trains of ancestors of whose existence there is no evidence; he can marshal hosts of equally imaginary foes; he can call up
continents, floods, and peculiar atmospheres, he can dry up oceans, split
islands, and parcel out eternity at will; surely with these advantages he
must be a dull fellow if he cannot scheme some series of animals and circumstances explaining our assumed difficulty quite naturally. Feeling the difficulty of dealing with adversaries who command so huge a domain of fancy,
we will abandon these arguments, and trust to those which at least cannot
be assailed by mere efforts of imagination. Our arguments as to the efficiency
of natural selection may be summed up as follows:—

We must distinguish several kinds of conceivable variation in individuals.

First, We have the ordinary variations peculiar to each individual. The
effect of the struggle for life will be to keep the stock in full vigour by
selecting the animals which in the main are strongest. When circumstances
alter, one special organ may become eminently advantageous, and then
natural selection will improve that organ. But this efficiency is limited
to the cases in which the same variation occurs in enormous numbers of
individuals, and in which the organ improved is already used by the mass
of the species. This case does not apply to the appearance of new organs
or habits.

Secondly, We have abnormal variations called sports, which may be supposed to introduce new organs or habits in rare individuals. This case

must be again subdivided; we may suppose the offspring of the sports to be intermediate between their ancestor and the original tribe. In this case the sport will be swamped by numbers, and after a few generations its peculiarity will be obliterated. Or, we may suppose the offspring of the sport faithfully to reproduce the advantageous peculiarity undiminished. In this case the new variety will supplant the old species; but this theory implies a succession of phenomena so different from those of the ordinary variations which we see daily, that it might be termed a theory of successive creations; it does not express the Darwinian theory, and is no more dependent on the theory of natural selection than the universally admitted fact that a new strong race, not intermarrying with an old weak race, will surely supplant it. So much may be conceded.

Lapse of Time.–Darwin says with candour that he 'who does not admit how incomprehensibly vast have been the past periods of time,' may at once close his volume, admitting thereby that an indefinite, if not infinite time is required by his theory. Few will on this point be inclined to differ from the ingenious author. We are fairly certain that a thousand years has made no very great change in plants or animals living in a state of nature. The mind cannot conceive a multiplier vast enough to convert this trifling change by accumulation into differences commensurate with those between a butterfly and an elephant, or even between a horse and a hippopotamus. A believer in Darwin can only say to himself, Some little change does take place every thousand years; these changes accumulate, and if there be no limit to the continuance of the process, I must admit that in course of time any conceivable differences may be produced. He cannot think that a thousandfold the difference produced in a thousand years would suffice, according to our present observation, to breed even a dog from a cat. He may perhaps think that by careful selection, continued for this million years, man might do quite as much as this; but he will readily admit that natural selection does take a much longer time, and that a million years must by the true believer be looked upon as a minute. Geology lends her aid to convince him that countless ages have elapsed, each bearing countless generations of beings, and each differing in its physical conditions very little from the age we are personally acquainted with. This view of past time is, we believe, wholly erroneous. So far as this world is concerned, past ages are far from countless; the ages to come are numbered; no one age has resembled its predecessor, nor will any future time repeat the past. The estimates of geologists must yield before more accurate methods of computation, and these show that our world cannot have been habitable for more than an infinitely insufficient period for the execution of the Darwinian transmutation.

Before the grounds of these assertions are explained, let us shortly consider the geological evidence. It is clear that denudation and deposition of vast masses of matter have occurred while the globe was habitable. The present rate of deposit and denudation is very imperfectly known, but it is nevertheless sufficiently considerable to account for all the effects we know of, provided sufficient time be granted. Any estimate of the time occupied in depositing or denuding a thousand feet of any given formation, even on this hypothesis of constancy of action, must be very vague. Darwin makes the denudation of the Weald occupy 300,000,000 years, by supposing that a cliff 500 feet high was taken away one inch per century. Many people will admit that a strong current washing the base of such a cliff as this, might get on at least a hundredfold faster, perhaps a thousandfold; and on the other hand, we may admit, that for aught geology can show, the denudation of the Weald may have occupied a few million times more years than the number Darwin arrives at. The whole calculation savours a good deal of that known among engineers as 'guess at the half and multiply by two.'

But again, what are the reasons for assuming uniformity of action, for believing that currents were no stronger, storms no more violent, alternations of temperature no more severe in past ages than at present? These reasons, stated shortly, are that the simple continuance of actions we are acquainted with would produce all the known results, that we are not justified in assuming any alteration in the rate of violence of those actions without direct evidence, that the presence of fossils and the fineness of the ancient deposits show directly that things of old went on much as now. This last reason, apparently the strongest, is really the weakest; the deposits would assuredly take place in still waters, and we may fairly believe that still waters then resembled still waters now. The sufficiency of present actions is an excellent argument in the absence of all proof of change, but falls to utter worthlessness in presence of the direct evidence of change. We will try to explain the nature of the evidence, which does prove not only that the violence of all natural changes has decreased, but also that it is decreasing, and must continue to decrease.

Perpetual motion is popularly recognised as a delusion; yet perpetual *motion* is no mechanical absurdity, but in given conditions is a mechanical necessity. Set a mass in motion and it must continue to move for ever, unless stopped by something else. This something else takes up the motion in some other form, and continues it till the whole or part is again transmitted to other matter; in this sense perpetual motion is inevitable. But this is not the popular meaning of 'perpetual motion,' which represents a vague idea that a watch will not go unless it is wound up. Put into more accurate form, it means that no finite construction of physical materials can continue to *do work* for an infinite time; or in other words, one part of the construction cannot continue to part with its energy and another part to receive it for ever, nor can the action be perpetually reversed. All motion we can produce in this world is accompanied by the performance of a certain amount of work in the form of overcoming friction, and this involves a redistribution of energy. No continual motion can therefore be produced by any finite chemical, mechanical, or other physical construction. In this case, what is true on a small scale is equally true on a large scale. Looking on the sun and planets as a certain complex physical combination, differing in degree but not in kind from those we can produce in the workshop by using similar materials subject to the same laws, we at once admit that if there be no resistance, the planets may continue to revolve round the sun for ever, and may have done so from infinite time. Under these circumstances, neither the sun nor planets gain or lose a particle of energy in the process. Perpetual motion is, therefore, in this case quite conceivable. But when we find the sun raising huge masses of water daily from the sea to the skies, lifting yearly endless vegetation from the earth, setting breeze and hurricane in motion, dragging the huge tidal wave round and round our earth; performing, in fine, the great bulk of the endless labour of this world and of other worlds, so that the energy of the sun is continually being given away; then, we may say this continual work cannot go on for ever. This would be precisely the perpetual motion we are for ever ridiculing as an exploded delusion, and yet how many persons will read these lines, to whom it has occurred that the physical work done in the world requires a motive power, that no physical motive power is infinite or indefinite, that the heat of the sun, and the sum of all chemical and other physical affinities in the world, is just as surely limited in its power of doing work as a given number of tons of coal in the boiler of a steam-engine. Most readers will allow that the power man can extract from a ton of coals is limited, but perhaps not one reader in a thousand will at first admit that the power of the sun and that of the chemical affinities of bodies on the earth is equally limited.

There is a loose idea that our perpetual motions are impossible because we cannot avoid friction, and that friction entails somehow a loss of power, but that nature either works without friction, or that in the general system, friction entails no loss, and so her perpetual motions are possible; but nature no more works without friction than we can, and friction entails a loss of available power in all cases. When the rain falls, it feels the friction as much as drops from Hero's fountain; when the tide rolls round the world it rubs upon the sea-floor, even as a ball of mercury rubs on the artificial inclined planes used by ingenious inventors of impossibilities; when the breeze plays among the leaves, friction occurs according to the same laws as when artificial fans are driven through the air. Every chemical action in nature is as finite as the combustion of oxygen and carbon. The stone which, loosened by the rain, falls down a mountain-side, will no more raise itself to its first height, than the most ingeniously devised counter-poise of mechanism will raise an equal weight an equal distance. How comes it then that the finite nature of natural actions has not been as generally recognised as the finite nature of the so-called artificial combina-tions? Simply because, till very lately, it was impossible to follow the com-plete cycle of natural operations in the same manner as the complete cycle of any mechanical operations could be followed. All the pressures and resistances of the machine were calculable; we knew not so much as if there were analogous pressures and resistances in nature's mechanism. The establishment of the doctrine of conservation of energy, showing a numerical equivalence between the various forms of physical energy exhibited by *vis viva*, heat, chemical affinity, electricity, light, elasticity, and gravitation, has enabled us to examine the complete series of any given actions in nature, even as the successive actions of a train of wheels in a mill can be studied. There is no missing link; there is no unseen gearing, by which, in our ignorance, we might assume that the last wheel of the set somehow managed to drive the first. We have experimentally proved one law,—that the total quantity of energy in the universe is constant, meaning by energy something perfectly intelligible and measurable, equivalent in all cases to the product of a mass into the square of a velocity, sometimes latent, that is to say, producing or undergoing no change; at other times in action, that is to say, in the act of producing or undergoing change, not a change in amount, but a change of distribution. First, the hand about to throw a ball, next, the ball in motion, lastly, the heated wall struck by the ball, contain the greater part of the energy of the construction; but, from first to last, the sum of the energies contained by the hand, the ball, and the wall, is constant. At first sight, this constancy, in virtue of which no energy is ever lost, but simply transferred from mass to mass, might seem to favour the notion of a possible eternity of change, in which the earlier and later states of the universe would differ in no essential feature. It is to Professor Sir W. Thomson of Glasglow* that we owe the demonstration of the fallacy of this conception, and the establishment of the contrary doctrine of a continual dissipation of energy, by which the available power to produce change in any finite quantity of matter diminishes at every change of the distribution of energy. A simple illustration of the meaning of this doctrine is afforded by an unequally heated bar of iron. Let one end be hot and the other cold. The total quantity of heat (representing one form of energy) contained by the bar is mensurable and finite, and the bar contains within itself the elements of change,—the heated end may become cooler and the cold end warmer. So long as any two parts differ in temperature, change may occur; but so soon as all parts of the bar are at one temperature, the bar *quoad* heat can produce no change in itself, and yet if we conceive radiation or conduction from the surface to have been prevented, the bar will contain the same total energy as before. In the first condition, it had the power of doing work, and if it had not been a simple bar, but a more complex arrangement of materials of which the two parts had been at different temperatures, this difference might have been used to set wheels going, or to produce a thermo-electric current; but gradually the wheels would have been stopped by friction producing heat once more, the thermo-electric current would have died out, producing heat in its turn, and the final quantity of heat in the system would have been the same as before. Its distribution only, as in the simple case, would have been different. At first, great differences in the distribution existed; at last, the distribution was absolutely uniform; and in that condition, the system could suffer no alteration until affected by some other body in a different condition, outside itself. Every change in the distribution of energy depends on a difference between bodies, and every change tends, on the whole, to diminish this difference, and so render the total future possible change less in amount. Heat is the great agent in this gradual decay. No sooner does energy take this form than it is rapidly dissipated, *i.e.,* distributed among a large number of bodies, which assume a nearly equal temperature; once energy has undergone this transformation, it is practically lost. The equivalent of the energy is there; but it can produce no change until some fresh body, at a very different temperature, is presented to it. Thus it is that friction is looked upon as the grand enemy of so-called perpetual motion;

* [Later Lord Kelvin, one of Darwin's most powerful adversaries; see the editor's comments following this review.]

it is the commonest mode by which *vis viva* is converted into heat; and we all practically know, that once the energy of our coal, boiling water, steam, piston, fly-wheel, rolling mills, gets into this form, it is simply conducted away, and is lost to us for ever; just so, when the chemical or other energies of nature, contained, say, in our planetary system, once assume the form of heat, they are in a fair way to be lost for all available purposes. They will produce a greater or less amount of change according to circumstances. The greater the difference of the temperature produced between the surrounding objects, the greater the physical changes they will effect, but the degradation is in all cases inevitable. Finally, the sun's rays take the form of heat, whether they raise water or vegetation, or do any other work, and in this form the energy quits the earth radiated into distant space. Nor would this gradual degradation be altered if space were bounded and the planets enclosed in a perfect non-conducting sphere. Everything inside that sphere would gradually become equally hot, and when this consummation was reached no further change would be possible. We might say (only we should not be alive) that the total energy of the system was the same as before, but practically the universe would contain mere changeless death, and to this condition the material universe tends, for the conclusion is not altered even by an unlimited extension of space. Moreover, the rate at which the planetary system is thus dying is perfectly mensurable, if not yet perfectly measured. An estimate of the total loss of heat from the sun is an estimate of the rate at which he is approaching the condition of surrounding space, after reaching which he will radiate no more. We intercept a few of his rays, and can measure the rate of his radiation very accurately; we know that his mass contains many of the materials our earth is formed of, and we know the capacity for heat and other forms of energy which those materials are capable of and so can estimate the total possible energy contained in the sun's mass. Knowing thus approximately, how much he has, and how fast he is losing it, we can, or Professor Thomson* can, calculate how long it will be before he will cool down to any given temperature. Nor is it possible to assume that, *per contra,* he is receiving energy to an unlimited extent in other ways. He may be supplied with heat and fuel by absorbing certain planetary bodies, but the supply is limited, and the limit is known and taken into account in the calculation, and we are assured that the sun will be too cold for our or Darwin's purposes before many millions of years—a long time, but far enough from countless ages; quite similarly past countless ages are inconceivable, inasmuch as the heat required by the sun to have allowed him to cool from time immemorial, would be such as to turn him into mere vapour, which would extend over the whole planetary system, and evaporate us entirely. It has been thought necessary to give the foregoing sketch of the necessary gradual running down of the heavenly mechanism, to show that this reasoning concerning the sun's heat does not depend on any one special fact, or sets of facts, about heat, but is the mere accidental form of decay, which in some shape is inevitable, and the very essential condition of action. There is a kind of vague idea, when the sun is said to be limited in its heating powers, that somehow chemistry or electricity, etc., may reverse all that; but it has been explained that every one of these agencies is subject to the same law; they can never twice produce the same change in its entirety. Every change is a decay, meaning by change a change in the distribution of energy.

Another method by which the rate of decay of our planetary system can be measured, is afforded by the distribution of heat in the earth. If a man were to find a hot ball of iron suspended in the air, and were carefully to ascertain the distribution in the ball, he would be able to determine whether the ball was being heated or cooled at the time. If he found the outside hotter than the inside, he would conclude that in some way the ball was receiving heat from outside; if he found the inside hotter than the outside, he would conclude that the ball was cooling, and had therefore been hotter before he found it than when he found it. So far mere common sense would guide him, but with the aid of mathematics and some physical knowledge of the properties of iron and air, he would go much further, and be able to calculate how hot the ball must have been at any given moment, if it had not been interfered with. Thus he would be able to say, the ball must have been hung up less than say five hours ago, for at that time the heat of the ball would have been such, if left in its present position, that the metal would be fused, and so could not hang where he saw it. Precisely analogous reasoning holds with respect to the earth; it is such a ball; it is hotter inside than outside. The distribution of the heat near its surface is approximately known. The properties of the matter of which it is composed are approximately known, and hence an approximate calculation can be made of the period of time within which it must have been hot enough to fuse the materials of which it is composed, provided it has occupied its present position, or a similar position, in space. The data for this calculation are still very imperfect, but the result of analogous calculation applied to the sun, as worked out by Professor Sir W. Thomson, is five hundred million years, and the results

derived from the observed temperatures of the earth are of the same order
of magnitude. This calculation is a mere approximation. A better knowledge
of the distribution of heat in the interior of the globe may modify materially
our estimates. A better knowledge of the conducting powers of rocks, etc.,
for heat, and their distribution in the earth, may modify it to a less degree,
but unless our information be wholly erroneous as to the gradual increase
of temperature as we descend towards the centre of the earth, the main
result of the calculation, that the centre is gradually cooling, and if uninter-
fered with must, with a limited time, have been in a state of complete
fusion, cannot be overthrown. Not only is the time limited, but it is limited
to periods utterly inadequate for the production of species according to
Darwin's views. We have seen a lecture-room full of people titter when
told that the world would not, without supernatural interference, remain
habitable for more than one hundred million years. This period was to
those people ridiculously beyond anything in which they could take an
interest. Yet a thousand years is an historical period well within our
grasp,—as a Darwinian or geological unit it is almost uselessly small. Darwin
would probably admit that more than a thousand times this period, or
a million years, would be no long time to ask for the production of a
species differing only slightly from the parent stock. We doubt whether
a thousand times more change than we have any reason to believe has
taken place in wild animals in historic times, would produce a cat from
a dog, or either from a common ancestor. If this be so, how preposterously
inadequate are a few hundred times this unit for the action of the Darwinian theory!

But it may be said they are equally inadequate for the geological formations which we know of, and therefore your calculations are wrong. Let
us see what conclusions the application of the general theory of the gradual
dissipation of energy would lead to, as regards these geological formations.
We may perhaps find the solution of the difficulty in reconciling the results
of the calculation of the rate of secular cooling, with the results deduced
from the denudation or deposition of strata in the following consideration.
If there have been a gradual and continual dissipation of energy, there
will on the whole have been a gradual decrease in the violence or rapidity
of all physical changes. When the gunpowder in a gun is just lighted,
the energy applied in a small mass produces rapid and violent changes;
as the ball rushes through the air it gradually loses speed; when it strikes
rapid changes again occur, but not so rapid as at starting. Part of the
energy is slowly being diffused through the air; part is being slowly con-

ducted as heat from the interior to the exterior of the gun, only a residue shatters the rampart, and that residue, soon changing into heat, is finally diffused at a gradually decreasing rate into surrounding matter. Follow any self-contained change, and a similar gradual diminution on the whole will be observed. There are periods of greater and less activity, but the activity on the whole diminishes. Even so must it have been, and so will it be, with our earth. Extremes tend to diminish; high places become lower, low places higher, by denudation. Conduction is continually endeavouring to reduce extremes of heat and cold; as the sun's heat diminishes so will the violence of storms; as inequalities of surface diminish, so will the variations of climate. As the external crust consolidates, so will the effect of internal fire diminish. As internal stores of fuel are consumed, or other stores of chemical energy used up, the convulsions or gradual changes they can produce must diminish; on every side, and from whatever cause changes are due, we see the tendency to their gradual diminution of intensity or rapidity. To say that things must or can always have gone on at the present rate is a sheer absurdity, exactly equivalent to saying that a boiler fire once lighted will keep a steam-engine going for ever at a constant rate; to say all changes that have occurred, or will occur, since creation, have been due to the same causes as those now in action; and further, that those causes have not varied in intensity according to any other laws than they are now varying, is, we believe, a correct scientific statement, but then we contend that those causes must and do hourly diminish in intensity, and have since the beginning diminished in intensity, and will diminish, till further sensible change ceases, and a dead monotony is the final physical result of the mechanical laws which matter obeys.

Once this is granted, the calculations as to the length of geological periods, from the present rates of denudation and deposit, are blown to the winds. They are rough, very rough, at best. The present assumed rates are little better than guesses; but even were these really known, they could by no means be simply made use of in a rule-of-three sum, as has generally been done. The rates of denudation and deposition have been gradually, on the whole, slower and slower, as the time of fusion has become more and more remote. There has been no age of cataclysm, in one sense, no time, when the physical laws were other than they now are, but the results were as different as the rates of a steam-engine driven with a boiler first heated to 1500 degrees Fahrenheit, and gradually cooling to 200.

A counter argument is used, to the effect that our argument cannot be correct, since plants grew quietly, and fine deposits were formed in the earliest geological times. But, in truth, this fact in no way invalidates our argument. Plants grow just as quietly on the slope of Vesuvius, with a few feet between them and molten lava, as they do in a Kentish lane; but they occasionally experience the difference of the situation. The law according to which a melted mass cools would allow vegetation to exist and animals to walk unharmed over an incredibly thin crust. There would be occasional disturbances; but we see that a few feet of soil are a sufficient barrier between molten lava and the roots of the vine; each tendril grows not the less slowly and delicately because it is liable in a year or two to be swallowed up by the stream of lava. Yet no one will advance the proposition that changes on the surface of a volcano are going on at the same rate as elsewhere. Even so in the primeval world, barely crusted over, with great extremes of climate, violent storms, earthquakes, and a general rapid tendency to change, tender plants may have grown, and deep oceans may have covered depths of perfect stillness, interrupted occasionally by huge disturbances. Violent currents or storms in some regions do not preclude temperate climates in others, and after all the evidence of tranquillity is very slight. There are coarse deposits as well as fine ones; now a varying current sifts a deposit better than a thousand sieves, the large stones fall first in a rapid torrent, then the gravel in a rapid stream, then the coarse sand, and finally, the fine silt cannot get deposited till it meets with still water. And still water might assuredly exist at the bottom of oceans, the surface of which was traversed by storms and waves of an intensity unknown to us. The soundings in deep seas invariably produce samples of almost intangible ooze. All coarser materials are deposited before they reach regions of such deathlike stillness, and this would always be so. As to the plants, they may have grown within a yard of red-hot gneiss.

Another class of objections to the line of argument pursued consists in the suggestion that it is impossible to prove that since the creation things always have been as they are. Thus, one man says,—'Ah, but the world and planetary system may have passed through a warm region of space, and then your deductions from the radiation of heat into space go for nothing; or, a fresh supply of heat and fuel may have been supplied by regular arrivals of comets or other fourgons; or the sun and centre of the earth may be composed of materials utterly dissimilar to any we are acquainted with, capable of evolving heat from a limited space at a rate which we have no example of, leaving coal or gunpowder at an infinite distance behind them. Or it may please the Creator to continue creating energy in the form of heat at the centre of the sun and earth; or the mathematical laws of cooling and radiation, and conservation of energy and dissipation of energy may be actually erroneous, since man is, after all, fallible.' Well, we suppose all these things *may* be true, but we decline to allow them the slightest weight in the argument, until some reason can be shown for believing that any one of them *is* true.

To resume the arguments in this chapter—Darwin's theory requires countless ages, during which the earth shall have been habitable, and he claims geological evidence as showing an inconceivably great lapse of time, and as not being in contradiction with inconceivably greater periods than are even geologically indicated,—periods of rest between formations, and periods anterior to our so-called first formations, during which the rudimentary organs of the early fossils became degraded from their primeval uses. In answer, it is shown that a general physical law obtains, irreconcilable with the persistence of active change at a constant rate; in any portion of the universe, however large, only a certain capacity for change exists, so that every change which occurs renders the possibility of future change less, and, on the whole, the rapidity or violence of changes tends to diminish. Not only would this law gradually entail in the future the death of all beings and cessation of all change in the planetary system, and in the past point to a state of previous violence equally inconsistent with life, if no energy were lost by the system, but this gradual decay from a previous state of violence is rendered far more rapid by the continual loss of energy going on by means of radiation. From this general conception pointing either to a beginning, or to the equally inconceivable idea of infinite energy in finite materials, we pass to the practical application of the law to the sun and earth, showing that their present state proves that they cannot remain for ever adapted to living beings, and that living beings can have existed on the earth only for a definite time, since in distant periods the earth must have been in fusion, and the sun must have been mere hot gas, or a group of distant meteors, so as to have been incapable of fulfilling its present functions as the comparatively small centre of the system. From the earth we have no very safe calculation of past time, but the sun gives five hundred million years as the time separating us from a condition inconsistent with life. We next argue that the time occupied in the arrangement of the geological formations need not have been longer than is fully consistent with this view, since the gradual dissipation of energy must have resulted in a gradual diminution of violence of all kinds, so that calculations of the time occupied by denudations or deposits based on the simple division of the total mass of a deposit, or denudation

by the annual action now observed, are fallacious, and that even as the
early geologists erred in attempting to compress all action into six thousand
years, so later geologists have outstepped all bounds in their figures, by
assuming that the world has always gone on much as it now does, and
that the planetary system contains an inexhaustible motive power, by which
the vast labour of the system has been, and can be maintained for ever.
We have endeavoured to meet the main objections to these views, and
conclude, that countless ages cannot be granted to the expounder of any
theory of living beings, but that the age of the inhabited world is proved
to have been limited to a period wholly inconsistent with Darwin's views.

Difficulty of Classification.–It appears that it is difficult to classify animals
or plants, arranging them in groups as genera, species, and varieties; that
the line of demarcation is by no means clear between species and sub-species,
between sub-species and well-marked varieties, or between lesser varieties
and individual differences; that these lines of demarcation, as drawn by
different naturalists, vary much, being sometimes made to depend on this,
sometimes on that organ, rather arbitrarily. This difficulty chiefly seems
to have led men to devise theories of transmutation of species, and is the
very starting point of Darwin's theory, which depicts the differences between
various individuals of any one species as identical in nature with the differences between individuals of various species, and supposes all these differences, varying in degree only, to have been produced by the same causes;
so that the subdivision into groups is, in this view, to a great extent arbitrary,
but may be considered rational if the words variations, varieties, sub-species,
species, and genera, be used to signify or be considered to express that
the individuals included in these smaller or greater groups, have had a
common ancestor very lately, some time since, within the later geological
ages, or before the primary rocks. The common terms, explained by Darwin's principles, signify, in fact, the more or less close blood-relationship
of the individuals. This, if it could be established, would undoubtedly afford
a less arbitrary principle of classification than pitching on some one organ
and dragging into a given class all creatures that had this organ in any
degree similar. The application of the new doctrine might offer some difficulty, as it does not clearly appear what would be regarded as the sign
of more or less immediate descent from a common ancestor, and perhaps
each classifier would have pet marks by which to decide the question,
in which case the new principle would not be of much practical use;
yet if the theory were really true, in time the marks of common ancestry
would probably come to be known with some accuracy, and meanwhile

the theory would give an aim and meaning to classification, which otherwise might be looked upon as simply a convenient form of catalogue.

If the arguments already urged are true, these descents from common ancestors are wholly imaginary. 'How, then,' say the supporters of transmutation, 'do you account for our difficulty in distinguishing, a priori, varieties from species? The first, we know by experience, have descended from a common ancestor; the second you declare have not, and yet neither outward inspection nor dissection will enable us to distinguish a variety from what you call a species. Is not this strange, if there be an essential difference?'

No, it is not strange. There is nothing either wonderful or peculiar to organized beings, in the difficulty experienced in classification, and we have no reason to expect that the differences between beings which have had no common ancestor should be obviously greater than those occurring in the descendants of a given stock. Whatever origin species may have had, whether due to separate creation or some yet undiscovered process, we ought to expect a close approximation between these species, and difficulty in arranging them as groups. We find this difficulty in all classification, and the difficulty increases as the number of objects to be classified increases. Thus the chemist began by separating metals from metalloids, and found no difficulty in placing copper and iron in one category, and sulphur and phosphorus in the other. Now-a-days, there is or has been a doubt, whether hydrogen gas be a metal or no. It probably ought to be so classed. Some physical properties of tellurium would lead to its classification as a metal; its chemical properties are those of a metalloid. Acids and bases were once very intelligible headings to large groups of substances. Now-a-days there are just as finely drawn distinctions as to what is an acid, and what a base, as eager discussions which substance in a compound plays the part of acid or base, as there can possibly be about the line of demarcation between animal or vegetable life, and any of the characteristics used to determine the group that shall claim a given shell or plant. Nay, some chemists are just as eager to abandon the old terms altogether, as Darwin to abolish species. His most advanced disciple will hardly contend that metals and metalloids are the descendants of organic beings, which, in the struggle for life, have gradually lost all their organs; yet is it less strange that inorganic substances should be hard to class, than that organic beings, with their infinitely greater complexity, should be difficult to arrange in neat, well-defined groups? In the early days of chemistry, a theory might well have been started, perhaps was started, that all metals were alloys of a couple of unknown substances. Each newly discovered metal would have appeared to occupy an intermediate place between old metals. Alloys similarly occupied an intermediate place between the metals composing them; why might not all metals be simply sets of alloys, of which the elements were not yet discovered? An alloy can no more be distinguished by its outward appearance than a hybrid can. Alloys differ as much from one another, and from metals, as metals do one from another, and a whole set of Darwinian arguments might be used to prove all metals alloys. It is only of late, by a knowledge of complicated electrical and other properties, that we could feel a certainty that metals were not alloys.

Other examples may be given, and will hereafter be given, of analogous difficulties of classification; but let us at once examine what expectations we might naturally form, *a priori,* as to the probable ease or difficulty in classifying plants and animals, however these may have originated. Are not animals and plants combinations, more or less complex, of a limited number of elementary parts? The number of possible combinations of a given number of elements is limited, however numerous these elements may be. The limits to the possible number of combinations become more and more restricted, as we burden these combinations with laws more and more complicated,—insisting, for instance, that the elements shall only be combined in groups of threes or fives, or in triple groups of five each, or in *n* groups, consisting respectively of *a, b, c, d, . . . n* elements arranged each in a given order. But what conceivable complexity of algebraic arrangement can approach the complexity of the laws which regulate the construction of an organic being out of inorganic elements? Let the chemist tell us the laws of combination of each substance found in an organized being. Let us next attempt to conceive the complexity of the conditions required to arrange these combinations in a given order, so as to constitute an eating, breathing, moving, feeling, self-reproducing thing. When our mind has recoiled baffled, let us consider whether it is not probable, nay certain, that there should be a limit to the possible number of combinations, called animals or vegetables, produced out of a few simple elements, and grouped· under the above inconceivably complex laws. Next, we may ask whether, as in the mathematical permutations, combinations, and arrangements, the complete set of possible organized beings will not necessarily form a continuous series of combinations, each resembling its neighbour, even as the letters of the alphabet grouped say in all possible sets of five each, might be arranged so as to form a continuous series of groups, or sets of series, according as one kind of resemblance or another be chosen to guide us in the arrangement. It is clear that the number of combinations or animals

will be immeasurably greater when these combinations are allowed to resemble each other very closely, than when a condition is introduced, that
given marked differences shall exist between them. Thus, there are upwards
of 7,890,000 words or combinations of five letters in the English alphabet.
These are reduced to 26 when we insert a condition that no two combinations shall begin with the same letter, and to 5 when we stipulate that
no two shall contain a single letter alike. Thus we may expect, if the
analogy be admitted, to find varieties of a given species, apparently, though
not really, infinite in number, since the difference between these varieties
is very small, whereas we may expect that the number of well-marked
possible species will be limited, and only subject to increase by the insertion
of fresh terms or combinations, intermediate between those already existing.
Viewed in this light, a species is the expression of one class of combination;
the individuals express the varieties of which that class is capable.

It may be objected that the number of elements in an organized being
is so great, as practically to render the number of possible combinations
infinite; but unless infinite divisibility of matter be assumed, this objection
will not hold, inasmuch as the number of elements or parts in the germ
or seed of a given animal or plant appears far from infinite. Yet it is
certain that differences between one species and another, one variety and
another, one individual and another, exist in these minute bodies, containing
very simple and uniform substances if analysed chemically. Probably, even
fettered by these conditions, the number of possible animals or plants is
inconceivably greater than the number which exist or have existed; but
the greater the number, the more they must necessarily resemble one
another.

It may perhaps be thought irreverent to hold an opinion that the Creator
could not create animals of any shape and fashion whatever; undoubtedly
we may conceive all rules and all laws as entirely self-imposed by him,
as possibly quite different or non-existent elsewhere; but what we mean
is this, that just as with the existing chemical laws of the world, the number
of possible chemical combinations of a particular kind is limited, and not
even the Creator could make more without altering the laws he has himself
imposed, even so, if we imagine animals created or existing under some
definite law, the number of species, and of possible varieties of one species,
will be limited; and these varieties and species being definite arrangements
of organic compounds, will as certainly be capable of arrangement in series
as inorganic chemical compounds are. These views no more imply a limit
to the power of God than the statement that the three angles of a triangle
are necessarily equal to two right angles.

It is assumed that all existing substances or beings of which we have any scientific knowledge exist under definite laws. Under any laws there will be a limit to the possible number of combinations of a limited number of elements. The limit will apply to size, strength, length of life, and every other quality. Between any extremes the number of combinations called animals or species can only be increased by filling in gaps which exist between previously existing animals, or between these and the possible limits, and therefore whatever the general laws of organization may be, they must produce results similar to those we observe, and which lead to difficulty in classification, and to the similarity between one species or variety and another. Turning the argument, we might say that the observed facts simply prove that organisms exist and were created under definite laws, and surely no one will be disposed to deny this. Darwin assumes one law, namely, that every being is descended from a common ancestor (which, by the way, implies that every being shall be capable of producing a descendant like any other being), and he seems to think this the only law which would account for the close similarity of species, whereas any law may be expected to produce the same results. We observe that animals eat, breathe, move, have senses, are born, and die, and yet we are expected to feel surprise that combinations, which are all contrived to perform the same functions, resemble one another. It is the apparent variety that is astounding, not the similarity. Some will perhaps think it absurd to say that the number of combinations are limited. They will state that no two men ever were or will be exactly alike, no two leaves in any past or future forest; it is not clear how they could find this out, or how they could prove it. But as already explained, we quite admit that by allowing closer and closer similarity, the number of combinations of a fixed number of elements may be enormously increased. We may fairly doubt the identity of any two of the higher animals, remembering the large number of elements of which they consist, but perhaps two identical foraminiferae have existed. As an idle speculation suggested by the above views, we might consider whether it would be possible that two parts of any two animals should be identical, without their being wholly identical, looking on each animal as one possible combination, in which no part could vary without altering all the others. It would be difficult to ascertain this by experiment.

It is very curious to see how man's contrivances, intended to fulfill some common purpose, fall into series, presenting the difficulty complained of by naturalists in classifying birds and beasts, or chemists in arranging compounds. It is this difficulty which produces litigation under the Patent Laws. Is or is not this machine comprised among those forming the subject of the patent? At first sight nothing can be more different than the drawing in the patent and the machine produced in court, and yet counsel and witnesses shall prove to the satisfaction of judge, jury, and one party to the suit, that the essential part, the important organ, is the same in both cases. The case will often hinge on the question, What is the important organ? Just the question which Darwin asks; and quite as difficult to answer about a patented machine as about an organic being.

This difficulty results from the action of man's mind contriving machines to produce a common result according to definite laws, the laws of mechanics. An instance of this is afforded by the various forms of bridge. Nothing would appear more distinct than the three forms of suspension-bridge, girder, and arch; the types of which are furnished by a suspended rope, a balk of wood, and a stone arch; yet if we substitute an iron-plate girder of approved form for the wooden balk, and then a framed or lattice girder for the plate-iron girder, we shall see that the girder occupies an intermediate place between the two extremes, combining both the characteristics of the suspension and arched rib,—the upper plates and a set of diagonal strutts being compressed like the stones of an arch, the lower plates and a set of diagonal ties being extended like a suspended rope. Curve the top plates, as is often done, and the resemblance to an arch increases, yet every member of the girder remains. Weaken the bracing, leaving top and bottom plates as before, the bridge is now an arched bridge with the abutments tied together. Weaken the ties gradually, and you gradually approach nearer and nearer to the common arch with the usual abutments. Quite similarly the girder can be transformed into a suspension-bridge by gradual steps, so that none can say when the girder ends and the suspension-bridge begins. Nay, take the common framed or lattice girder, do not alter its shape in any way, but support it, first, on flat stones, like a girder, then wedge it between sloping abutments like an arch, and lastly, hang it up between short sloping links like those of a suspension-bridge, attached to the upper corners at the end,—you will so alter the strains in the three cases that in order to bear the same load, the relative parts of the framework must be altered in their proportions in three distinct ways, resembling in the arrangement of the strongest parts, first a girder, next an arch, and finally a suspension-bridge. Yet the outline might remain the same, and not a single member be removed.

Thus we see, that though in three distinct and extreme cases it is easy to give distinctive names with clear characteristics, it is very difficult as the varieties multiply to draw distinct lines between them. Shall the distribution of strains be the important point? Then one and the same piece of framework will have to be included under each of three heads, according to the manner in which it is suspended or supported. Shall form be the important point? We may construct a ribbed arch of string, of a form exactly similar to many compressed arches, we may support this from below, and yet the whole arch shall be in tension, and bear a considerable load. Shall the mode of support be the important point? It would be an odd conclusion to arrive at, that any stiff beam hung up in a particular way was a suspension-bridge. Nor is this difficulty simply a sophistical one invented for the occasion; the illustration was suggested by a practical difficulty met with in drawing up a patent; and in ordinary engineering practice, one man will call a certain bridge a stiffened arch, while another calls it a girder of peculiar form; a third man calls a bridge a strengthened girder, which a fourth says differs in no practical way from a suspension-bridge. Here, as in the case of animals or vegetables, when the varieties are few, classification is comparatively easy; as they are multiplied it becomes difficult; and when all the conceivable combinations are inserted it becomes impossible. Nor must it be supposed that this is due to the suggestion of one form by another in a way somewhat analogous to descent by animal reproduction. The facts would be the same however the bridges were designed. There are only certain ways in which a stream can be bridged; the extreme cases are easily perceived, and ingenuity can then only fill in an indefinite number of intermediate varieties. The possible varieties are not created by man, they are found out, laid bare. Which are laid bare will frequently depend on suggestion or association of ideas, so that groups of closely analogous forms are discovered about the same time; but we may *a priori* assert that whatever is discovered will lie between the known extremes, and will render the task of classification, if attempted, more and more difficult.

Legal difficulties furnish another illustration. Does a particular case fall within a particular statute? is it ruled by this or that precedent? The number of statutes or groups is limited; the number of possible combinations of events almost unlimited. Hence, as before, the uncertainty [as to] which group a special combination shall be classed within. Yet new combinations, being doubtful cases, are so, precisely because they are intermediate between others already known.

It might almost be urged that all the difficulties of reasoning, and all differences of opinion, might be reduced to difficulties of classification, that is to say, of determining whether a given minor is really included in a certain major proposition; and of discovering the major proposition or genus we are in want of. As trivial instances, take the docketing of letters or making catalogues of books. How difficult it is to devise headings, and how difficult afterwards to know under what head to place your book. The most arbitrary rule is the only one which has a chance of being carried out with absolute certainty.

Yet while these difficulties meet us wherever we turn, in chemistry, in mechanics, law, or mere catalogues of heterogeneous objects, we are asked to feel surprise that we cannot docket off creation into neat rectangular pigeon-holes, and we are offered a special theory of transmutation, limited to organic beings, to account for a fact of almost universal occurrence.

To resume this argument:–Attention has been drawn to the fact, that when a complete set of combinations of certain elements is formed according to a given law, they will necessarily be limited in number, and form a certain sequence, passing from one extreme to the other by successive steps.

Organized beings may be regarded as combinations, either of the elementary substances used to compose them, or of the parts recurring in many beings; for instance, of breathing organs, apparatus for causing blood to circulate, organs of sense, reproduction, etc., in animals. The conclusion is drawn that we can feel no reasonable surprise at finding that species should form a graduated series which it is difficult to group as genera, or that varieties should be hard to group into various distinct species.

Nor is it surprising that newly discovered species and varieties should almost invariably occupy an intermediate position between some already known, since the number of varieties of one species, or the number of possible species, can only be indefinitely increased by admitting varieties or species possessing indefinitely small differences one from another.

We observe that these peculiarities require no theory of transmutation, but only that the combination of the parts, however effected, should have been made in accordance with some law, as we have every reason to expect they would be.

In illustration of this conclusion, cases of difficult classification are pointed out containing nothing analogous to reproduction, and where no struggle for life occurs.

Observed Facts supposed to support Darwin's Views.–The chief arguments used to establish the theory rest on conjecture. Beasts may have varied; variation may have accumulated; they may have become permanent; continents may have arisen or sunk, and seas and winds been so arranged as to dispose of animals just as we find them, now spreading a race widely, now confining it to one Galapagos island. There may be records of infinitely more animals than we know of in geological formations yet unexplored. Myriads of species differing little from those we know to have been preserved, may actually not have been preserved at all. There may have been an inhabited world for ages before the earliest known geological strata. The world may indeed have been inhabited for an indefinite time; even the geological observations may perhaps give a most insufficient idea of the enormous times which separated one formation from another; the peculiarities of hybrids may result from accidental differences between the parents, not from what have been called specific differences.

We are asked to believe all these maybe's happening on an enormous scale, in order that we may believe the final Darwinian 'maybe,' as to the origin of species. The general form of his argument is as follows:–All these things may have been, therefore my theory is possible, and since my theory is a possible one, all those hypotheses which it requires are rendered probable. There is little direct evidence that any of these maybe's actually *have been*.

In this essay an attempt has been made to show that many of these assumed possibilities are actually impossibilities, or at the best have not occurred in this world, although it is proverbially somewhat difficult to prove a negative.

Let us now consider what direct evidence Darwin brings forward to prove that animals really are descended from a common ancestor. As direct evidence we may admit the possession of webbed feet by unplumed birds; the stripes observed on some kinds of horses and hybrids of horses, resembling not their parents, but other species of the genus; the generative variability of abnormal organs; the greater tendency to vary of widely diffused and widely ranging species, certain peculiarities of distribution. All these facts are consistent with Darwin's theory, and if it could be shown that they could not possibly have occurred except in consequence of natural selection, they would prove the truth of this theory. It would, however, clearly be impossible to prove that in no other way could these phenomena have been produced, and Darwin makes no attempt to prove this. He only says he cannot imagine why unplumed birds should have webbed feet, unless in consequence of their direct descent from web-footed ancestors who lived in the water; that he thinks it would in some way be derogatory to the Creator to let hybrids have stripes on their legs, unless some ancestors of theirs had stripes on his leg. He cannot imagine why abnormal organs and widely diffused genera should vary more than others, unless his views

be true; and he says he cannot account for the peculiarities of distribution
in any way but one. It is perhaps hardly necessary to combat these arguments, and to show that our inability to account for certain phenomena,
in any way but one, is no proof of the truth of the explanation given,
but simply is a confession of our ignorance. When a man says a glowworm
must be on fire, and in answer to our doubts challenges us to say how
it can give out light unless it be on fire, we do not admit his challenge
as any proof of his assertion, and indeed we allow it no weight whatever
as against positive proof we have that the glowworm is not on fire. We
conceive Darwin's theory to be in exactly the same case; its untruth can,
as we think, be proved, and his or our own inability to explain a few
isolated facts consistent with his views would simply prove his and our
ignorance of the true explanation. But although unable to give any certainly
true explanation of the above phenomena, it is possible to suggest explanations perhaps as plausible as the Darwinian theory, and though the fresh
suggestions may very probably not be correct, they may serve to show that
at least more than one conceivable explanation may be given.

It is a familiar fact that certain complexions go with certain tempera-
ments, that roughly something of a man's character may be told from
the shape of his head, his nose, or perhaps from most parts of his body.
We find certain colours almost always accompanying certain forms and
tempers of horses. There is a connexion between the shape of the hand
and the foot, and so forth. No horse has the head of a cart-horse and
the hind-quarters of a racer; so that, in general, if we know the shape
of most parts of a man or horse, we can make a good guess at the probable
shape of the remainder. All this shows that there is a certain correlation
of parts, leading us to expect that when the heads of two birds are very
much alike, their feet will not be very different. From the assumption
of a limited number of possible combinations or animals, it would naturally
follow that the combination of elements producing a bird having a head
very similar to that of a goose, could not fail to produce a foot also somewhat similar. According to this view, we might expect most animals to
have a good many superfluities of a minor kind, resulting necessarily from
the combination required to produce the essential or important organs.
Surely, then, it is not very strange that an animal intermediate by birth
between a horse and ass should resemble a quagga, which results from
a combination intermediate between the horse and ass combination. The
quagga is in general appearance intermediate between the horse and ass,
therefore, *a priori,* we may expect that in general appearance a hybrid

between the horse and the ass will resemble the quagga, and if in general it does resemble a quagga, we may expect that owing to the correlation of parts it will resemble the quagga in some special particulars. It is difficult to suppose that every stripe on a zebra or quagga, or cross down a donkey's back, is useful to it. It seems possible, even probable, that these things are the unavoidable consequences of the elementary combination which will produce the quagga, or a beast like it. Darwin himself appears to admit that correlation will or may produce results which are not themselves useful to the animal; thus how can we suppose that the beauty of feathers which are either never uncovered, or very rarely so, can be of any advantage to a bird? Nevertheless those concealed parts are often very beautiful, and the beauty of the markings on these parts must be supposed due to correlation. The exposed end of a peacock's feather could not be so gloriously coloured without beautiful colours even in the unexposed parts. According to the view already explained, the combination producing the one was impossible unless it included the other. The same idea may perhaps furnish the clue to the variability of abnormal organs and widely diffused species, the abnormal organ may with some plausibility be looked upon as the rare combination difficult to effect, and only possible under very special circumstances. There is little difficulty in believing that it would more probably vary with varying circumstances than a simple and ordinary combination. It is easy to produce two common wine-glasses which differ in no apparent manner; two Venice goblets could hardly be blown alike. It is not meant here to predicate ease or difficulty of the action of omnipotence; but just as mechanical laws allow one form to be reproduced with certainty, so the occult laws of reproduction may allow certain simpler combinations to be produced with much greater certainty than the more complex combinations. The variability of widely diffused species might be explained in a similar way. These may be looked on as the simple combinations of which many may exist similar one to the other, whereas the complex combinations may only be possible within comparatively narrow limits, inside which one organ may indeed be variable, though the main combination is the only possible one of its kind.

We by no means wish to assert that we know the above suggestions to be the true explanations of the facts. We merely wish to show that other explanations than those given by Darwin are conceivable, although this is indeed not required by our argument, since, if his main assumptions can be proved false, his theory will derive no benefit from the few facts which may be allowed to be consistent with its truth.

The peculiarities of geographical distribution seem very difficult of explanation on any theory. Darwin calls in alternately winds, tides, birds, beasts, all animated nature, as the diffusers of species, and then a good many of the same agencies as impenetrable barriers. There are some impenetrable barriers between the Galapagos Islands, but not between New Zealand and South America. Continents are created to join Australia and the Cape of Good Hope, while a sea as broad as the British Channel is elsewhere a valid line of demarcation. With these facilities of hypothesis there seems to be no particular reason why many theories should not be true. However an animal may have been produced, it must have been produced somewhere, and it must either have spread very widely, or not have spread, and Darwin can give good reason for both results. If produced according to any law at all, it would seem probable that groups of similar animals would be produced in given places. Or we might suppose that all animals having been created anywhere or everywhere, those have been extinguished which were not suited to such climate; nor would it be an answer to say that the climate, for instance, of Australia, is less suitable now to marsupials than to other animals introduced from Europe, because we may suppose that this was not so when the race began; but in truth it is hard to believe any of the suppositions, nor can we just now invent any better, and this peculiarity of distribution, namely, that all the products of a given continent have a kind of family resemblance, is the sole argument brought forward by Darwin which seems to us to lend any countenance to the theory of a common origin and the transmutation of species.

Our main arguments are now completed. Something might be said as to the alleged imperfection of the geological records. It is certain that, when compared with the total number of animals which have lived, they must be very imperfect; but still we observe that of many species of beings thousands and even millions of specimens have been preserved. If Darwin's theory be true, the number of varieties differing one from another a very little must have been indefinitely great, so great indeed as probably far to exceed the number of individuals which have existed of any one variety. If this be true, it would be more probable that no two specimens preserved as fossils should be of one variety than that we should find a great many specimens collected from a very few varieties, provided, of course, the chances of preservation are equal for all individuals. But this assumption may be denied, and some may think it probable that the conditions favourable to preservation only recur rarely, at remote periods, and never last long enough to show a gradual unbroken change. It would rather seem probable that fragments, at least, of perfect series would be preserved of those beings which lead similar lives favourable to their preservation as fossils. Have any fragments of these Darwinian series been found where the individuals merge from one variety insensibly to another?

It is really strange that vast numbers of perfectly similar specimens should be found, the chances against their perpetuation as fossils are so great; but it is also very strange that the specimens should be so exactly alike as they are, if, in fact, they came and vanished by a gradual change. It is, however, not worth while to insist much on this argument, which by suitable hypotheses might be answered, as by saying, that the changes were often quick, taking only a few myriad ages, and that then a species was permanent for a vastly longer time, and that if we have not anywhere a gradual change clearly recorded, the steps from variety to variety are gradually being diminished as more specimens are discovered. These answers do not seem to us sufficient, but the point is hardly worth contesting, when other arguments directly disproving the possibility of the assumed change have been advanced.

These arguments are cumulative. If it be true that no species can vary beyond defined limits, it matters little whether natural selection would be efficient in producing definite variations. If natural selections, though it does select the stronger average animals, and under peculiar circumstances may develop special organs already useful, can never select new imperfect organs such as are produced in sports, then, even though eternity were granted, and no limit assigned to the possible changes of animals, Darwin's cannot be the true explanation of the manner in which change has been brought about. Lastly, even if no limit be drawn to the possible difference between offspring and their progenitors, and if natural selection were admitted to be an efficient cause capable of building up even new senses, even then, unless time, vast time, be granted, the changes which might have been produced by the gradual selection of peculiar offspring have not really been so produced. Any one of the main pleas of our argument, if established, is fatal to Darwin's theory. What then shall we say if we believe that experiment has shown a sharp limit to the variation of every species, that natural selection is powerless to perpetuate new organs even should they appear, that countless ages of a habitable globe are rigidly proven impossible by the physical laws which forbid the assumption of infinite power in a finite mass? What can we believe but that Darwin's theory is an ingenious and plausible speculation, to which future physiologists will look back with the kind of admiration we bestow on the atoms of Lucretius, or the crystal spheres of Eudoxus, containing like these some faint half-truths, marking at once the ignorance of the age and the ability of the philosopher. Surely the time is past when a theory unsupported by evidence is received as probable, because in our ignorance we know not why it should be false, though we cannot show it to be true. Yet we have heard grave men gravely urge, that because Darwin's theory was the most plausible known, it should be believed. Others seriously allege that it is more consonant with a lofty idea of the Creator's action to suppose that he produced beings by natural selection, rather than by the finikin process of making each separate little race by the exercise of Almighty power. The argument such as it is, means simply that the user of it thinks that this is how he personally would act if possessed of almighty power and knowledge, but his speculations as to his probable feelings and actions, after such a great change of circumstances, are not worth much. If we are told that our experience shows that God works by laws, then we answer, 'Why the special Darwinian law?' A plausible theory should not be accepted while unproven; and if the arguments of this essay be admitted, Darwin's theory of the origin of species is not only without sufficient support from evidence, but is proved false by a cumulative proof.

Comments on Jenkin

Henry Charles Fleeming Jenkin was a professor of engineering at Glasgow and worked with Lord Kelvin in laying the transatlantic cable. His paper represents the best efforts of a physical scientist to evaluate evolutionary theory. The nature and extent of Jenkin's contribution to Darwin's thought have been some matter of dispute. Because of Darwin's remark that Jenkin's arguments concerning variation had convinced him and because of the appearance of Jenkin's review just prior to the extensively revised fifth edition of the *Origin,* some authors have concluded that it was Jenkin who convinced Darwin that natural selection by chance variation was inadequate and caused him to rely more heavily than he had previously on Lamarckian and other mechanisms. Peter Vorzimmer (1963) has shown that this interpretation cannot be correct, since Darwin did not read Jenkin's paper until after it was too late to make any extensive revisions in the fifth edition. In addition, Darwin understood Jenkin's arguments to re-enforce his own beliefs concerning the importance of individual differences rather than refuting them. A fair estimate of Jenkin's contribution to Darwin's views on inheritance is that it was comparable to Malthus's contribution to his conception of natural selection. As Gavin de Beer has shown, Darwin had worked out all aspects of evolution by natural selection before reading Malthus, just as Vorzimmer shows that Darwin had already considered every one of Jenkin's arguments before reading Jenkin. Both Malthus and Jenkin functioned as coagulants. They caused Darwin to bring together various lines of investigation and to see their immediate interrelations more clearly than he had before.

The difficulties which Darwin was having over inheritance were mainly conceptual. Much of the necessary data was available, but for Darwin to have properly understood it would have required him to do the type of ordering and organizing that he did most poorly. To focus on a few isolated cases of a 3 to 1 ratio in the transmission of very special kinds of properties in a peculiar kind of plant and to develop a theoretical mechanism for them was not in Darwin's makeup. He would have been able to bring forth hundreds of apparent counterexamples to such a simple Mendelian mechanism of heredity.

The conceptual difficulties which confronted Darwin in his work on heredity are revealed in the difficulty he had in communicating his views to Wallace. The necessary terminological distinctions had not been made yet. The terms "individual differences," "single variations," and "blending inheritance" were not adequate for the job that had to be done. When naturalists spoke of blending inheritance, sometimes they meant that the observable characters of the organism blended, sometimes that the material basis of heredity (pangenes, gemmules, etc.) blended. The two phenomena were thought to be associated more closely than they actually were.

When the blending of observable characters was discussed, several different phenomena were intended: the intermediate expression of two different parental characters in the offspring (a white-flowered plant producing a descendant with pink flowers when crossed with a red-flowered plant); offspring being a mixture of unblended parental characters (a child could have his father's eyes, hair, and walk but his mother's smile, skin color, and disposition); and the statistical contribution made by a variation within a population or a succession of populations. Variations could be large (the addition of an extra digit) or small (a slightly longer tail or the appearance of slight bony protruberances on the skull of a ruminant). Large variations were frequently called saltations or sports.

The most important distinction, however, and the distinction not seen clearly by any of the participants was between those variations in observable characters which were due to an alteration in the material basis of heredity

and those that had some other cause. The concepts that were missing were
that of genotype and phenotype, reaction norm, and mutation. Two individuals can have the same genotype but differing phenotypes, and vice
versa. Each genotype has a reaction norm within which phenotypic characters can vary, depending on the environment. Mutation, a substantive
change in genetic makeup, results in a new reaction norm. In addition,
there are various mechanisms which result in the same genetic units when
passed on in rearranged states to produce new characters.

For Darwin the important difference was between single variations and
individual differences. By "single variation" Darwin meant those new or
altered characters which appear in a *single* individual in a population.
They could be large or small, but since the noticeable ones were large,
Darwin usually thought of them as being large. By "individual differences"
Darwin meant those characters which occurred frequently and simultaneously in a population. For example, twenty litters of cubs might be
produced in a wolf pack. In each litter, one or two cubs would be larger
than the rest and would tend to have a better chance of surviving, mating,
and passing on their size. Darwin always looked upon these changes as
small. The only cases of large alterations which occurred frequently and
simultaneously throughout a population were reversions, and they were
hardly the stuff of evolution.

What Darwin did not know, of course, was that most of his individual
differences were not due to substantive changes in the hereditary material
but to such mechanisms as recombination. This explains his failure to find
his kind of changes (individual differences) in asexual species. Although
asexual reproduction does allow of some nonmutational variation, it is not
nearly as extensive as that permitted by sexual reproduction.

Jenkin raised two major objections to the unlimited variation required
by evolutionary theory—the swamping effect and the existence of spheres
of limitation for each species. Contrary to Vorzimmer's reading, Jenkin
does not seem to limit his swamping effect argument to single variations,
that is, to saltations or sports. Jenkin began his discussion of the swamping
effect with a reference to "slight advantages" and ends it with reference
to "slight and imperceptible variations." Jenkin's argument is that *on the
supposition of blending inheritance,* even the harshest selection could not
result in the replacement of an established character by the single occurrence
of a new character. For example, a white man might be introduced into
a population of 100 blacks. Assume that the white man kills all the black
males and then mates with all of the black females—sexual selection at

its most extreme. The next generation would be medium brown in color. Now assume that the white man kills all of his sons and former wives and then mates with his daughters. He repeats this process for the second and third generations until he finally dies. Each individual in the first generation would be 50 percent white, in the second 75 percent, the third 86½ per cent, and so on. After his death, the shade of skin then prevalent would remain unaltered until the island received another visitor, whether white or black. Thus, even in the most extreme case, a white man could not turn a nation of blacks white. In less extreme instances, he would have even less effect.

What is right about this example is that *new* genetical material is being introduced into the population. What is wrong with it is the assumption of blending inheritance. The lesson which Jenkin brought home to Darwin was "how rarely single variations whether, slight or strongly-marked, could be perpetuated" (Darwin, 1872, p. 96). Thus, Darwin was forced to rely even more heavily on the simultaneous and recurrent appearance of numerous individual differences, the kinds of differences exhibited by the various members of a single brood or litter. For example, all of the members of a native tribe are not uniformly black. On a scale of 10, the mean shade might be 9, with a single point variation in either direction. Assuming that lighter skin had a slightly higher selective value than darker skin, more of the lighter-skinned individuals would survive to reproduce. In the next generation, a higher percentage of the tribe with skin at the 8 end of the scale would survive than at the 10 end. Hence, the mean shade might now be 8.5. *If the extremes of variation remained one point in either direction,* the lightest members of this generation would be 7.5 and the darkest 9.5. Thus, even on the supposition of blending inheritance, if a character continues to vary in each generation "in all directions" and one extreme is selected for, evolution is sure to occur.

Jenkin seized upon the weakest assumption in the preceding example. A character may vary within limits in one generation, one extreme may be selected for, and the mean would then be shifted in the next generation. It does not follow, however, that the *limits of variation* would thereby be shifted. Jenkin's metaphor was that of a sphere of limitation. If a limit of variation existed for each character, the percentage of individuals in successive generations which possessed the advantageous extreme might increase without resulting in evolution. Like the decreasing velocity of a cannon ball, the ease with which new varieties are produced might decrease as the limits of these various spheres were approached. The experience of horticulturists and animal breeders was on Jenkin's side. Initially, great changes could be produced by selective breeding but gradually decreased until even the most rigorous selection would produce no change. Even so, Darwin felt that the evidence was not irrevocably against him. We *know* why cannon balls slow down. No one knew of a mechanism in heredity analogous to friction which would assure a limit to variation. Thus, Darwin felt justified in retaining his belief in unlimited variation.

As is usually the case in disputes of this kind, both sides were partly right and partly wrong. Jenkin was right in suggesting that the individual differences among the offspring of a single set of parents typically have limits—reaction norms in contemporary terminology. As long as the genetic pool remains unchanged, considerable but not unlimited variation is possible. Darwin was right in thinking that slight variations were the raw material of evolution and that unlimited variation was possible. On occasion a phenotypic variation occurs which looks no different from any of the other individual differences observed among the offspring but which is the result of a mutation, thus resulting in a new reaction norm, a new sphere of limitation. For example, two wolf cubs might be produced of equally large size, one due to the possession of all the genes which its parents possessed that contributed to an increase in size, the other due to no special combination of its parent's genes but to the appearance of a new gene controlling size. Phenotypically, both cubs would be identical but genotypically they would be different. Of course, neither Darwin nor Jenkin saw the dispute in this light.

Both of the preceding arguments and their respective examples assumed blending inheritance. If a white mated with a black, the offspring would be roughly intermediate in skin color. Some might be slightly darker, some slightly lighter, but none completely white or completely black. Both Darwin and Jenkin were aware of characters which were transmitted unchanged and which appeared again unchanged in later generations, but the former were looked upon as anomalous rarities and the latter as reversions. Lest he leave Darwin any avenue of escape, Jenkin also entertained the possibility of an organism's being able to pass on its characters indefinitely and undiluted to its descendants. On such a hypothesis, evolution could occur at a rapid pace. Jenkin's comments on this suggestion presaged the bias and misunderstanding that were to surround the rediscovery of Mendel's laws. "But this theory of the origin of species is surely not the Darwinian theory; it simply amounts to the hypothesis that, from time to time, an animal is born differing appreciably from its progenitors, and possessing the power of transmitting the difference to its descendants. What is this but stating that from time to time, a new species is created?"

Jenkin is assuming that unblending characters are always sports. Evolution by saltation would run counter to evolutionary theory as Darwin set it out, but not to the version proposed by Huxley. On either interpretation, however, the process would not be "creation" as the term was used at the time of the *Origin,* though Owen argued that this is what *he* meant by "creation." In actual fact, however, most of the unblending new characters are small and beautifully fit the needs of Darwin's theory of evolution. Darwin's mistake was in thinking that typical, single differences were the raw material of evolution, when only the rare, genetically based, individual differences were operative.

Darwin does not mention the second half of Jenkin's paper, which deals with the physical objections to evolutionary theory. Numerous physicists attacked evolutionary theory because of its incompatibility with physics as it was then known. Though the desire to reconcile conflicting scientific theories was justified, the arrogance with which physical scientists assumed that when in conflict the biological theories must be at fault was not. Although Jenkin's paper is easily the best of its kind, the genre is none too good. The ignorance of biology and the undisguised contempt of Haughton were typical. Darwin himself was very suspicious of mathematicians and physicists. He advised his fellow naturalists to be slow in admitting the conclusions of astronomers and to be cautious in trusting mathematicians, (*More Letters,* 1:134–135).

Lyell had argued for uniformitarianism in geology. If the processes now observed to effect deposition, denudation, and the like (including floods, volcanic eruptions, and such) operated at the same speed and with the same force in previous periods of time as they were then, a rough estimate of the age of the various strata could be calculated. These geological calculations provided Darwin with the great expanses of time he needed for evolution to occur. For example, he estimated that the denudation of the Weald must have required close to three hundred million years (Darwin, 1859, p. 267), a hundred million years more than Haughton had allowed for all geological time! Although current calculations show that Darwin's estimate was not far off, it came under such sustained and scornful attack that he deleted it from later editions of the *Origin.*

Almost to the man, physicists argued against Lyell's uniformitarianism and Darwin's special applications of it (see, for example, Thomson, 1865, Tait, 1869, and Thomson and Tait, 1867). Both Kelvin and Tait argued that uniformitarianism was an absurdity. The whole heavenly mechanism was running down, tending toward what is now called increased entropy. Specifically, they established the duration of time which Lyell and Darwin had available for their purposes by the underground temperature of the earth, the tidal retardation of the earth's rotation, and the rate of diminution of the sun's heat. At the outside, the earth could not have been habitable for more than ten or fifteen million years, a bit shy of the three hundred million years Darwin postulated for a single denudation. Thus, if the conclusions of physics were to square with the data of geology, the processes resulting in the formation of geological strata and fossils must have operated at a greater speed and intensity in earlier times than they do at present, and they will operate even more slowly in the future.

Lyell, Huxley, and Hooker all replied that physicists should let them be (Huxley, 1869; Hooker, 1868). Naturalists and geologists had their own disciplines, and according to their calculations, the earth was many times older than the physicists thought. Both sides agreed that the findings of physics, geology, and biology would eventually have to be made compatible, but they disagreed that absolute precedence had to be given to the calculations of the physicists. In this instance, geologists and naturalists, as untrained as their logical faculties might be, were correct and the physicists and mathematicians were mistaken.*

* See also Colvin and Ewing (1887).

St. George Jackson Mivart (1827–1900)
Chauncey Wright (1830–1875)

A mathematician, Mr. Wright, has been writing on the geometry of
bee-cells in the United States in consequence of my book; but I can hardly
understand his paper.—C. Darwin to W. H. Miller, Down, December 1,
1859 (*More Letters*, 1:124)

I hope to God there will be nothing disagreeable to you in Vol. II.
[of *The Descent of Man,* concerning sexual selection], and that I have
spoken fairly of your views; I am fearful on this head, because I have
just read (but not with sufficient care) Mivart's book, and I feel *absolutely
certain* that he meant to be fair (but he was stimulated by theological
fervour); yet I do not think he has been quite fair . . . Mivart is savage
or contemptuous about my "moral sense," and so probably will you be.
I am extremely pleased that he agrees with my position, *as far as animal
nature is concerned,* of man in the series; or if anything, thinks I have
erred in making him too distinct.—C. Darwin to A. R. Wallace, Down,
January 30, 1871 (*Life and Letters,* 2:315–316)

I send . . . revised proofs of an article which will be published in the
July number of the 'North American Review,' sending it in the hope that
it will interest or even be of greater value to you. Mr. Mivart's book of
which this article is substantially a review, seems to me a very good background from which to present the considerations which I have endeavoured
to set forth in the article, in defense and illustration of the theory of Natural
Selection. My special purpose has been to contribute to the theory by placing
it in its proper relations to philosophical inquiries in general.—C. Wright
to C. Darwin, June 21, 1871 (*Life and Letters,* 2:323)

I send by this post a review by Chauncey Wright, as I much want
your opinion of it as soon as you can send it. I consider you an incomparably better critic than I am. The article, though not very clearly written, and

poor in parts from want of knowledge, seems to me admirable. Mivart's book is producing a great effect against Natural Selection, and more especially against me. Therefore if you think the article even somewhat good I will write and get permission to publish it as a shilling pamphlet, together with the MS. additions (enclosed), for which there was not room at the end of the review . . .

I am now at work at a new and cheap edition of the 'Origin,' and shall answer several points in Mivart's book, and introduce a new chapter for this purpose; but I treat the subject so much more concretely, and I dare say less philosophically, than Wright, that we shall not interfere with each other. You will think me a bigot when I say, after studying Mivart, I was never before in my life so convinced of the *general* (i.e. not in detail) truth of the views in the 'Origin.' I grieve to see the omission of the words by Mivart, detected by Wright. I complained to Mivart that in two cases he quotes only the commencement of sentences by me, and thus modifies my meaning; but I never supposed he would have omitted words. There are other cases of what I consider unfair treatment. I conclude with sorrow that though he means to be honourable he is so bigoted that he cannot act fairly . . .—C. Darwin to A. R. Wallace, Down, July 9, 1871 (*Life and Letters*, 2:323–324)

There is a most cutting review of me in the 'Quarterly'; I have only read a few pages. The skill and style make me think of Mivart. I shall soon be viewed as the most despicable of men. This 'Quarterly Review' tempts me to republish Ch. Wright, even if not read by any one, just to show some one will say a word against Mivart, and that his (i.e. Mivart's) remarks ought not to be swallowed without some reflection . . . God knows whether my strength and spirit will last out to write a chapter versus Mivart and others; I do so hate controversy and feel I shall do it so badly.—C. Darwin to A. R. Wallace, Down, July 12, 1871 (*Life and Letters*, 2:326)

I have hardly ever in my life read an article which has given me so much satisfaction as the review which you have been so kind as to send me. I agree to almost everything which you say. Your memory must be wonderfully accurate, for you know my works as well as I do myself, and your power of grasping other men's thoughts is something quite surprising; and this, as far as my experience goes, is a very rare quality. As I read on I perceived how you have acquired this power, viz. by thoroughly analyzing each word.—C. Darwin to C. Wright, Down, July 14, 1871 (*Life and Letters*, 2:324–325)

Many thanks for your article in the 'North American Review,' which I have read with great interest. Nothing can be clearer than the way in which you discuss the permanence or fixity of species. It never occurred to me to suppose that any one looked at the case as it seems Mr. Mivart does. Had I read his answer to you, perhaps I should have perceived this; but I have resolved to waste no more time in reading reviews of my works or on Evolution, excepting when I hear that they are good and contain new matter . . . It is pretty clear that Mr. Mivart has come to the end of his tether on this subject.—C. Darwin to C. Wright, Down, June 3, 1872 (*Life and Letters,* 2:342–343)

I am preparing a new and cheap edition of the *Origin,* and shall introduce a new chapter on gradation, and on the uses of initial commencements of useful structures; for this, I observe, has produced the greatest effect on most persons. Every one of his [Mivart's] cases, as it seems to me, can be answered in a fairly satisfactory manner. He is very unfair, and never says what he must have known could be said on my side. He ignores the effect of use, and what I have said in all my later books and editions on the direct effects of the conditions of life and so-called spontaneous variation. I send you by this post a very clever, but ill-written review from N. America by a friend of Asa Gray, which I have republished.

I am glad to hear about Huxley. You never read such strong letters Mivart wrote to me about respect towards me, begging that I would call on him, etc., etc.; yet in the *Q. Review* he shows the greatest scorn and animosity towards me, and with uncommon cleverness says all that is most disagreeable. He makes me the most arrogant, odious beast that ever lived. I cannot understand him; I suppose that accursed religious bigotry is at the root of it. Of course he is quite at liberty to scorn and hate me, but why take such trouble to express something more than friendship? It has mortified me a good deal.—C. Darwin to J. D. Hooker, Down, September 16, 1871 (*More Letters,* 1:332–333)

What a wonderful man you are to grapple with those old metaphisico-divinity books. It quite delights me that you are going to some extent to answer and attack Mivart. His book, as you say, has produced a great effect; yesterday I perceived the reverberations from it, even from Italy. It was this that made me ask Chauncey Wright to publish at my expense his article, which seems to me very clever, though ill-written. He has not knowledge enough to grapple with Mivart in detail. I think there can be no shadow of doubt that he is the author of the article in the 'Quarterly Review.'—C. Darwin to T. H. Huxley, Down, September 21, 1871 (*Life and Letters,* 2:327)

It was very good of you to send the proof-sheets, for I was *very* anxious to read your article. I have been delighted with it. How you do smash

Mivart's theology . . . There are scores of splendid passages, and vivid flashes of wit. I have been a good deal more than merely pleased by the concluding part of your review; and all the more, as I own I felt mortified by the accusation of bigotry, arrogance, &c., in the 'Quarterly Review.' But I assure you, he may write his worst, and he will never mortify me again.—C. Darwin to T. H. Huxley, Down, September 30, 1871 (*Life and Letters*, 2:328–329)

This leads me to the remark that I have always been treated honestly by my reviewers, passing over those without scientific knowledge as not worthy of notice. My views have often been grossly misrepresented, bitterly opposed and ridiculed, but this has been generally done, as I believe, in good faith. I must, however, except Mr. Mivart, who as an American expressed it in a letter has acted towards me "like a pettifogger", or as Huxley has said "like and Old Bailey lawyer."—C. Darwin, *Autobiography*, pp. 125–126

Darwin's Descent of Man*

[ST. GEORGE JACKSON MIVART]

In Mr. Darwin's last work we possess at length a complete and thorough exposition of his matured views. He gives us the results of the patient labour of many years' unremitting investigation and of the application of a powerful and acute intellect, combined with an extraordinarily active imagination, to an unequalled collection of most varied, interesting, and important biological data. In his earlier writings a certain reticence veiled, though it did not hide, his ultimate conclusions as to the origin of our own species; but now all possibility of misunderstanding or of a repetition of former disclaimers on the part of any disciple is at an end, and the entire and naked truth as to the logical consequences of Darwinism is displayed with a frankness which we had a right to expect from the distinguished author. What was but obscurely hinted in the 'Origin of Species' is here fully and fairly stated in all its bearings and without disguise. Mr. Darwin has, in fact, 'crowned the edifice,' and the long looked for and anxiously awaited detailed statement of his views as to the human race is now unreservedly put before us.

We rise from the careful perusal of this book with mingled feelings of admiration and disappointment. The author's style is clear and attractive—clearer than in his earlier works—and his desire to avoid every kind

* From the *Quarterly Review* July 1871, 131:47–90; American edition, 131:25–48.

of conscious misrepresentation is as conspicuous as ever. The number of interesting facts brought forward is as surprising as is the ingenuity often displayed in his manipulation of them. Under these circumstances it is a most painful task to point out grave defects and serious shortcomings. Mr. Darwin, however, seems in his recent work even more than in his earlier productions to challenge criticism, and to have thrown out ideas and suggestions with a distinct view to their subsequent modification by others. It is but an act of fairness to call attention to this:

> False facts, says Mr. Darwin, are highly injurious to the progress of science, for they often endure long; but false views, if supported by some evidence, do little harm, for every one takes a salutary pleasure in proving their falseness; and when this is done, one path toward error is closed and the road to truth is often at the same time opened.—*Descent of Man,* vol. ii. p. 385.

Although we are unable to agree entirely with Mr. Darwin in this remark, it none the less contains an undoubted truth. We cannot agree, because we feel that a false theory which keenly solicits the imagination, put forward by a writer widely and deservedly esteemed, and which reposes on a multitude of facts difficult to verify, skilfully interwoven, and exceedingly hard to unravel, is likely to be very prejudicial to science. Nevertheless, science cannot make progress without the action of two distinct classes of thinkers: the first consisting of men of creative genius, who strike out brilliant hypotheses, and who may be spoken of as 'theorizers' in the good sense of the word; the second, of men possessed of the critical faculty, and who test, mould into shape, perfect or destroy, the hypotheses thrown out by the former class.

Obviously important as it is that there should be such theorizers, it is also most important that criticism should clearly point out when a theory is really proved, when it is but probable, and when it is a mere arbitrary hypothesis. This is all the more necessary if, as may often and very easily happen, from being repeatedly spoken of, and being connected with celebrated, and influential names, it is likely to be taken for very much more than it is really worth.

The necessity of caution in respect to this is clearly shown by Mr. Darwin's present work, in which 'sexual selection,' from being again and again referred to as if it had been proved to be a *vera causa,* may readily be accepted as such by the uninstructed or careless reader. For many persons, at first violently opposed through ignorance or prejudice to Mr. Darwin's

views, are now, with scarcely less ignorance and prejudice, as strongly inclined in their favour.

Mr. Darwin's recent work, supplementing and completing, as it does,
his earlier publications, offers a good opportunity for reviewing his whole
position. We shall thus be better able to estimate the value of his convictions
regarding the special subject of his present inquiry. We shall first call
attention to his earlier statements, in order that we may see whether he
has modified his views, and, if so, how far and with what results. If he
has, even by his own showing and admission, been overhasty and seriously
mistaken previously, we must be the more careful how we commit ourselves
to his guidance now. We shall endeavour to show that Mr. Darwin's convictions have undergone grave modifications, and that the opinions adopted
by him now are quite distinct from, and even subversive of, the views
he originally put forth. The assignment of the law of 'natural selection'
to a subordinate position is virtually an abandonment of the Darwinian
theory; for the one distinguishing feature of that theory was the all-suffi-
ciency of 'natural selection.' Not the less, however, ought we to feel grateful
to Mr. Darwin for bringing forward that theory, and for forcing on men's
minds, by his learning, acuteness, zeal, perseverance, firmness, and candour,
a recognition of the probability, if not more, of evolution and of the certainty of the action of 'natural selection.' For though the 'survival of the
fittest' is a truth which readily presents itself to any one who considers
the subject, and though its converse, the destruction of the least fit, was
recognized thousands of years ago, yet to Mr. Darwin, and (through Mr.
Wallace's reticence) to Mr. Darwin alone, is due the credit of having
first brought it prominently forward and demonstrated its truth in a volume
which will doubtless form a landmark in the domain of zoological science.

We find even in the third edition of his 'Origin of Species' the following
passages:–'Natural selection can act only by taking advantage of slight
successive variations; she can never take a leap, but must advance by
short and slow steps' (p. 214). Again he says:–'If it could be demonstrated
that any complex organ existed, which could not possibly have been formed
by numerous, successive, slight modifications, my theory would absolutely
break down. But I can find out no such case' (p. 208). He adds:–

> Every detail of structure in every living creature (making some little
> allowance for the direct action of physical conditions) may be viewed
> either as having been of special use to some ancestral form, or as being
> now of special use to the descendants of this form—either directly, or
> indirectly though the complex laws of growth, [and] if it could be proved

that any part of the structure of any one species had been formed for
the exclusive good of another species, it would annihilate my theory, for
such could not have been produced through natural selection (p. 220).

It is almost impossible for Mr. Darwin to have used words by which
more thoroughly to stake the whole of his theory on the non-existence
or non-action of causes of any moment other than natural selection. For
why should such a phenomenon 'annihilate his theory'? Because the very
essence of his theory, as originally stated, is to recognize only the conservation of minute variations directly beneficial to the creature presenting them,
by enabling it to obtain food, escape enemies, and propagate its kind.
But once more he says:–

> We have seen that species at any one period are not indefinitely vari-
> able, and are not linked together by a multitude of intermediate grada-
> tions, partly because the process of natural selection will always be very
> slow, and will act, at any one time, only on a very few forms; and
> partly because the very process of natural selection almost implies the
> continual supplanting and extinction of preceding and intermediate
> gradations—P. 223.

Such are Mr. Darwin's earlier statements. At present we read as follows:–

> I now admit, after reading the essay by Nägeli on plants, and the
> remarks by various authors with respect to animals, more especially those
> recently made by Professor Broca, that in the earlier editions of my
> "Origin of Species" I probably attributed too much to the action of
> natural selection or the survival of the fittest. . . . I had not formely
> sufficiently considered the existence of many structures which appear
> to be, as far as we can judge, neither beneficial nor injurious; and this
> I believe to be one of the greatest oversights as yet detected in my
> work.–('Descent of Man,' vol i. p. 152.)

A still more remarkable admission is that in which he says, after referring
to the action of both natural and sexual selection:–

> An unexplained residuum of change, perhaps a large one, must be
> left to the assumed action of those *unknown agencies,* which occasionally
> induce strongly marked and abrupt deviations of structure in our domestic
> productions.—vol. i. p. 154.

But perhaps the most glaring contradiction is presented by the following
passage:–

> No doubt man, as well as every other animal, presents structures, which
> as far as we can judge with our little knowledge, are not now of any
> service to him, nor have been so during any former period of his existence,

either in relation to his general conditions of life, or of one sex to the other. Such structures cannot be accounted for by any form of selection, or by the inherited effects of the use and disuse of parts. We know, however, that many strange and strongly marked peculiarities of structure occasionally appear in our domesticated productions; and if the unknown causes which produce them were to act more uniformly, they would probably become common to all the individuals of the species.—vol. ii. p. 387.

Mr. Darwin, indeed, seems now to admit the existence of internal, innate powers, for he goes on to say:–

We may hope hereafter to understand something about the causes of such occasional modifications, especially through the study of monstrosities. . . . In the greater number of cases we can only say that the cause of each slight variation and of each monstrosity lies much more *in the nature or constitution of the organism*[1] than in the nature of the surrounding conditions; though new and changed conditions certainly play an important part in exciting organic changes of all kinds.

Also, in a note (vol. i. p. 223), he speaks of 'incidental results of certain unknown differences in the constitution of the reproductive system.'

Thus, then, it is admitted by our author that we may have 'abrupt, strongly marked' changes, 'neither beneficial nor injurious' to the creatures possessing them, produced 'by unknown agencies' lying deep in 'the nature or constitution of the organism,' and which, if acting uniformly, would 'probably' modify similarly 'all the individuals of a species.' If this is not an abandonment of 'natural selection,' it would be difficult to select terms more calculated to express it. But Mr. Darwin's admissions of error do not stop here. In the fifth edition of his 'Origin of Species' (p. 104) he says, 'Until reading an able and valuable article in the "North British Review" (1867), I did not appreciate how rarely single variations, whether slight or strongly marked, could be perpetuated.' Again: he was formerly 'inclined to lay much stress on the principle of protection, as accounting for the less bright colours of female birds' ('Descent of Man,' vol. ii. p. 198); but now he speaks as if the correctness of his old conception of such colours being due to protection was unlikely. 'Is it probable,' he asks, 'that the head of the female chaffinch, the crimson on the breast of the female bullfinch,—the green of the female chaffinch,—the crest of the female golden-crested wren, have all been rendered less bright by the slow process of selection for the sake of protection? *I cannot think so*' (vol. ii. p. 176).

1. The italics in the quotations from Mr. Darwin's book in this article are, in almost all cases, our's and not the author's [St. G. J. M.].

Once more Mr. Darwin shows us (vol. i. p. 125) how he has been over-hasty in attributing the development of certain structures to reversion. He remarks, 'in my "Variations of Animals under Domestication" (vol. ii. p. 57) I attributed the not very rare cases of supernumerary mammae in women to reversion.' 'But Professor Preyer states that *mammae erraticae* have been known to occur in other situations, even on the back; so that the force of my argument is greatly weakened or perhaps quite destroyed.'

Finally, we have a postscript at the beginning of the second volume of the 'Descent of Man' which contains an avowal more remarkable than even the passages already cited. He therein declares:—

> I have fallen into a serious and unfortunate error, in relation to the sexual differences of animals, in attempting to explain what seemed to me a singular coincidence in the late period of life at which the necessary variations have arisen in many cases, and the late period at which sexual selection acts. The explanation given is wholly erroneous, as I have discovered by working out an illustration in figures.

While willingly paying a just tribute of esteem to the candour which dictated these several admissions, it would be idle to dissemble, and disingenuous not to declare, the amount of distrust with which such repeated over-hasty conclusions and erroneous calculations inspire us. When their Author comes before us anew, as he now does, with opinions and conclusions still more startling, and calculated in a yet greater degree to disturb convictions reposing upon the general consent of the majority of cultivated minds, we may well pause before we trust ourselves unreservedly to a guidance which thus again and again declares its own reiterated fallibility. Mr. Darwin's conclusions may be correct, but we feel we have now indeed a right to demand that they shall be proved before we assent to them; and that since what Mr. Darwin before declared '*must* be,' he now admits not only to be unnecessary but untrue, we may justly regard with extreme distrust the numerous statements and calculations which, in the 'Descent of Man,' are avowedly recommended by a mere '*may* be.' This is the more necessary, as the Author, starting at first with an avowed hypothesis, constantly asserts it as an undoubted fact, and claims for it, somewhat in the spirit of a theologian, that it should be received as an article of faith. Thus the formidable objection to Mr. Darwin's theory, that the great break in the organic chain between man and his nearest allies, cannot be bridged over by any extinct or living species, is answered simply by an appeal 'to a *belief* in the general principle of evolution' (vol. i. p. 200), or by a confident statement that 'we have *every reason to believe* that breaks in the series are simply the result of many forms having become extinct' (vol. i. 187). So, in like manner, we are assured that 'the early progenitors of man were, *no doubt,* once covered with hair, both sexes having beards; their ears were pointed and capable of movement; and their bodies were provided with a tail, having the proper muscles' (vol. i. p. 206). And, finally, we are told, with a dogmatism little worthy of a philosopher, that, '*unless we wilfully close our eyes,*' we must recognize our parentage (vol. i. p. 213).

These are hard words; and, even at the risk of being accused of wilful blindness, we shall now proceed, with an unbiassed and unprejudiced mind, to examine carefully the arguments upon which Mr. Darwin's theory rests. Must we acknowledge that 'man with all his noble qualities, with sympathy which feels for the most debased, with benevolence which extends not only to other men but to the humblest living creature, with his god-like intellect which has penetrated into the movements and constitution of the solar system,' must we acknowledge that man 'with all these exalted powers' is descended from an Ascidian? Is this a scientific truth resting on scientific evidence, or is it to be classed with the speculations of a bygone age?

With regard to the Origin of Man, Mr. Darwin considers that both 'natural selection' and 'sexual selection' have acted. We need not on the present occasion discuss the action of natural selection; but it will be necessary to consider that of 'sexual selection' at some length. It plays a very important part in the 'descent of man' according to Mr. Darwin's views. He maintains that we owe to it our power of song and our hairlessness of body, and that also to it is due the formation and conservation of the various races and varieties of the human species. In this matter then we fear we shall have to make some demand upon our readers' patience. 'Sexual selection' is the corner-stone of Mr. Darwin's theory. It occupies three-fourths of his two volumes; and unless he has clearly established this point, the whole fabric falls to the ground. It is impossible, therefore, to review the book without entering fully into the subject, even at the risk of touching upon some points which, for obvious reasons, we should have preferred to pass over in silence . . .*

We now come to the consideration of a subject of great importance—namely, that of man's mental powers. Are they, as Mr. Darwin again and again affirms that they are,[2] different only in degree and not in kind from

* [Mivart's discussion of sexual selection has been deleted.]
2. 'There is no fundamental difference between man and the higher mammals in their mental faculties.'—*Descent of Man,* vol. i. p. 35.

the mental powers of brutes? As is so often the case in discussions, the error to be combated is in an implied negation. Mr. Darwin implies and seems to assume that when two things have certain characters in common there can be no fundamental difference between them.

To avoid ambiguity and obscurity, it may be well here to state plainly certain very elementary matters. The ordinary antecedents and concomitants of distinctly felt sensations may exist, with all their physical consequences, in the total absence of intellectual cognizance, as is shown by the well-known fact, that when through fracture of the spine the lower limbs of a man are utterly deprived of the power of feeling, the foot may nevertheless withdraw itself from tickling just as if a sensation was consciously felt. Amongst lower animals, a decapitated frog will join its hind feet together to push away an irritating object just as an uninjured animal will do. Here we have coadjusted actions resulting from stimuli which normally produce sensation, but occurring under conditions in which cerebral action does not take place. Did it take place we should have sensations, but by no means necessarily intellectual action.

'Sensation' is not 'thought,' and no amount of the former would constitute the most rudimentary condition of the latter, though sensations supply the conditions for the existence of 'thought' or 'knowledge.'

Altogether, we may clearly distinguish at least six kinds of action to which the nervous system ministers:—

I. That in which impressions received result in appropriate movements without the intervention of sensation or thought, as in the cases of injury above given. (This is the reflex action of the nervous system.)

II. That in which stimuli from without result in sensations through the agency of which their due effects are wrought out. (Sensation.)

III. That in which impressions received result in sensations which give rise to the observation of sensible objects.—Sensible perception.

IV. That in which sensations and perceptions continue to coalesce, agglutinate, and combine in more or less complex aggregations according to the laws of the association of sensible perceptions.—Association.

The above four groups contain only indeliberate operations, consisting, as they do at the best, but of mere *presentative* sensible ideas in no way implying any reflective or *representative* faculty. Such actions minister to and form *Instinct*. Besides these, we may distinguish two other kinds of mental action, namely:—

V. That in which sensations and sensible perceptions are reflected on by thought and recognised as our own and we ourselves recognised by ourselves as affected and perceiving.—Self-consciousness.

VI. That in which we reflect upon our sensations or perceptions, and ask what they are and why they are.—Reason.

These two latter kinds of action are deliberate operations, performed, as they are, by means of representative ideas implying the use of a *reflective representative* faculty. Such actions distinguish the *intellect* or rational faculty. Now, we assert that possession in perfection of all the first four (*presentative*) kinds of action by no means implies the possession of the last two (*representative*) kinds. All persons, we think, must admit the truth of the following proposition:–

Two faculties are distinct, not in degree but in *kind*, if we may possess the one in perfection without that fact implying that we possess the other also. Still more will this be the case if the two faculties tend to increase in an inverse ratio. Yet this is the distinction between the *instinctive* and the *intellectual* parts of man's nature.

As to animals, we fully admit that they may possess all the first four groups of actions—that they may have, so to speak, mental images of sensible objects combined in all degrees of complexity, as governed by the laws of association. We deny to them, on the other hand, the possession of the last two kinds of mental action. We deny them, that is, the power of reflecting on their own existence or of enquiring into the nature of objects and their causes. We deny that they know that they know or know themselves in knowing. In other words, we deny them *reason*. The possession of the presentative faculty, as above explained, in no way implies that of the reflective faculty; nor does any amount of direct operation imply the power of asking the reflective question before mentioned, as to 'what' and 'why.'

According to our definition, then, given above, the faculties of men and those of other animals differ in kind; and brutes low in the scale supply us with a good example in support of this distinctness; for it is in animals generally admitted to be wanting in reason—such as insects (*e.g.* the ant and the bee)—that we have the very summit and perfection of instinct made known to us.

We will shortly examine Mr. Darwin's arguments, and see if he can bring forward a single instance of brute action implying the existence in it of the representative reflective power. Before doing so, however, one or two points as to the conditions of the controversy must be noticed.

In the first place, the position which we maintain is the one in possession—that which is commended to us by our intuitions, by ethical considerations, and by religious teaching universally. The *onus probandi* should surely therefore rest with him who, attacking the accepted position, maintains the essential similarity and fundamental identity of powers the effects of which are so glaringly diverse. Yet Mr. Darwin quietly assumes the whole point in dispute, by asserting identity of *intuition* where there is identity of *sensation* (vol. i. p. 36), which, of course, implies that there is no mental power whatever except sensation. For if the existence of another faculty were allowed by him, it is plain that the action of that other faculty might modify the effects of mere sensation in any being possessed of such additional faculty.

Secondly, it must be remembered that it is a law in all reasoning that where known causes are sufficient to account for any phenomena we shall not gratuitously call in additional causes. If, as we believe to be the case, there is no need whatever to call in the *representative* faculty as an explanation of brute mental action;—if the phenomena brutes exhibit can be accounted for by the *presentative* faculty—that is, by the presence of sensible perceptions and emotions together with the reflex and co-ordinating powers of the nervous system;—then to ascribe to them the possession of reason is thoroughly gratuitous.

Thirdly, in addition to the argument that brutes have not intellect because their actions can be accounted for without the exercise of that faculty, we have other and positive arguments in opposition to Mr. Darwin's view of their mental powers. These arguments are based upon the absence in brutes of articulate and rational speech, of true concerted action and of educability, in the human sense of the word. We have besides, what may be called an experimental proof in the same direction. For if the germs of a rational nature existed in brutes, such germs would certainly ere this have so developed as to have produced unmistakeably rational phenomena, considering the prodigious lapse of time passed since the entombment of the earliest known fossils. To this question we will return later.

We shall perhaps be met by the assertion that many men may also be taken to be irrational animals, so little do the phenomena they exhibit exceed in dignity and importance the phenomena presented by certain brutes. But, in reply, it is to be remarked that we can only consider men who are truly men—not idiots, and that all *men*, however degraded their social condition, have self-consciousness properly so called, possess the gift of articulate and rational speech, are capable of true concerted action, and have a perception of the existence of right and wrong. On the other hand, no brute has the faculty of articulate, rational speech: most persons will also admit that brutes are not capable of truly concerted action, and we contend most confidently that they have no self-consciousness properly so called, and no perception of the difference between truth and falsehood and right and wrong.

Let us now consider Mr. Darwin's facts in favour of an opposite conclusion.

1st. His testimony drawn from his own experience and information regarding the lowest races of men.

2nd. The anecdotes he narrates in favour of the intelligence of brutes.

In the first place, we have to thank our author for very distinct and unqualified statements as to the substantial unity of men's mental powers. Thus he tells us:–

> The Fuegians rank amongst the lowest barbarians; but I was continually struck with surprise how closely the three natives on board H.M.S. "Beagle," who had lived some years in England, and could talk a little English, resembled us in disposition, and in most of our mental qualities.—vol. i. p. 34.

Again he adds:–

> The American aborigines, Negroes and Europeans differ as much from each other in mind as any three races that can be named; yet I was incessantly struck, whilst living with the Fuegians on board the "Beagle," with the many little traits of character, showing how similar their minds were to ours; and so it was with a full-blooded negro with whom I happened once to be intimate.—vol. i. p. 232.

Again:–

> Differences of this kind (mental) between the highest men of the highest races and the lowest savages, are connected by the finest gradations (vol. i. p. 35).

Mr. Darwin then, plainly tells us that all the essential mental characters of civilised man are found in the very lowest races of men, though in a less completely developed state; while, in comparing their mental powers with those of brutes, he says, 'No doubt the difference in this respect is enormous' (vol. i. p. 34). As if, however, to diminish the force of this admission, he remarks, what no one would dream of disputing, that there are psychical phenomena common to men and to other animals. He says of man that

> He uses in common with the lower animals inarticulate cries to express his meaning, aided by gestures and the movements of the muscles of

the face. This especially holds good with the more simple and vivid feelings, which are *but little connected with the higher intelligence.* Our cries of pain, fear, surprise, anger, together with their appropriate actions, and the murmur of a mother to her beloved child, are more expressive than any words.—vol. i. p. 54.

But, inasmuch as it is admitted on all hands that man *is* an animal, and therefore has all the four lower faculties enumerated in our list, as well as the two higher ones, the fact that he makes use of common instinctive actions in no way diminishes the force of the distinction between him and brutes as regards the representative, reflective faculties. It rather follows as a matter of course from his animality that he should manifest phenomena common to him and to brutes. That man has a common nature with them is perfectly compatible with his having, besides, a superior nature and faculties of which no brute has any rudiment or vestige. Indeed, all the arguments and objections in Mr. Darwin's second chapter may be met by the fact that man, being an animal, has corresponding faculties, whence arises a certain external conformity with other animals as to the modes of expressing some mental modifications. In the overlooking of this possibility of coexistence of two natures lies that error of negation to which we before alluded. Here, as in other parts of the book, we may say there are two quantities a and $a + x$, and Mr. Darwin, seeing the two a's but neglecting the x, represents the quantities as equal.

We will now notice the anecdotes narrated by Mr. Darwin in support of the rationality of brutes. Before doing so, however, we must remark that our author's statements, given on the authority (sometimes secondhand authority) of others, afford little evidence of careful criticism. This is the more noteworthy when we consider the conscientious care and pains which he bestows on all the phenomena which he examines himself.

Thus for example, we are told on the authority of Brehm that—

An eagle seized a young cercopithecus, which clinging to a branch, was not at once carried off; it cried loudly for assistance, upon which other members of the troop, with much uproar rushed to the rescue, surrounded the eagle, and pulled out so many feathers that he no longer thought of his prey, but only how to escape.—vol. i. p. 76.

We confess we wish that Mr. Darwin had himself witnessed this episode. Perhaps, however, he has seen other facts sufficiently similar to render this one credible. In the absence of really trustworthy evidence we should, however, be inclined to doubt the fact of a young cercopithecus, unexpectedly seized, being able, by clinging, to resist the action of an eagle's wings.

, We are surprised that Mr. Darwin should have accepted the following tale without suspicion:–

> One female baboon had so capacious a heart that she not only adopted young monkeys of other species, but stole young dogs and cats which she continually carried about. Her kindness, however, did not go so far as to share her food with her adopted offspring, at which Brehm was surprised, as his monkeys always divided everything quite fairly with their own young ones. An adopted kitten scratched the above-mention affectionate baboon, *who certainly had a fine intellect,* for she was much astonished at being scratched, and immediately examined the kitten's feet, and without more ado bit off the claws. (!!)—vol. i. p. 41.

Has Mr. Darwin ever tested this alleged fact? Would it be possible for a baboon to bite off the claws of a kitten without keeping the feet perfectly straight?

Again we have an anecdote on only second-hand authority (namely a quotation by Brehm or Schimper) to the following effect:–

> In Abyssinia, when the baboons belonging to one species (*C. gelada*) descend in troops from the mountains to plunder the fields, they some-times encounter troops of another species (*C. hamadryas*), and then a fight ensues. The Geladas roll down great stones, which the Hamadryas try to avoid, and then both species, making a great uproar, rush furiously against each other. Brehm, when accompanying the Duke of Coburg-Gotha, aided in an attack with fire-arms on a troop of baboons in the pass of Mensa in Abyssinia. The baboons in return rolled so many stones down the mountain, some as large as a man's head, that the attackers had to beat a hasty retreat; and the pass was actually for a time closed against the caravan. It deserves notice that these baboons thus acted in concert.—vol. i. p. 51.

Now, if every statement of fact here given be absolutely correct, it in no way even tends to invalidate the distinction we have drawn between 'instinct' and 'reason'; but the positive assertion that the brutes 'acted in concert,' when the evidence proves nothing more than that their actions were simultaneous, shows a strong bias on the part of the narrator. A flock of sheep will simultaneously turn round and stare and stamp at an intruder; but this is not 'concerted action,' which means that actions are not only simultaneous, but are so in consequence of a reciprocal understanding and convention between the various agents. It may be added that if any brutes were capable of such really *concerted* action the effects would soon make themselves known to us so forcibly as to prevent the possibility of mistake.

We now come to Mr. Darwin's instances of brute rationality. In the first place he tells us: –

> I had a dog who was savage and averse to all strangers, and I purposely tried his memory after an absence of five years and two days. I went near the stable where he lived, and shouted to him in my old manner; he showed no joy, but instantly followed me out walking and obeyed me exactly as if I had parted with him only half an hour before. A train of old associations, dormant during five years, had thus been instantaneously awakened in his mind.—vol. i. p. 45.

No doubt! but this is not 'reason.' Indeed we could hardly have a better instance of the mere action of associated sensible impressions. What is there here which implies more than memory, impressions of sensible objects and their association? Had there been reason there would have been signs of joy and wonder, though such signs would not alone prove reason to exist. It is evident that Mr. Darwin's own mode of explanation is the sufficient one—namely, by a train of associated sensible impressions. Mr. Darwin surely cannot think that there is in this case any evidence of the dog's having put to himself those questions, which, under the circumstances, a rational being would put. Mr. Darwin also tells us how a monkey-trainer gave up in despair the education of monkeys, of which the attention was easily distracted from his teaching, while 'a monkey which carefully attended to him could always be trained.' But 'attention' does not imply 'reason.' The anecdote only shows that some monkeys are more easily impressed and more retentive of impressions than others.

Again, we are told, as an instance of *reason,* that 'Rengger sometimes put a live wasp in paper so that the monkeys in hastily unfolding it got stung; after this had once happened, they always first held the packet to their ears to detect any movement within.' But here again we have no need to call in the aid of 'reason.' The monkeys had had the group of sensations 'folded paper' associated with the other groups—'noise and movement' and 'stung fingers.' The second time they experience the group of sensations 'folded paper,' the succeeding sensations (in this instance only too keenly associated) are forcibly recalled, and with the recollection of the sensation of hearing, the hand goes to the ear. Yet Mr. Darwin considers this unimportant instance of such significance that he goes on to say: –

> Any one who is not convinced by such facts as these, and by what he may observe with his own dogs, that animals can reason, would not

be convinced by anything I could add. Nevertheless, I will give one
case with respect to dogs, as it rests on two distinct observers, and can
hardly depend on the modification of any instinct. Mr. Colquhoun
winged two wild ducks, which fell on the opposite side of a stream;
his retriever tried to bring over both at once, but could not succeed;
she then, though never before known to ruffle a feather, deliberately
killed one, brought over the other, and returned for the dead bird.
Colonel Hutchinson relates that two partridges were shot at once, one
being killed and the other wounded; the latter ran away, and was caught
by the retriever, who on her return came across the dead bird; she
stopped, evidently greatly puzzled, and after one or two trials, finding
that she could not take it up without permitting the escape of the winged
bird, she considered a moment, then deliberately murdered it by giving
it a severe crunch, and afterwards brought away both together. This
was the only known instance of her having wilfully injured any game.

Mr. Darwin adds:

> Here we have reason, though not quite perfect, for the retriever might
> have brought the wounded bird first and then returned for the dead
> one, as in the case of the two wild ducks.—vol. i. pp. 47, 48.

Here we reply we have nothing of the kind, and to bring 'reason' into
play is gratuitous. The circumstances can be perfectly explained (and on
Mr. Darwin's own principles) as evidences of the revival of an old instinct.
The ancestors of sporting dogs of course killed their prey, and that trained
dogs do not kill it is simply due to man's action, which has suppressed
the instinct by education, and which continually thus keeps it under control.
It is indubitable that the old tendency *must* be latent, and that a small
interruption in the normal retrieving process, such as occurred in the cases
cited, would probably be sufficient to revive that old tendency and call
the obsolete habit into exercise.

But perhaps the most surprising instance of groundless inference is presented in the following passage:—

> My dog, a full grown and very sensible animal, was lying on the
> lawn during a hot and still day; but at a little distance a slight breeze
> occasionally moved an open parasol, which would have been wholly dis-
> regarded by the dog, had any one stood near it. As it was, every time
> that the parasol slightly moved, the dog growled fiercely and barked.
> He must, I think, have reasoned to himself in a rapid and unconscious
> manner, that movement without any apparent cause indicated the pres-
> ence of some strange living agent, and no stranger had a right to be
> on his territory.—vol. i. p. 67.

The consequences deduced from this trivial incident are amazing. Probably, however, Mr. Darwin does not mean what he says; but, on the face of it, we have a brute credited with the abstract ideas 'movement,' 'causation,' and the notions logically arranged and classified in subordinate genera—'agent,' 'living agent,' 'strange living agent.' He also attributes to it the notion of 'a right' of 'territorial limitation,' and the relation of such 'limited territory' and 'personal ownership.' It may safely be affirmed that if a dog could so reason in one instance he would in others, and would give much more unequivocal proofs for Mr. Darwin to bring forward.

Mr. Darwin, however, speaks of reasoning in an 'unconscious manner,' so that he cannot really mean any process of reasoning at all but, if so, his case is in no way apposite. Even an insect can be startled, and will exhibit as much evidence of rationality as is afforded by the growl of a dog; and all that is really necessary to explain such a phenomenon exists in an oyster, or even in the much talked-of Ascidian.

Thus, then, it appears that, even in Mr. Darwin's especially-selected instances, there is not a tittle of evidence tending, however slightly, to show that any brute possesses the representative reflective faculties. But if, as we assert, brute animals are destitute of such higher faculties, it may well be that those lower faculties which they have (and which we more or less share with them) are highly developed, and their senses possess a degree of keenness and quickness inconceivable to us. Their[3] minds being entirely occupied with such lower faculties, and having, so to speak, nothing else to occupy them, their sensible impressions may become interwoven and connected to a far greater extent than in us. Indeed, in the absence of free will, the laws of this association of ideas obtain supreme command over the minds of brutes: the brute being entirely immersed, as it were, in his presentative faculties.

There yet remain two matters for consideration, which tend to prove the fundamental difference which exists between the mental powers of man and brutes: 1. The mental equality between animals of very different grades of structure, and their non-progressiveness. 2. The question of articulate speech.

Considering the vast antiquity of the great animal groups,[4] it is, indeed,

3. The words 'mind,' 'mental,' 'intelligence,' &c., are here made use of in reference to the psychical faculties of brutes, in conformity to popular usage, and not as strictly appropriate.

4. Mr. Darwin (vol. i. p. 360) refers to Dr. Scudder's discovery of 'a fossil insect in the Devonian formation of New Brunswick, furnished with the well-known tympanum or stridulating apparatus of the male Locustidae.'

remarkable how little advance in mental capacity has been achieved even by the highest brutes. This is made especially evident by Mr. Darwin's own assertions as to the capacities of lowly animals. Thus he tells us that—

Mr. Gardner, whilst watching a shore-crab (Gelasimus) making its burrow, threw some shells toward the hole. One rolled in, and three other shells remained within a few inches of the mouth. In about five minutes the crab brought out the shell which had fallen in, and carried it away to the distance of a foot; it then saw the three other shells lying near, and *evidently thinking* that they might likewise roll in, carried them to the spot where it had laid the first.—vol. i. p. 334.

Mr. Darwin adds or quotes the astonishing remark, 'It would, I think, be difficult to distinguish this act from one performed by man by the aid of reason.' Again, he tells us:—

Mr. Lonsdale informs me that he placed a pair of land-shells (*Helix pomatia*), one of which was weakly, into a small and ill-provided garden. After a short time the strong and healthy individual disappeared, and was traced by its track of slime over a wall into an adjoining well-stocked garden. Mr. Lonsdale concluded that it had deserted its sickly mate; but after an absence of twenty-four hours it returned, and apparently communicated the result of its successful exploration, for both then started along the same track and disappeared over the wall.—vol. i. p. 326.

Whatever may be the real value of the statements quoted, they harmonize with a matter which is incontestable. We refer to the fact that the intelligence of brutes, be they high or be they low, is essentially one in kind, there being a singular parity between animals belonging to groups widely different in type of structure and in degree of development.

Apart from the small modifications which experience occasionally introduces into the habits of animals—as sometimes occurs after man has begun to frequent a newly-discovered island—it cannot be denied that, looking broadly over the whole animal kingdom, there is no evidence of advance in mental power on the part of brutes. This absence of progression in animal intelligence is a very important consideration, and it is one which does not seem to be adverted to by Mr. Darwin, though the facts detailed by him are exceedingly suggestive of it.

When we speak of this absence of progression we do not, of course, mean to deny that the dog is superior in mental activity to the fish, or the jackdaw to the toad. But we mean that, considering the vast period of time that must (on Mr. Darwin's theory) have elapsed for the evolution of an Orang from an Ascidian, and considering how beneficial increased intelligence must be to all in the struggle for life, it is inconceivable (on

Mr. Darwin's principles only) that a mental advance should not have taken place greater in degree, more generally diffused, and more in proportion to the grade of the various animals than we find to be actually the case. For in what respect is the intelligence of the ape superior to that of the dog or of the elephant? It cannot be said that there is one point in which its psychical nature approximates to man more than that of those four-footed beasts. But, again, where is the great superiority of a dog or an ape over a bird? The falcon trained to hawking is at least as remarkable an instance of the power of education as the trained dog. The tricks which birds can be taught to perform are as complex and wonderful as those acted by the mammal. The phenomena of nidification, and some of those now brought forward by Mr. Darwin as to courtship, are fully comparable with analogous phenomena of quasi-intelligence in any beast.

This, however, is but a small part of the argument. For let us descend to the invertebrata, and what do we find?—a restriction of their quasi-mental faculties proportioned to their constantly inferior type of structure? By no means. We find, e.g., in ants, phenomena which simulate those of an intelligence such as ours far more than do any phenomena exhibited by the highest beasts. Ants display a complete and complex political organization, classes of beings socially distinct, war resulting in the capture of slaves, and the appropriation and maintenance of domestic animals (Aphides) analogous to our milk-giving cattle.

Mr. Darwin truthfully remarks on the great difference in these respects between such creatures as ants and bees, and singularly inert members of the same class—such as the scale insect or coccus. But can it be pretended that the action of natural and sexual selection has alone produced these phenomena in certain insects, and failed to produce them in any other mere animals even of the very highest class? If these phenomena are due to a power and faculty similar in kind to human intelligence, and which power is latent and capable of evolution in all animals, then it is certain that this power must have been evolved in other instances also, and that we should see varying degrees of it in many, and notably in the highest brutes as well as in man. If, on the other hand, the faculties of brutes are different in kind from human intelligence, there can be no reason whatever why animals most closely approaching man in physical structure should resemble him in psychical nature also.

This reflection leads us to the difference which exists between men and brutes as regards the faculty of articulate speech. Mr. Darwin remarks that of the distinctively human characters this has 'justly been considered as one of the chief' (vol. i. p. 53). We cannot agree in this. Some brutes can articulate, and it is quite conceivable that brutes might (though as a fact they do not) so associate certain sensations and gratifications with certain articulate sounds as, in a certain sense, to speak. This, however, would in no way even tend to bridge over the gulf which exists between the representative reflective faculties and the merely presentative ones. Articulate signs of sensible impressions would be fundamentally as distinct as mere gestures are from truly rational speech.

Mr. Darwin evades the question about language by in one place (vol. i. p. 54) attributing that faculty in man to his having acquired a higher intellectual nature; and in another (vol. ii. p. 391), by ascribing his higher intellectual nature to his having acquired that faculty.

Our author's attempts to bridge over the chasm which separates instinctive cries from rational speech are remarkable examples of groundless speculation. Thus he ventures to say–

> That primeval man, or rather some early progenitor of man, *probably* (the italics are ours) used his voice largely, as does one of the gibbon-apes at the present day, in producing true musical cadences, that is in singing; we may conclude from a widely-spread analogy that this power would have been especially exerted during the courship of the sexes, serving to express various emotions, as love, jealousy, triumph, and serving as a challenge to their rivals. The imitation by articulate sounds of musical cries *might* have given rise to words expressive of various complex emotions.

And again:

> It does not appear *altogether incredible,* that some unusually wise ape-like animal should have thought of imitating the growl of a beast of prey, so as to indicate to his fellow monkeys the nature of the expected danger. And this would have been a first step in the formation of a language.—vol. i. p. 56.

But the question is, not whether it is incredible, but whether there are any data whatever to warrant such a supposition. Mr. Darwin brings forward none: we suspect none could be brought forward.

It is not, however, emotional expressions or manifestations of sensible impressions, in whatever way exhibited, which have to be accounted for, but the enunciation of distinct deliberate judgments as to 'the what,' 'the how,' and 'the why,' by definite articulate sounds; and for these Mr. Darwin not only does not account but he does not adduce anything even tending to account for them. Altogether we may fairly conclude, from the complete failure of Mr. Darwin to establish identity of kind between the mental faculties of man and of brutes, that identity cannot be established; as we are not likely for many years to meet with a naturalist so competent to collect and marshal facts in support of such identity, if any such facts there are. The old barrier, then, between 'presentative instinct' and 'representative reason' remains still unimpaired, and, as we believe, insurmountable.

We now pass to another question, which is of even greater consequence than that of man's intellectual powers. Mr. Darwin does not hesitate to declare that even the 'moral sense' is a mere result of the development of brutal instincts. He maintains, 'the first foundation or origin of the moral sense lies in the social instincts, including sympathy; and these instincts no doubt were primarily gained, as in the case of the lower animals, through natural selection' (vol. ii. p. 394).

Everything, however, depends upon what we mean by the 'moral sense.' It is a patent fact that there does exist a perception of the qualities 'right' and 'wrong' attaching to certain actions. However arising, men have a consciousness of an absolute and immutable rule *legitimately* claiming obedience with an authority necessarily supreme and absolute—in other words, intellectual judgments are formed which imply the existence of an ethical ideal in the judging mind.

It is the existence of this power which has to be accounted for; neither its application nor even its validity have to be considered. Yet instances of difference of opinion respecting the moral value of particular concrete actions are often brought forward as if they could disprove the *existence* of moral intuition. Such instances are utterly beside the question. It is amply sufficient for our purpose if it be conceded that developed reason dictates to us that certain modes of action, abstractedly considered, are intrinsically wrong; and this we believe to be indisputable.

It is equally beside the question to show that the existence of mutually beneficial acts and of altruistic habits can be explained by 'natural selection.' No amount of benevolent habits tend even in the remotest degree to account for the intellectual perception of 'right' and 'duty.' Such habits may make the doing of beneficial acts pleasant, and their omission painful' but such feelings have essentially nothing whatever to do with the perception of 'right' and 'wrong,' nor will the faintest incipient stage of the perception be accounted for by the strongest development of such sympathetic feelings. Liking to do acts which happen to be good, is one thing; seeing that actions are good, whether we or others like them or not, is quite another.

Mr. Darwin's account of the moral sense is very different from the above. It may be expressed most briefly by saying that it is the prevalence of more enduring instincts over less persistent ones—the former being social instincts, the latter personal ones. He tells us:—

> As man cannot prevent old impressions continually repassing through his mind, he will be compelled to compare the weaker impressions of, for instance, past hunger, or of vengeance satisfied or danger avoided at the cost of other men, with the instinct of sympathy and goodwill to his fellows, which is still present and ever in some degree active in his mind. He will then feel in his imagination that a stronger instinct has yielded to one which now seems comparatively weak; and then that sense of dissatisfaction will inevitably be felt with which man is endowed, like every other animal in order that his instincts may be obeyed.—vol. i. p. 90.

Mr. Darwin means by 'the moral sense' an instinct, and adds, truly enough, that 'the very essence of an instinct is, that it is followed independently of reason' (vol. i. p. 100). But the very essence of moral action is that it is *not* followed independently of reason.

Having stated our wide divergence from Mr. Darwin with respect to what the term 'moral sense' denotes, we might be dispensed from criticising instances which must from our point of view be irrelevant, as Mr. Darwin would probably admit. Nevertheless, let us examine a few of these instances, and see if we can discover in them any justification of the views he propounds.

As illustrations of the development of self-reproach for the neglect of some good action, he observes:—

> A young pointer when it first scents game, apparently cannot help pointing. A squirrel in a cage who pats the nuts which it cannot eat, as if to bury them in the ground, can hardly be thought to act thus either from pleasure or pain. Hence the common assumption that men must be impelled to every action by experiencing some pleasure or pain may be erroneous. Although a habit may be blindly and implicitly followed, independently of any pleasure or pain felt at the moment, yet if it be forcibly and abruptly checked, a vague sense of dissatisfaction is generally experienced; and this is especially true in regard to persons of feeble intellect.— vol. i. p. 80.

Now, passing over the question whether in the 'pointing' and 'patting' referred to there may not be some agreeable sensations, we contend that such instincts have nothing to do with 'morality,' from their blind nature, such blindness simply *ipso facto* eliminating every vestige of morality from an action.

Mr. Darwin certainly exaggerates the force and extent of social sympathetic feelings. Mr. Mill admits that they are 'often wanting;' but Mr. Darwin claims the conscious possession of such feelings for all, and quotes Hume as saying that the view of the happiness of others 'communicates a secret joy,' while the appearance of their misery 'throws a melancholy damp over the imagination.'[5] One might wish that this remark were universally true, but unfortunately some men take pleasure in the pain of others; and Larochefoucauld even ventured on the now well-known saying, 'that there is something in the misfortunes of our best friends not unpleasant to us.' But our feeling that the sufferings of others are pleasant or unpleasant has nothing to do with the question, which refers to the *judgment* whether the indulging of such feelings is 'right' or 'wrong.'

If the 'social instinct' were the real basis of the moral sense, the fact that society approved of anything would be recognised as the supreme sanction of it. Not only, however, is this not so, not only do we judge as to whether society in certain cases is right or wrong, but we demand a reason why we should obey society at all; we demand a rational basis and justification for social claims, if we happen to have a somewhat inquiring turn of mind. We shall be sure avowedly or secretly to despise and neglect the performance of acts which we do not happen to desire, and which have not an intellectual sanction.

The only passage in which our author seems as if about to meet the real question at issue is very disappointing, as the difficulty is merely evaded. He remarks, 'I am aware that some persons maintain that actions performed impulsively do not come under the dominion of the moral sense, and cannot be called moral' (vol. i. p. 87). This is not a correct statement of the intuitive view, and the difficulty is evaded thus: 'But it appears scarcely possible to draw any clear line of distinction of this kind, though the distinction may be real!' It seems to us, however, that there is no difficulty at all in drawing a line between a judgment as to an action being right or wrong and every other kind of mental act. Mr. Darwin goes on to say:—

Moreover, an action repeatedly performed by us, will at last be done without deliberation or hesitation, and can then hardly be distinguished from an instinct; yet surely no one will pretend that an action thus done ceases to be moral. On the contrary, we all feel that an act cannot

5. [David Hume] 'Enquiry concerning the Principles of Morals,' 1751 ed., p. 132.

be considered as perfect, or as performed in the most noble manner, unless it is done impulsively, without deliberation or effort, in the same manner as by a man in whom the requisite qualities are innate.—vol. i. p. 88.

To this must be replied, in one sense, 'Yes;' in another, 'No.' An action which has ceased to be directly or indirectly deliberate has ceased to be moral as a distinct act, but it is moral as the continuation of those preceding deliberate acts through which the good habit was originally formed, and the rapidity with which the will is directed in the case supposed may indicate the number and constancy of antecedent meritorious volitions. Mr. Darwin seems to see this more or less, as he adds: 'He who is forced to overcome his fear or want of sympathy before he acts, deserves, however, in one way higher credit than the man whose innate disposition leads him to a good act without effort.'

As an illustration of the genesis of remorse, we have the case

of a temporary though for the time strongly persistent instinct conquering another instinct which is usually dominant over all others. [Swallows] at the proper season seem all day long to be impressed with the desire to migrate; their habits change; they become restless, are noisy, and congregate in flocks. Whilst the mother-bird is feeding or brooding over her nestlings, the maternal instinct is probably stronger than the migratory; but the instinct which is more persistent gains the victory, and at last, at a moment when her young ones are not in sight, she takes flight and deserts them. When arrived at the end of her long journey, and the migratory instinct ceases to act, what an agony of remorse each bird would feel, if, from being endowed with great mental activity, she could not prevent the image continually passing before her mind of her young ones perishing in the bleak north from cold and hunger. vol. i. p. 90.

Let us suppose she does suffer 'agony,' that feeling would be nothing to the purpose. What is requisite is that she shall judge that she *ought not* to have left them. To make clear our point, let us imagine a man formerly entangled in ties of affection which in justice to another his conscience has induced him to sever. The image of distress his act of severance has caused may occasion him keen emotional suffering for years, accompanied by a clear perception that his act has been right. Again, let us suppose another case: The struggling father of a family becomes aware that the property on which he lives really belongs to another, and he relinquishes it. He may continue to judge that he has done a proper action, whilst tortured by the trials in which his act of justice has involved him. To assert that these acts are merely instinctive would be absurdly false. In the cases supposed, obedience is paid to a clear intellectual perception and against the very strongest instincts.

That we have not misrepresented Mr. Darwin's exposition of 'conscience' is manifest. He says that if a man has gratified a passing instinct, to the neglect of an enduring instinct, he 'will then feel dissatisfied with himself, and will resolve with more or less force to act differently for the future. This is conscience; for conscience looks backwards and judges past actions, inducing that kind of dissatisfaction, which if weak we call regret, and if severe remorse' (vol. i. p. 91). 'Conscience' certainly 'looks back and judges,' but not all that 'looks back and judges' is 'conscience.' A judgment of conscience is one of a particular kind, namely a judgment according to the standard of moral worth. But for this, a *gourmand,* looking back and judging that a particular sauce had occasioned him dyspepsia, would, in the dissatisfaction arising from his having eaten the wrong dish at dinner, exercise his conscience!

Indeed, elsewhere (vol. i. p. 103) Mr. Darwin speaks of 'the standard of morality rising higher and higher,' though he nowhere explains what he means either by the 'standard' or by the 'higher;' and, indeed, it is very difficult to understand what can possibly be meant by this 'rising of the standard,' if the 'standard' is from first to last pleasure and profit.

We find, again, the singular remark:–'If any desire or instinct leading to an action opposed to the good of others, still appears to a man, when recalled to mind, as strong as or stronger than his social instinct, he will feel no keen regret at having followed it' (vol. i. p. 92).

Mr. Darwin is continually mistaking a merely beneficial action for a moral one; but, as before said, it is one thing to *act well* and quite another to be a moral agent. A dog or even a fruit-tree may act well, but neither is a moral agent. Of course, all the instances he brings forward with regard to animals are not in point, on account of this misconception of the problem to be solved. He gives, however, some examples which tell strongly against his own view. Thus, he remarks of the *Law of Honour*—'The breach of this law, even when the breach is known to be strictly accordant with true morality, has caused many a man more agony than a real crime. We recognise the same influence in the sense of burning shame which most of us have felt, even after the interval of years, when calling to mind some accidental breach of a trifling, though fixed, rule of etiquette' (vol. i. p 92). This is most true; some trifling breach of good manners may indeed occasion us pain; but this may be unaccompanied by a judgment that we are morally blameworthy. It is judgment, and not feeling, which has to do with right and wrong. But a yet better example might be given. What quality can have been more universally useful to social communities than courage? It has always been, and is still, greatly admired and highly appreciated, and is especially adapted, both directly and indirectly, to enable its possessors to become the fathers of succeeding generations. If the social instinct were the basis of the moral sense, it is infallibly certain that courage must have come to be regarded as supremely 'good,' and cowardice to be deserving of the deepest moral condemnation. And yet what is the fact? A coward feels probably self-contempt and that he has incurred the contempt of his associates, but he does not feel 'wicked.' He is painfully conscious of his defective organization, but he knows that an organization, however defective, cannot, in itself, constitute moral demerit. Similarly, we, the observers, despise, avoid, or hate a coward; but we can clearly understand that a coward may be a more virtuous man than another who abounds in animal courage.

The better still to show how completely distinct are the conceptions 'enduring or strong instincts' and 'virtuous desires' on the one hand, and 'transient or weak impulses' and 'vicious inclinations' on the other, let us substitute in the following passage for the words which Mr. Darwin, on his own principles, illegitimately introduces, others which accord with those principles, and we shall see how such substitution eliminates every element of morality from the passage:

> Looking to future generations, there is no cause to fear that the social instincts will grow weaker, and we may expect that enduring (virtuous) habits will grow stronger, becoming perhaps fixed by inheritance. In this case the struggle between our stronger (higher) and weaker (lower) impulses will be less severe, and the strong (virtue) will be triumphant (vol. i. p. 104).

As to past generations, Mr. Darwin tells us (vol. i. p. 166) that at all times throughout the world tribes have supplanted other tribes; and as social acts are an element in their success, sociality must have been intensified, and this because 'an increase in the number of well-endowed men will certainly give an immense advantage to one tribe over another.' No doubt! but this only explains an augmentation of mutually beneficial actions. It does not in the least even tend to explain how the moral judgment was first formed.

Having thus examined Mr. Darwin's theory of Sexual Selection, and his comparison of the mental powers of man (including their moral application) with those of the lower animals, we have a few remarks to make upon his mode of conducting his argument.

In the first place we must repeat what we have already said as to his singular dogmatism, and in the second place we must complain of the way in which he positively affirms again and again the existence of the very things which have to be proved. Thus, to take for instance the theory of the descent of man from some inferior form, he says:–'the grounds upon which this conclusion rests *will never be shaken*' (vol. ii. p. 385), and 'the possession of exalted mental powers is *no* insuperable objection to this conclusion' (vol. i. p. 107). Speaking of sympathy, he boldly remarks,–'this instinct *no doubt* was originally acquired like all the other social instincts through natural selection' (vol. i. p. 164); and 'the fundamental social instincts *were* originally thus gained' (vol. i. p. 173).

Again, as to the stridulating organs of insects, he says:–'No one who admits the agency of natural selection, will dispute that these musical instruments have been acquired through sexual selection.' Speaking of the peculiarities of humming-birds and pigeons, Mr. Darwin observes, 'the *sole* difference between these cases is, that in one the result is due to man's selection, whilst in the other, as with humming-birds, birds of paradise, &c., it is due to sexual selection,–that is, to the selection by the females of the more beautiful males' (vol. ii. p. 78). Of birds, the males of which are brilliant, but the hens are only slightly so, he remarks: 'these cases *are almost certainly* due to characters primarily acquired by the male, having been transferred, in a greater or less degree, to the female' (vol. ii. p. 128). 'The colours of the males may *safely* be attributed to sexual selection' (vol. ii. p. 194). As to certain species of birds in which the males alone are black, we are told, 'there can *hardly be a doubt,* that blackness in these cases has been a sexually selected character' (vol. ii. p. 226). The following, again, is far too positive a statement:–'Other characters proper to the males of the lower animals, such as bright colours, and various ornaments, *have been* acquired by the more attractive males having been preferred by the females. There are, however, exceptional cases, in which the males, instead of having been selected, *have been* the selectors' (vol. ii. p. 371).

It is very rarely that Mr. Darwin fails in courtesy to his opponents; and we were therefore surprised at the tone of the following passage (vol. ii. p. 386):–'He who is not content to look, *like a savage,* at the phenomena of nature as disconnected, *cannot* any longer believe that man is the work of a separate act of creation. He will be *forced* to admit' the contrary.

What justifies Mr. Darwin in his assumption that to suppose the soul of man to have been especially created, is to regard the phenomena of nature as disconnected?

In connexion with this assumption of superiority on Mr. Darwin's part, we may notice another matter of less importance, but which tends to produce the same effect on the minds of his readers. We allude to the terms of panegyric with which he introduces the names or opinions of every disciple of evolutionism, while writers of equal eminence, who have not adopted Mr. Darwin's views, are quoted, for the most part, without any commendation. Thus we read of our 'great anatomist and philosopher, Prof. Huxley,'–of 'our great philosopher, Herbert Spencer,'–of 'the remarkable work of Mr. Galton,'–of 'the admirable treatises of Sir Charles Lyell and Sir John Lubbock,'–and so on. We do not grudge these gentlemen such honorific mention, which some of them well deserve, but the repetition produces an unpleasant effect; and we venture to question the good taste on Mr. Darwin's part, in thus speaking of the adherents to his own views, when we do not remember, for example, a word of praise bestowed upon Prof. Owen in the numerous quotations which our author has made from his works.

Secondly, as an instance of Mr. Darwin's practice of begging the question at issue, we may quote the following assertion:–'any animal whatever, endowed with well-marked social *instincts,* would inevitably acquire a moral sense or *conscience,* as soon as its intellectual powers had become as well developed, or nearly as well developed, as in man' (vol. i. p. 71). This is either a monstrous assumption or a mere truism; it is a truism, for of course, any creature with the intellect of a man would perceive the qualities men's intellect is capable of perceiving, and, amongst them—moral worth.

Mr. Darwin in a passage before quoted (vol. i. p. 86) slips in the whole of absolute morality, by employing the phrase 'appreciation of justice.' Again (vol. i. p. 168), when he speaks of aiding the needy, he remarks:–'Nor could we check our sympathy, if so urged by hard reason, without deterioration in the *noblest* part of our nature.' How noblest? According to Mr. Darwin, a virtuous instinct is a strong and permanent one. There can be, according to his views, no other elements of quality than intensity and duration. Mr. Darwin, in fact, thus silently and unconsciously introduces the moral element into his 'social instinct,' and then, of course, has no difficulty in finding in this latter what he had previously put there. This, however, is quite illegitimate, as he makes the social instinct synonymous with the gregariousness of brutes. In such gregariousness, how-ever, there is no moral element, because the mental powers of brutes are not equal to forming reflective, deliberate, representative judgments.

The word 'social' is ambiguous, as gregarious animals may metaphorically be called social, and man's social relations may be regarded both bene-ficentially and morally. Having first used 'social' in the former sense, it is subsequently applied in the latter; and it is thus that the really moral conception is silently and illegitimately introduced.

We may now sum up our judgment of Mr. Darwin's work on the 'Descent of Man'—of its execution and tendency, of what it fails to accomplish, and of what it has successfully attained.

Although the style of the work is, as we have said, fascinating, nevertheless we think that the author is somewhat encumbered with the multitude of his facts, which at times he seems hardly able to group and handle so effectively as might be expected from his special talent. Nor does he appear to have maturely reflected over the data he has so industriously collected. Moreover, we are surprised to find so accurate an observer receiving as facts many statements of a very questionable nature, as we have already pointed out, and frequently on second-hand authority. The reasoning also is inconclusive, the author having allowed himself constantly to be carried away by the warmth and fertility of his imagination. In fact, Mr. Darwin's power of reasoning seems to be in an inverse ratio to his power of observa-tion. He now strangely exaggerates the action of 'sexual selection,' as previ-ously he exaggerated the effects of the 'survival of the fittest.' On the whole, we are convinced that by the present work the cause of 'natural selection' has been rather injured than promoted; and we confess to a feeling of surprise that the case put before us is not stronger, since we had anticipated the production of far more telling and significant details from Mr. Darwin's biological treasurehouse.

A great part of the work may be dismissed as beside the point—as a mere elaborate and profuse statement of the obvious fact, which no one denies, that man is an animal, and has all the essential properties of a highly organised one. Along with this truth, however, we find the assump-tion that he is *no more* than an animal—an assumption which is necessarily implied in Mr. Darwin's distinct assertion that there is no difference of *kind,* but merely one of *degree,* between man's mental faculties and those of brutes.

We have endeavoured to show that this is distinctly untrue. We maintain that while there is no need to abandon the received position that man is truly an animal, he is yet the only rational one known to us, and that his rationality constitutes a fundamental distinction—one of *kind* and not of *degree.* The estimate we have formed of man's position differs therefore most widely from that of Mr. Darwin.

Mr. Darwin's remarks, before referred to . . . concerning the difference between the instincts of the coccus (or scale insect) and those of the ant— and the bearing of that difference on their zoological position (as both are members of the class insecta) and on that of man—exhibit clearly his misapprehension as to the true significance of man's mental powers.

For in the first place zoological classification is morphological. That is to say, it is a classification based upon form and structure—upon the number and shape of the several parts of animals, and not at all upon what those parts *do,* the consideration of which belongs to physiology. This being the case we not only may, but *should,* in the field of zoology, neglect all questions of diversities of instinct or mental power, equally with every other power, as is evidenced by the location of the bat and the porpoise in the same class, mammalia, and the parrot and the tortoise in the same larger group, Sauropsida.

Looking, therefore, at man with regard to his bodily structure, we not only may, but *should,* reckon him as a member of the class mammalia, and even (we believe) consider him as the representative of a mere family of the first order of that class. But all men are not zoologists; and even zoologists must, outside their science, consider man in his totality and not merely from the point of view of anatomy.

If then we are right in our confident assertion that man's mental faculties are different *in kind* from those of brutes, and if he is, as we maintain, the only rational animal; then is man, as a whole, to be spoken of by preference from the point of view of his animality, or from the point of view of his rationality? Surely from the latter, and, if so, we must consider not structure, but action.

Now Mr. Darwin seems to concede[6] that a difference in kind *would* justify the placing of man in a distinct kingdom, inasmuch as he says a difference in degree does not so justify; and we have no hesitation in affirming (with Mr. Darwin) that between the instinctive powers of the coccus and the ant there *is* but a difference of degree, and that, therefore, they do belong to the same kingdom; but we contend it is quite otherwise with man. Mr. Darwin doubtless admits that all the wonderful actions of ants are mere modifications of instinct. But if it were not so—if the

6. *Descent of Man,* vol. i. p. 186.

piercing of tunnels beneath rivers, &c., were evidence of their possession of reason, then far from agreeing with Mr. Darwin, we should say that ants also are rational animals, and that, while considered from the anatomical stand-point they would be insects, from that of their rationality they would rank together with man in a kingdom apart of 'rational animals.' Really, however, there is no tittle of evidence that ants possess the reflective, self-conscious, deliberate faculty; while the perfection of their instincts is a most powerful argument against the need of attributing a rudiment of rationality to any brute whatever.

We seem then to have Mr. Darwin on our side when we affirm that animals possessed of mental faculties distinct in kind should be placed in a kingdom apart. And man possesses such a distinction.

Is this, however, all that can be said for the dignity of his position? Is he merely one division of the visible universe co-ordinate with the animal, vegetable, and mineral kingdoms?

It would be so if he were intelligent and no more. If he could observe the facts of his own existence, investigate the co-existences and successions of phenomena, but all the time remain like the other parts of the visible universe a mere floating unit in the stream of time, incapable of one act of free self-determination or one voluntary moral aspiration after an ideal of absolute goodness. This, however, is far from being the case. Man is not merely an intellectual animal, but he is also a free moral agent, and, as such—and with the infinite future such freedom opens out before him— differs from all the rest of the visible universe by a distinction so profound that none of those which separate other visible beings is comparable with it. The gulf which lies between his being as a whole, and that of the highest brute, marks off vastly more than a mere kingdom of material beings; and man, so considered, differs far more from an elephant or a gorilla than do these from the dust of the earth on which they tread.

Thus, then, in our judgment the author of the 'Descent of Man' has utterly failed in the only part of his work which is really important. Mr. Darwin's errors are mainly due to a radically false metaphysical system in which he seems (like so many other physicists) to have become entangled. Without a sound philosophical basis, however, no satisfactory scientific superstructure can ever be reared; and if Mr. Darwin's failure should lead to an increase of philosophic culture on the part of physicists, we may therein find some consolation for the injurious effects which his work is likely to produce on too many of our half-educated classes. We sincerely trust Mr. Darwin may yet live to furnish us with another work, which, while enriching physical science, shall not, with needless opposition, set at naught the first principles of both philosophy and religion.

The Genesis of Species*

CHAUNCEY WRIGHT

It is now nearly twelve years since the discussion of that "mystery of mysteries," the origin of species, was reopened by the publication of the first edition of Mr. Darwin's most remarkable work. Again and again in the history of scientific debate this question had been discussed, and, after exciting a short-lived interest, had been condemned by cautious and conservative thinkers to the limbo of insoluble problems or to the realm of religious mystery. They had, therefore, sufficient grounds, a priori, for anticipating that a similar fate would attend this new revival of the question, and that, in a few years, no more would be heard of the matter; that the same condemnation awaited this movement which had overwhelmed the venturesome speculations of Lamarck and of the author of the "Vestiges of Creation." This not unnatural anticipation has been, however, most signally disappointed. Every year has increased the interest felt in the question, and at the present moment the list of publications which we place at the head of this article testifies to the firm hold which the subject has acquired in this short period on the speculative interests of all inquisitive minds. But what can we say has really been accomplished by this debate; and what reasons have we for believing that the judgment of conservative thinkers will not, in the main, be proved right after all, though present indications are against them? One permanent consequence, at least, will remain, in the great additions to our knowledge of natural history, and of general physiology, or theoretical biology, which the discussion has produced; though the greater part of this positive contribution to science. is still to be credited directly to Mr. Darwin's works, and even to his original researches. But, besides this, an advantage has been gained which cannot be too highly estimated. Orthodoxy has been won over to the doctrine of evolution. In asserting this result, however, we are obliged to make what will appear to many persons important qualifications and explanations. We do not mean that the heads of leading religious bodies, even in the most enlightened communities, are yet willing to withdraw the dogma that

---

* From *The North American Review* (July 1871), 113:63–103; reprinted as *Darwinism: Being an Examination of Mr. St. George Mivart's 'Genesis of Species'* (1871). London, John Murray, 46 pp, with an appendix on final causes.

the origin of species is a special religious mystery, or even to assent to
the hypothesis of evolution as a legitimate question for scientific inquiry.
We mean only, that many eminent students of science, who claim to be
orthodox, and who are certainly actuated as much by a spirit of reverence
as by scientific inquisitiveness, have found means of reconciling the general
doctrine of evolution with the dogmas they regard as essential to religion.
Even to those whose interest in the question is mainly scientific this result
is a welcome one, as opening the way for a freer discussion of subordinate
questions, less trammelled by the religious prejudices which have so often
been serious obstacles to the progress of scientific researches.

But again, in congratulating ourselves on this result, we are obliged to
limit it to the doctrine of evolution in its most general form, the theory
common to Lamarck's zoological philosophy, to the views of the author
of the "Vestiges of Creation," to the general conclusions of Mr. Darwin's
and Mr. Wallace's theory of Natural Selection, to Mr. Spencer's general
doctrine of evolution, and to a number of minor explanations of the processes by which races of animals and plants have been derived by descent
from different ancestral forms. What is no longer regarded with suspicion
as secretly hostile to religious beliefs by many truly religious thinkers is
that which is denoted in common by the various names "transmutation,"
"development," "derivation," "evolution," and "descent with modification."
These terms are synonymous in their primary and general signification,
but refer secondarily to various hypotheses of the processes of derivation.
But there is a choice among them on historical grounds, and with reference
to associations, which are of some importance from a theological point
of view. "Transmutation" and "development" are under ban. "Derivation"
is, perhaps, the most innocent word; though "evolution" will probably
prevail, since, spite of its etymological implication, it has lately become
most acceptable, not only to the theological critics of the theory, but to
its scientific advocates; although, from the neutral ground of experimental
science, "descent with modification" is the most pertinent and least excep-
tionable name.

While the general doctrine of evolution has thus been successfully re-
deemed from theological condemnation, this is not yet true of the subordinate hypothesis of Natural Selection, to the partial success of which this
change of opinion is, in great measure, due. It is, at first sight, a paradox
that the views most peculiar to the eminent naturalist, whose work has
been chiefly instrumental in effecting this change of opinion, should still
be rejected or regarded with suspicion by those who have nevertheless

been led by him to adopt the general hypothesis,–an hypothesis which his explanations have done so much to render credible. It would seem, at first sight, that Mr. Darwin has won a victory, not for himself, but for Lamarck. Transmutation, it would seem, has been accepted, but Natural Selection, its explanation, is still rejected by many converts to the general theory, both on religious and scientific grounds. But too much weight might easily be attributed to the deductive or explanatory part of the evidence, on which the doctrine of evolution has come to rest. In the half-century preceding the publication of the "Origin of Species," inductive evidence on the subject has accumulated, greatly outweighing all that was previously known; and the "Origin of Species" is not less remarkable as a compend and discussion of this evidence than for the ingenuity of its explanations. It is not, therefore, to what is now known as "Darwinism" that the prevalence of the doctrine of evolution is to be attributed, or only indirectly. Still, most of this effect is due to Mr. Darwin's work, and something undoubtedly to the indirect influence of reasonings that are regarded with distrust by those who accept their conclusions; for opinions are contagious, even where their reasons are resisted.

The most effective general criticism of the theory of Natural Selection which has yet appeared, or one which, at least, is likely to exert the greatest influence in overcoming the remaining prejudice against the general doctrine of evolution, is the work of Mr. St. George Mivart "On the Genesis of Species." Though, as we shall show in the course of this article, the work falls far short of what we might have expected from an author of Mr. Mivart's attainments as a naturalist, yet his position before the religious world, and his unquestionable familiarity with the theological bearings of his subject, will undoubtedly gain for him and for the doctrine of evolution a hearing and a credit, which the mere student of science might be denied. His work is mainly a critique of "Darwinism"; that is, of the theories peculiar to Mr. Darwin and the "Darwinians," as distinguished from the believers in the general doctrine of evolution which our author accepts. He also puts forward an hypothesis in opposition to Mr. Darwin's doctrine of the predominant influence of Natural Selection in the generation of organic species, and their relation to the conditions of their existence. On this hypothesis, called "Specific Genesis," an organism, though at any one time a fixed and determinate species, approximately adapted to surrounding conditions of existence, is potentially, and by innate potential combinations of organs and faculties, adapted to many other conditions of existence. It passes, according to the hypothesis, from one form to another of specific "manifestation," abruptly and discontinuously in conformity to the emergencies of its outward life; but in any condition to which it is tolerably adapted it retains a stable form, subject to variation only within determinate limits, like oscillations in a stable equilibrium. For this conception our author is indebted to Mr. Galton, who, in his work on "Hereditary Genius," "compares the development of species with a many-faceted spheroid tumbling over from one facet or stable equilibrium to another. The existence of internal conditions in animals," Mr. Mivart adds (p. 111), "corresponding with such facets is denied by pure Darwinians, but it is contended in this work that something may also be said for their existence." There are many facts of variation, numerous cases of abrupt changes in individuals both of natural and domesticated species, which, of course, no Darwinian or physiologist denies, and of which Natural Selection professes to offer no direct explanation. The causes of these phenomena, and their relations to external conditions of existence, are matters quite independent of the principle of Natural Selection, except so far as they may directly affect the animal's or plant's well-being, with the origin of which this principle is alone concerned. General physiology has classified some of these sudden variations under such names as "reversion" and "atavism," or returns more or less complete to ancestral forms. Others have been connected together under the law of "correlated or concomitant variations," changes that, when they take place, though not known to be physically dependent on each other, yet usually or often occur together. Some cases of this law have been referred to the higher, more fundamental laws of homological variations, or variations occurring together on account of the relationships of homology, or due to similarities and physical relations between parts of organisms, in tissues, organic connections, and modes of growth. Other variations are explained by the laws and causes that determine monstrous growths. Others again are quite inexplicable as yet, or cannot yet be referred to any general law or any known antecedents. These comprise, indeed, the most common cases. The almost universal prevalence of well-marked phenomena of variation in˜ species, the absolutely universal fact that no two individual organisms are exactly alike, and that the description of a species is necessarily abstract and in many respects by means of averages,–these facts have received no particular explanations, and might indeed be taken as ultimate facts or highest laws in themselves, were it not that in biological speculations such an assumption would be likely to be misunderstood, as denying the existence of any real determining causes and more ultimate laws, as well as denying any known antecedents or regularities

in such phenomena. No physical naturalist would for a moment be liable
to such a misunderstanding, but would, on the contrary, be more likely
to be off his guard against the possibility of it in minds otherwise trained
and habituated to a different kind of studies. Mr. Darwin has undoubtedly
erred in this respect. He has not in his works repeated with sufficient frequency his faith in the universality of the law of causation, in the phenomena of general physiology or theoretical biology, as well as in all the
rest of physical nature. He has not said often enough, it would appear,
that in referring any effect to "accident," he only means that its causes
are like particular phases of the weather, or like innumerable phenomena
in the concrete course of nature generally, which are quite beyond the
power of finite minds to anticipate or to account for in detail, though
none the less really determinate or due to regular causes. That he has
committed this error appears from the fact that his critic, Mr. Mivart,
has made the mistake, which nullifies nearly the whole of his criticism,
of supposing that "the theory of Natural Selection may (though it need
not) be taken in such a way as to lead men to regard the present organic
world as formed, so to speak, *accidentally,* beautiful and wonderful as
is confessedly the hap-hazard result" (p. 33). Mr. Mivart, like many another
writer, seems to forget the age of the world in which he lives and for
which he writes,–the age of "experimental philosophy," the very stand-point
of which, its fundamental assumption, is the universality of physical causation. This is so familiar to minds bred in physical studies, that they rarely
imagine that they may be mistaken for disciples of Democritus, or for
believers in "the fortuitous concourse of atoms," in the sense, at least,
which theology has attached to this phrase. If they assent to the truth
that may have been meant by the phrase, they would not for a moment
suppose that the atoms move fortuitously, but only that their conjunctions,
constituting the actual concrete orders of events, could not be anticipated
except by a knowledge of the natures and regular histories of each and
all of them,–such knowledge as belongs only to omniscience. The very hope
of experimental philosophy, its expectation of constructing the sciences into
a true philosophy of nature, is based on the induction, or, if you please,
the *a priori* presumption, that physical causation is universal; that the
constitution of nature is written in its actual manifestations, and needs
only to be deciphered by experimental and inductive research; that it is
not a latent invisible writing, to be brought out by the magic of mental
anticipation or metaphysical meditation. Or, as Bacon said, it is not by
the "anticipations of the mind," but by the "interpretation of nature,"

that natural philosophy is to be constituted; and this is to presume that
the order of nature is decipherable, or that causation is everywhere either
manifest or hidden, but never absent.

Mr. Mivart does not wholly reject the process of Natural Selection, or
disallow it as a real cause in nature, but he reduces it to "a subordinate
rôle" in his view of the derivation of species. It serves to perfect the imperfect adaptations and to meet within certain limits unfavorable changes
in the conditions of existence. The "accidents" which Natural Selection
acts upon are allowed to serve in a subordinate capacity and in subjection
to a foreordained, particular, divine order, or to act like other agencies
dependent on an evil principle, which are compelled to turn evil into good.
Indeed, the only difference on purely scientific grounds, and irrespective
of theological considerations, between Mr. Mivart's views and Mr. Darwin's
is in regard to the *extent* to which the process of Natural Selection has
been effective in the modifications of species. Mr. Darwin himself, from
the very nature of the process, has never supposed for it, as a cause, any
other than a co-ordinate place among other causes of change, though he
attributes to it a superintendent, directive, and controlling agency among
them. The student of the theory would gather quite a different impression
of the theory from Mr. Mivart's account of it, which attributes to "Darwinians" the absurd conception of this cause as acting "alone" to produce
the changes and stabilities of species; whereas, from the very nature of
the process, other causes of change, whether of a known or as yet unknown
nature, are presupposed by it. Even Mr. Galton's and our author's hypothetical "facets," or internal conditions of abrupt changes and successions
of stable equilibriums, might be among these causes, if there were any
good inductive grounds for supposing their existence. Reversional and correlated variations are, indeed, due to such internal conditions and to laws
of inheritance, which have been ascertained inductively as at least laws
of phenomena, but of which the causes, or the antecedent conditions in
the organism, are unknown. Mr. Darwin continually refers to variations
as arising from unknown causes, but these are always such, so far as observation can determine their relations to the organism's conditions of existence,
that they are far from accounting for, or bearing any relations to, the
adaptive characters of the organism. It is solely upon and with reference
to such adaptive characters that the process of Natural Selection has any
agency, or could be supposed to be effective. If Mr. Mivart had cited
anywhere in his book, as he has not, even a single instance of sudden
variation in a whole race, either in a state of nature or under domestication,

which is not referable by known physiological laws to the past history of the race on the theory of evolution, and had further shown that such a variation was an adaptive one, he might have weakened the arguments for the agency and extent of the process of Natural Selection. As it is, he has left them quite intact.

The only direct proofs which he adduces for his theory that adaptive as well as other combinations proceed from innate predeterminations wholly within the organism, are drawn from, or rather assumed in, a supposed analogy of the specific forms in organisms to those of crystals. As under different circumstances or in different media the same chemical substances or constituent substances assume different and distinct crystalline forms, so, he supposes, organisms are distinct manifestations of typical forms, one after another of which will appear under various external conditions. He quotes from Mr. J. J. Murphy, "Habit and Intelligence," that, "it needs no proof that in the case of spheres and crystals, the forms and structures are the effect and not the cause of the formative principle. Attraction, whether gravitative or capillary, produces the spherical form; the spherical form does not produce attraction. And crystalline polarities produce crystalline structure and form; crystalline structure and form do not produce polarities." And, by analogy, Mr. Murphy and our author infer that innate vital forces always produce specific vital forms, and that the vital forms themselves, or "accidental" variations of them, cannot modify the types of action in vital force. Now, although Mr. Murphy's propositions may need no proof, they will bear correction; and, clear as they appear to be, a better interpretation of the physical facts is needed for the purposes of tracing out analogy and avoiding paralogism. Strange as it may seem, Mr. Murphy's clear antitheses are not even partially true. No abstraction ever produced any other abstraction, much less a concrete thing. The abstract laws of attraction never produced any body, spherical or polyhedral. It was actual forces acting in definite ways that made the sphere or crystal; and the sizes, particular shapes, and positions of these bodies determined in part the action of these actual forces. It is the resultants of many actual attractions, dependent in turn on the actual products, that determine the spherical or crystalline forms. Moreover, in the case of crystals, neither these forces nor the abstract law of their action in producing definite angles reside in the finished bodies, but in the properties of the surrounding media, portions of whose constituents are changed into crystals, according to these properties and to other conditioning circumstances. So far as these bodies have any innate principle in them concerned in their own production, it is manifested in determining, not their general agreements but their particular differences in sizes, shapes, and positions. The particular position of a crystal that grows from some fixed base or nucleus, and the particular directions of its faces, may, perhaps, be said to be *innate;* that is, they were determined at the beginning of the particular crystal's growth. Finding, therefore, what Mr. Murphy and Mr. Mivart suppose to be innate to be really in the outward conditions of the crystal's growth, and what they would suppose to be superinduced to be all that is innate in it, we have really found the contrast in place of an analogy between a crystal and an organism. For, in organisms, no doubt, and as we may be readily convinced without resort to analogy, there is a great deal that is really innate, or dependent on actions in the organism, which diversities of external conditions modify very little, or affect at least in a very indeterminate manner, so far as observation has yet ascertained. External conditions are, nevertheless, essential factors in development, as well as in mere increase of growth. No animal or plant is developed, nor do its developments acquire any growth without very special external conditions. These are quite as essential to the production of an organism as a crystalline nucleus and fluid material are to the growth and particular form of a crystal; and as the general resemblances of the crystals of any species, the agreements in their angles, are results of the physical properties of their food and other surrounding conditions of their growth, so the general resemblances of animals or plants of any species, their agreements in specific characters, are doubtless due, in the main, to the properties of what is innate in them, yet not to any abstraction. This is sufficiently conspicuous not to "need any proof," and is denied by no Darwinian. The analogy is so close indeed between the internal determinations of growth in an organism and the external ones of crystals, that Mr. Darwin was led by it to invent his "provisional hypothesis of Pangenesis," or theory of gemmular reproduction. The gemmules in this theory being the perfect analogues of the hypothetical atoms of the chemical substances that are supposed to arrange themselves in crystalline forms, the theory rather gives probability to the chemical theory of atoms than borrows any from it. But we shall recur to this theory of Pangenesis further on.

General physiology, or physical and theoretical biology, are sciences in which, through the study of the laws of inheritance, and the direct and indirect effect of external conditions, we must arrive, if in any way, at a more and more definite knowledge of the causes of specific manifestations; and this is what Mr. Darwin's labors have undertaken to do, and have partially accomplished. Every step he has taken has been in strict conformity to the principles of method which the examples of inductive and experimental science have established. A stricter observance of these by Mr. Murphy and our author might have saved them from the mistake we have noticed, and from many others,–the "realism" of ascribing efficacy to an abstraction, making attraction and polarity produce structures and forms independently of the products and of the concrete matters and forces in them. A similar "realism" vitiates nearly all speculations in theoretical biology, which are not designedly, or even instinctively, as in Mr. Darwin's work, made to conform to the rigorous rules of experimental philosophy. These require us to assume no causes that are not true or phenomenally known, and known in some other way than in the effect to be explained; and to prove the sufficiency of those we do assume in some other way than by putting an abstract name or description of an effect for its cause, like using the words "attraction" and "polarity" to account for things the matters of which have *come together* in a *definite form*. It may seem strange to many readers to be told that Mr. Darwin, the most consummate speculative genius of our times, is no more a maker of hypotheses than Newton was, who, unable to discover the cause of the properties of gravitation, wrote the often-quoted but much misunderstood words, *"Hypotheses non fingo."* "For," he adds, "whatever is not deduced from the phenomena is to be called an hypothesis; and hypotheses, whether metaphysical or physical, whether of occult qualities or mechanical, have no place in experimental philosophy. In this philosophy particular propositions are inferred from the phenomena, and afterwards rendered general by induction. Thus it was that the impenetrability, the mobility, and the impulsive force of bodies, and the laws of motion and gravitation, were discovered. And to us it is enough that gravity does really exist and act according to the laws which we have explained, and abundantly serves to account for all the motions of the celestial bodies and of our sea." Thus, also, it is that the variability of organisms and the known laws of variation and inheritance, and of the influences of external conditions, and the law of Natural Selection, have been discovered. And though it is not enough that variability and selection do really exist and act according to laws which Mr. Darwin has explained (since the limits of their action and efficiency are still to be ascertained), yet it is enough for the present that Darwinians do not rest, like their opponents, contented with framing what Newton would have called, if he had lived after Kant, *"transcendental hypotheses,"* which have no place in experimental philosophy. It may be said that Mr. Darwin has invented the hypothesis of Pangenesis, against the rules of this philosophy; but so also did Newton invent the corpuscular theory of light, with a similar purpose and utility.

In determining the limits of the action of Natural Selection, and its sufficiency within these limits, the same demonstrative adequacy should not, for obvious reasons, be demanded as conditions of assenting to its highly probable truth, and Newton proved for his speculation. For the facts for this investigation are hopelessly wanting. Astronomy presents the anomaly, among the physical sciences, of being the only science that deals in the concrete with a few naturally isolated causes, which are separated from all other lines of causation in a way that in other physical sciences can only be imitated in the carefully guarded experiments of physical and chemical laboratories. The study of animals and plants under domestication is, indeed, a similar mode of isolating with a view to ascertaining the physical laws of life by inductive investigations. But the theory of Natural Selection, in its actual application to the phenomena of life and the origin of species, should not be compared to the theory of gravitation in astronomy, nor to the principles of physical science as they appear in the natures that are shut in by the experimental resources of the laboratory, but rather to these principles as they are actually working, and have been working, in the concrete courses of outward nature, in meteorology and physical geology. Still better, perhaps, at least for the purposes of illustration, we may compare the principle of Natural Selection to the fundamental laws of political economy, demonstrated and actually at work in the production of the values and the prices in the market of the wealth which human needs and efforts demand and supply. Who can tell from these principles what the market will be next week, or account for its prices of last week, even by the most ingenious use of hypotheses to supply the missing evidence? The empirical economist and statistician imagines that he can discover some other principles at work, some predetermined regularity in the market, some "innate" principles in it, to which the general laws of political economy are subordinated; and speculating on them, might risk his own wealth in trade, as the speculative "vitalist" might, if anything could be staked on a transcendental hypothesis. In the same way the empirical weather-philosopher thinks he can discern regularities in the weather, which the known principles of mechanical and chemical physics will not account for, and to which they are subordinate. This arises chiefly from his want of imagination, of a clear mental grasp of these principles, and of an adequate knowledge of the resources of legitimate hypothesis to supply the place of the unknown incidental causes through which these principles act. Such are also the sources of most of the difficulties which our author has found in the applications of the theory of Natural Selection.

His work is chiefly taken up with these difficulties. He does not so much insist on the probability of his own transcendental hypothesis, as endeavor to make way for it by discrediting the sufficiency of its rival; as if this could serve his purpose; as if experimental philosophy itself, without aid from "Darwinism," would not reject his metaphysical, occult, transcendental hypothesis of a specially predetermined and absolute fixity of species,–an hypothesis which multiplies species in an organism to meet emergencies,–the emergencies of theory,–much as the epicycles of Ptolemy had to be multiplied in the heavens. Ptolemy himself had the sagacity to believe that his was only a mathematical theory, a mode of representation, not a theory of causation; and to prize it only as representative of the facts of observation, or as "saving the appearances." Mr. Mivart's theory, on the other hand, is put forward as a theory of causation, not to save appearances, but to justify the hasty conclusion that they are real; the appearances, namely, of complete temporary fixity, alternating with abrupt changes, in the forms of life which are exhibited by the scanty records of geology and in present apparently unchanging natural species.

Before proceeding to a special consideration of our author's difficulties on the theory of Natural Selection, we will quote from Mr. Darwin's latest work, "The Descent of Man," his latest views of the extent of the action of this principle and its relations to the general theory of evolution. He says (Chapter IV) :–

> Thus a very large yet undefined extension may safely be given to the direct and indirect results of Natural Selection; but I now admit, after reading the essay by Nägeli on plants, and the remarks by various authors with respect to animals, more especially those recently made by Professor Broca, that in the earlier editions of my 'Origin of Species' I probably attributed too much to the action of Natural Selection, or the survival of the fittest. I have altered the fifth edition of the 'Origin' (the edition which Mr. Mivart reviews in his work), so as to confine my remarks to adaptive changes of structure. I had not formerly sufficiently considered the existence of many structures which appear to be, as far as we can judge, neither beneficial nor injurious: and this I believe to be one of the greatest oversights as yet detected in my work. I may be permitted to say, as some excuse, that I had two distinct objects in view: firstly, to show that species had not been separately created; and secondly, that Natural Selection had been the chief agent of change, though largely aided by the inherited effects of habit, and slightly

by the direct action of the surrounding conditions. Nevertheless, I was not able to annul the influence of my former belief, then widely prevalent, that each species had been purposely created; and this led to my tacitly assuming that every detail of structure, excepting rudiments, was of some special, though unrecognized, service. Any one with this assumption in his mind would naturally extend the action of Natural Selection, either during past or present times, too far. Some of those who admit the principle of evolution, but reject Natural Selection, seem to forget, when criticising my work, that I had the above two objects in view; hence, if I have erred in giving to Natural Selection great power, which I am far from admitting, or in having exaggerated its power, which is in itself probable, I have at least, as I hope, done good service in aiding to overthrow the dogma of separate creations.

In one other respect Mr. Darwin has modified his views of the action of Natural Selection, in consequence of a valuable criticism in the *North British Review* of June, 1867;* and our author† regards this modification as very important, and says of it that "this admission seems almost to amount to a change of front in the face of the enemy." It is not, as we shall see, an important modification at all, and does not change in any essential particular the theory as propounded in the first edition of the "Origin of Species," but our author's opinion of it has helped us to discover what, without this confirmation, seemed almost incredible,–how completely he has misapprehended, not merely the use of the theory in special applications, which is easily excusable, but also the nature of its general operation and of the causes employed by it; thus furnishing an additional illustration of what he says in his Introduction, that "few things are more remarkable than the way in which it (this theory) has been misunderstood." One other consideration has also been of aid to us. In his concluding chapter on "Theology and Evolution," in which he very ably shows, and on the most venerable authority, that there is no necessary conflict between the strictest orthodoxy and the theory of evolution, he remarks (and quotes Dr. Newman) on the narrowing effect of single lines of study. Not only inabilities may be produced by a one-sided pursuit, but "a positive distaste may grow up, which, in the intellectual order, may amount to a spontaneous and unreasoning disbelief in that which appears to be in opposition to the more familiar concept, and this at all times." This is, of course, meant to apply to those who, from want of knowledge, also lack ability and interest and even acquire a distaste for theological studies. But it also has other and equally important applications. Mr. Mivart, it would at first sight

---

* [See review by Fleeming Jenkin.]   † [Mivart.]

seem, being distinguished as a naturalist and also versed in theology, is
not trammelled by any such narrowness as to disable him from giving
just weight to both sides of the question he discusses. But what are the
two sides? Are they the view of the theologian and the naturalist? Not
at all. The debate is between the theologian and descriptive naturalist
on one side, or the theologian and the student of natural history in its
narrowest sense, that is, systematic biology; and on the other side the physical naturalist, physiologist, or theoretical biologist. Natural history and biology, or the general science of life, are very comprehensive terms, and comprise in their scope widely different lines of pursuit and a wide range
of abilities. In fact, the sciences of biology contain contrasts in the objects,
abilities, and interests of scientific pursuit almost as wide as that presented
by the physical sciences generally, and the sciences of direct observation,
description, and classification. The same contrast holds, indeed, even in
a science so limited in its material objects as astronomy. The genius of
the practical astronomer and observer is very different from that of the
physical astronomer and mathematician; though success in this science generally requires nowadays that some degree of both should be combined. So the
genius of the physiologist is different from that of the naturalist proper,
though in the study of comparative anatomy the observer has to exercise
some of the skill in analysis and in the use of hypotheses which are the
genius of the physical sciences in the search for unknown causes. We may,
perhaps, comprise all the forms of intellectual genius (excluding aesthetics)
under three chief classes, namely, first, the genius that pursues successfully
the researches for unknown causes by the skillful use of hypothesis and
experiment; secondly, that which, avoiding the use of hypotheses or preconceptions altogether and the delusive influence of names, brings together
in clear connections and contrasts in classification the objects of nature
in their broadest and realest relations of resemblance, and thirdly, that
genius which seeks with success for reasons and authorities in support of
cherished convictions.

That our author may have the last two forms of genius, even in a notable
degree, we readily admit; but that he has not the first to the degree needed
for an inquiry, which is essentially a branch of physical science, we propose
to show. We have already pointed out how his theological education, his
schooling against Democritus, has misled him in regard to the meaning
of "accidents" or accidental causes in physical science; as if to the physical
philosopher these could possibly be an absolute and distinct class, not included under the law of causation, "that every event must have a cause

or determinate antecedents," whether we can trace them out or not. The accidental causes of science are only "accidents" relatively to the intelligence of a man. Eclipses have the least of this character to the astronomer of all the phenomena of nature; yet to the savage they are the most terrible of monstrous accidents. The accidents of monstrous variation, or even of the small and limited variations normal in any race or species, are only accidents relatively to the intelligence of the naturalist, or to his knowledge of general physiology. An accident is what cannot be anticipated from what we know, or by any intelligence, perhaps, which is less than omniscient.

But this is not the most serious misconception of the accidental causes of science, which our author has fallen into. He utterly mistakes the particular class of accidents concerned in the process of Natural Selection, To make this clear, we will enumerate the classes of causes which are involved in this process. In the first place, there are the external conditions of an animal's or plant's life, comprising chiefly its relations to other organic beings, but partly its relations to inorganic nature, and determining its needs and some of the means of satisfying them. These conditions are consequences of the external courses of events or of the partial histories of organic and inorganic nature. In the second place, there are the general principles of the fitness of means to ends, or of supplies to needs. These comprise the best ascertained and most fundamental of all the principles of science, such as the laws of mechanical, optical, and acoustical science, by which we know how a leg, arm, or wing, a bony frame, a muscular or a vascular system, an eye or an ear, can be of *use*. In the third place, there are the causes introduced by Mr. Darwin to the attention of physiologists, as normal facts of organic nature, the little known phenomena of variation, and their relations to the laws of inheritance. There are several classes of these. The most important in the theory of Natural Selection are the diversities *always existing* in any race of animals or plants, called "individual differences," which *always* determine a better fitness of some individuals to the general conditions of the existence of a race than other less fortunate individuals have. The more than specific agreements in characters, which the best fitted individuals of a race must thus exhibit, ought, if possible, according to Cuvier's principles of zoölogy, to be included in the description of a species (as a norm or type which only the *best* exhibit), instead of the rough averages to which the naturalist really resorts in defining species by marks or characters that are variable. But probably such averages in variable characters are really close approximations to the characters of the best general adaptation; for variation being, so far as known, irrespective of adaptation, is as likely to exist to the same extent on one side of the norm of utility as on the other, or by excess as generally as by defect. Though variation is irrespective of utility, its limits are not. Too great a departure from the norm of utility must put an end to life and its successions. Utility therefore determines, along with the laws of inheritance, not only the middle line or safest way of a race, but also the bounding limits of its path of life; and so long as the conditions and principles of utility embodied in a form of life remain unchanged, they will, together with the laws of inheritance, maintain a race unchanged in its average characters. "Specific stability," therefore, for which theological and descriptive naturalists have speculated a transcendental cause, is even more readily and directly accounted for by the causes which the theory of Natural Selection regards than is specific change. But just as obviously it follows from these causes that a change in the conditions and resources of utility, not only may but must change the normal characters of a species, or else the race must perish. Again, a slow and gradual change in the conditions of existence must, on these principles, slowly change the middle line or safest way of life (the descriptive or graphic line); but always, of course, this change must be within the existing limits of variation, or the range of "individual differences." A change in these limits would then follow, or the range of "individual differences" would be extended, at least, so far as we know, in the direction of the change. That it is widened or extended to a greater range by rapid and important changes in conditions of existence, is a matter of observation in many races of animals and plants that have been long subject to domestication or to the capricious conditions imposed by human choice and care. This phenomenon is like what would happen if a roadway or path across a field were to become muddy or otherwise obstructed. The travelled way would swerve to one side, or be broadened, or abandoned, according to the nature and degree of the obstruction and to the resources of travel that remained. This class of variations, that is, "individual differences," constant and normal in a race, but having different ranges in different races, or in the same race under different circumstances, may be regarded as in no proper sense accidentally related to the advantages that come from them; or in no other sense that a tendril, or a tentacle, or a hand searching in the dark, is accidentally related to the object it succeeds in finding. And yet we say properly that it was by "accident" that a certain tendril was put forth so as to fulfil its function, and clasp the particular object by which it supports the vine; or that it was an accidental movement of the tentacle or hand that brought the object it has secured within its grasp. The search was, and continues to be, normal and general; it is the particular success only that is accidental; and this only in the sense that lines of causation, stretching backwards infinitely, and unrelated except in a first cause, or in the total order of nature, come together and by their concurrence produce it. Yet over even this concurrence "law" still presides, to the effect that for every such concurrence the same consequences follow.

But our author, with his mind filled with horror of "blind chance," and of "the fortuitous concourse of atoms," has entirely overlooked the class of accidental variations, on which, even in the earlier editions of the "Origin of Species," the theory of Natural Selection is based and has fixed his attention exclusively on another class, namely, abnormal or unusual variations, which Mr. Darwin at first supposed might also be of service in this process. The fault might, perhaps, be charged against Mr. Darwin for not sufficiently distinguishing the two classes, as well as overlooking, until it was pointed out by his critic in the "North British Review," before referred to, the fact that the latter class could be of no service; if it were not that our author's work is a review of the last edition of the "Origin of Species" and of the treatise on "Animals and Plants under Domestication," in both of which Mr. Darwin has emphatically distinguished these classes, and admitted that it is upon the first class only that Natural Selection can normally depend; though the second class of unusual and monstrous variations may give rise, by highly improbable though possible accidents, to changes in the characters of whole races. Mr. Mivart characterizes this admission by the words we have quoted, that "it seems almost to amount to a change of front in the face of the enemy"; of which it might have been enough to say, that the strategy of science is not the same as that of rhetorical disputation, and aims at cornering facts, not antagonists. But Mr. Mivart profits by it as a scholastic triumph over heresy, which he insists upon celebrating, rather than as a correction of his own misconceptions of the theory. He continues throughout his book to speak of the variations on which Natural Selection depends as if they were all of rare occurrence, like abrupt and monstrous variations, instead of being always present in a race; and also as having the additional disadvantage of being "individually slight," "minute," "insensible," "infinitesimal," "fortuitous," and "indefinite." These epithets are variously combined in different passages, but his favorite compendious formula is, "minute, fortuitous, and indefinite variations." When, however, he comes to consider the enormous time which such a process must have taken to produce the present forms of life, he brings to bear all his forces, and says (p. 154): "It is not easy to believe that less than two thousand million years would be required for the totality of animal development by no other means than minute, fortuitous, occasional, and intermitting variations in all conceivable directions." This exceeds very much–by some two hundred-fold–the length of time Sir William Thomson allows for the continuance of life on the earth. It is difficult to see how, with such uncertain "fortuitous, occasional, and intermitting" elements, our author could have succeed in making any calculations at all. On the probability of the correctness of Sir William Thomson's physical arguments "the author of this book cannot presume to advance an opinion; but," he adds (p. 150), "the fact that they have not been refuted pleads strongly in their favor when we consider how much they tell against the theory of Mr. Darwin." He can, it appears, judge of them on his own side.

For the descriptive epithets which our author applies to the variations on which he supposes Natural Selection to depend he has the following authority. He says (p. 35): "Now it is distinctly enunciated by Mr. Darwin that the spontaneous variations upon which his theory depends are individually slight, minute, and insensible. He says (*Animals and Plants under Domestication,* Vol II. p. 192): 'Slight individual differences, however, suffice for the work, and are probably the sole differences which are effective in the production of new species.' " After what we have said as to the real nature of the differences from which nature selects, it might be, perhaps, unnecessary to explain what ought at least to have been known to a naturalist, that by "individual differences" is meant the differences between the individuals of a race of animals or plants; that the slightness of them is only relative to the differences between the characters of species, and that they may be very considerable in themselves, or their effects, or even to the *eye* of the naturalist. How the expression "slight individual differences" could have got translated in our author's mind into "individually slight, minute, and insensible" ones, has no natural explanation. But this is not the only instance of such an unfathomable translation in our author's treatment of the theory of Natural Selection. Two others occur on page 133. In the first he says: "Mr. Darwin abundantly demonstrates the variability of dogs, horses, fowls, and pigeons, but he none the less shows the *very small* extent to which the goose, the peacock, and the guinea-fowl have varied. Mr. Darwin attempts to explain this fact as regards the goose by the animal being valued only for food and feathers, and from no pleasure having been felt in it on other accounts. He adds, however, at the end, the striking remark, which concedes the whole position, 'but the goose seems to have *a singularly inflexible organization.*'" The translation is begun in the author's italics, and completed a few pages further on (p. 141), where, recurring to this subject, he says: "We have seen that Mr. Darwin himself implicitly admits the principle of specific stability in asserting the singular inflexibility of the organization of the goose." This is what is called in scholastic logic, *Fallacia a dicto secundum quid ad dictum simpliciter.* The obvious meaning, both from the contexts and the evidence, of the expression "singularly inflexible," is that the goose has been much less changed by domestication than other domestic birds. But this *relative* inflexibility is understood by our author as an admission of an *absolute* one, in spite of the evidence that geese have varied from the wild type, and have individual differencer, and even differences of breeds, which are sufficiently conspicuous, even to the eye of a goose. The next instance of our author's translations (p. 133) is still more remarkable. He continues: "This is not the only place in which such expressions are used. He (Mr. Darwin) elsewhere makes use of phrases which quite harmonize with the conception of a normal specific constancy, but varying greatly and suddenly at intervals. Thus he speaks of a *whole organism seeming to have become plastic and tending to depart from the parental type* ('Origin of Species,' 5th edit., 1869, p. 13)." The italics are Mr. Mivart's. The passage from which these words are quoted (though they are not put in quotation-marks) is this: "It is well worth while carefully to study the several treatises on some of our old cultivated plants, as on the hyacinth, potato, even the dahlia, etc.; and it is really surprising to note the endless points in structure and constitution in which the varieties and sub-varieties differ slightly from each other. The whole organization seems to have become plastic, and tends to depart *in a slight degree* from that of the parental type." The words that we have italicized in this quotation are omitted by our author, though essential to the point on which he cites Mr. Darwin's authority, namely, as to the organism "varying greatly and suddenly at intervals." Logic has no adequate name for this fallacy; but there is another in our author's understanding of the passage which is very familiar,—the fallacy of ambiguous terms. Mr. Darwin obviously uses the word "plastic" in its secondary signification as the name of that which is "capable of being moulded, modelled, or fashioned to the purpose, as clay." But our author quite as obviously understands it in its primary signification as the name of anything "having the power to give form." But this is a natural enough misunderstanding, since in scholastic philosophy the primary signification of "plastic" is the prevailing one.*

That the highest products of nature are not the results of the mere forces of inheritance, and do not come from the birth of latent powers and structures, seems to be the lesson of the obscure discourse in which Jesus endeavored to instruct Nicodemus the Pharisee. How is it that a man can be born again, acquire powers and characters that are not developments of what is already innate in him? How is it possible when he is old to acquire new innate principles, or to enter a second time into his mother's womb and be born? The reply does not suggest our author's hypothesis of a life turning over upon a new "facet," or a new set of latent inherited powers. Only the symbols, water and the Spirit, which Christians have ever since worshipped, are given in reply; but the remarkable illustration of the accidentality of nature is added, which has been almost equally though independently admired. "Marvel not that I said unto thee, Ye must be born again. The wind bloweth where it listeth, and thou hearest the sound thereof, but canst not tell whence it cometh and whither it goeth; so is every one that is born of the Spirit." The highest products of nature are the outcome of its total and apparently accidental orders; or are born of water and the Spirit, which symbolize creative power. To this the Pharisee replied: "How can these things be?" And the answer is still more significant: "Art thou a master of Israel and knowest not these things?" We bring natural evidences, "and ye receive not our witness. If I have told you earthly (natural) things, and ye believe not, how shall ye believe if I tell you heavenly (supernatural) things?" The bearing of our subject upon the doctrine of Final Causes in natural history has been much discussed and is of considerable importance to our author's theory and criticism. But we propose, not only to distinguish between this branch of theology and the theories of inductive science on one hand, but still more emphatically, on the other hand, between it and the Christian faith in divine superintendency, which is very liable to be confounded with it. The Christian faith is that even the fall of a sparrow is included in this agency, and that as men are of more value than many sparrows, so much more is their security. So far from weakening this faith by showing the connection between value and security, science and the theory of Natural Selection have confirmed it. The very agencies that give values to life secure them

* [At this point Wright launches into a discussion of the possible origins of various organs e.g., the giraffe's long neck, mammary glands, and the rattle of rattlesnakes.]

by planting them most broadly in the immutable grounds of utility. But Natural Theology has sought by Platonic, not Christian, imaginations to discover, not the relations of security to value, but something *worthy* to be the source of the value considered as *absolute*, some particular worthy source of each valued end. This is the motive of that speculation of Final Causes which Bacon condemned as sterile and corruption to philosophy, interfering, as it does, with the study of the facts of nature, or of what *is,* by preconceptions, necessarily imperfect as to what *ought to be;* and by deductions from assumed *ends,* thought worthy to be the purposes of nature. The naturalists who "take care not to ascribe to God any intention," sin rather against the spirit of Platonism than that of Christianity, while obeying the precepts of experimental philosophy. Though, as our author says, in speaking of the moral sense and the impossibility, as he thinks, that the accumulations of small repugnances could give rise to the strength of its abhorrence and reprobation; though, as he says, "no stream can rise higher than its source"; while fully admitting the truth of this, we would still ask, Where is its source? Surely not in the little fountains that Platonic explorers go in search of, *a priori,* which would soon run dry but for the rains of heaven, the water and the vapor of the distilling atmosphere. Out of this come also the almost weightless snow-flakes, which combined in masses of great gravity, fall in the avalanche. The results of moralizing Platonism should not be confounded with the simple Christian faith in Divine superintendence. The often-quoted belief of Professor Gray, "that variation has been led along certain beneficial lines, like a stream along definite lines of irrigation," might be interpreted to agree with either view. The lines on which variations are generally useful are lines of search, and their particular successes, dependent, it is true, on no theological or absolute accidents, may be regarded as being lines of beneficial variations, seeing that they have resulted through laws of nature and principles of utility in higher living forms, or even in continuing definite forms of life on the earth. But thousands of movements of variation, or efforts of search, have not succeeded to one that has. These are not continued along evil lines, since thousands of forms have perished in consequence of them for every one that has survived.

The growth of a tree is a good illustration of this process, and more closely resembles the action of selection in nature generally than might at first sight appear; for its branches are selected growths, a few out of many thousands that have begun in buds; and this rigorous selection has been effected by the accidents that have determined superior relations in

surviving growths to their supplies of nutriment in the trunk and in exposure
to light and air. This exposure (as great as is consistent with secure connection with the sources of sap) seems actually to be sought, and the form
of the tree to be the result of some foresight in it. But the real seeking
process is budding, and the geometrical regularity of the production of
buds in twigs has little or nothing to do with the ultimate selected results,
the distributions of the branches, which are different for each individual
tree. Even if the determinate variations really existed,–the "facets" of stable
equilibrium in life, which our author supposes,–and were arranged with
geometrical regularity on their spheroid of potential forms, as leaves and
buds are in the twig, they would probably have as little to do with determining the ultimate diversities of life under the action of the selection
which our author admits as phyllotaxy has to do with the branching of
trees. But phyllotaxy, also, has its utility. Its orders are the best for packing
of the incipient leaves in the bud, and the best for the exposure to light
and air of the developed leaves of the stem. But here its utility ends, except
so far as its arrangements also present the greatest diversity of finite elements, within the smallest limits, for the subsequent choice of successful
growths; being *the nearest approaches that finite regularity could make*
to "indefinite variations in all conceivable directions." The general resemblance of trees of a given kind depends on no formative principle other
than physical and physiological properties in the woody tissue, and is related
chiefly to the tenacity, flexibility, and vascularity of this tissue, the degrees
of which might almost be inferred from the general form of the tree. It
cannot be doubted, in the case of the tree, that this tentative though regular
budding has been of service to the production of the tree's growth, and
that the particular growths which have survived and become the bases
of future growths were determined by a beneficial though accidental order
of events under the total orders of the powers concerned in the tree's
development. But if a rigorous selection had not continued in this growth,
no proper branching would have resulted. The tree would have grown
like a cabbage. Hence it is to selection, and not to variation,–or rather
to the *causes* of selection, and not to those of variation,–that species or well-
marked and widely separated forms of life are due. If we could study the
past and present forms of life, not only in different continents, which we
may compare to different individual trees of the same kind, or better,
perhaps, to different main branches from the same trunk and roots,
but could also study the past and present forms of life in different
planets, then diversities in the general outlines would probably be seen

similar to those which distinguish different kinds of trees, as the oak, the elm, and the pine; dependent, as in these trees, on differences in the physical and physiological properties of living matters in the different planets,–supposing the planets, of course, to be capable of sustaining life, like the earth, or, at least, to have been so at some period in the history of the solar system. We might find that these general outlines of life in other planets resemble elms or oaks, and are not pyramidal in form like the pine, with a "crowning" animal like man to lead their growths. For man, for aught we know or could guess (but for the highly probable accidents of nature, which blight the topmost terminal bud and give ascendency to some lateral one), except for these accidents, man *may* have always been the crown of earthly creation, or always "man," if you choose so to name and define the creature who, though once an ascidian (when the ascidian was the highest form of life), *may* have been the *best* of the ascidians. This would, perhaps, add nothing to the present value of the race, but it might satisfy the Platonic demand that the race, though not derived from a source quite worthy of it, yet should come from the *best* in nature.

We are thus led to the final problem, at present an apparently insoluble mystery, of the origin of the first forms of life on the earth. On this Mr. Darwin uses the figurative language of religious mystery, and speaks "of life with its several powers being originally breathed by the Creator into a few forms or into one." For this expression our author takes him to task, though really it could mean no more than if the gravitative properties of bodies were referred directly to the agency of a First Cause, in which the philosopher professed to believe; at the same time expressing his unwillingness to make hypotheses, that is, transcendental hypotheses, concerning occult modes of action. But life is, indeed, divine, and there is grandeur in the view, as Mr. Darwin says, which derives from so simple yet mysterious an origin, and "from the war of nature, from famine and death, the most exalted object which we are capable of conceiving, namely, the production of the higher animals." Our author, however, is much more "advanced" than Mr. Darwin on the question of the origin of life or archigenesis, and the possiblilty of it as a continuous and present operation of nature. He admits what is commonly called "spontaneous generation," believing it, however, to be not what in theology is understood by "spontaneous," but only a sudden production of life by chemical synthesis out of inorganic elements. The absence of decisive evidence on this point does not deter him, but the fact that the doctrine can be reconciled to the strictest

orthodoxy, and accords well with our author's theory of sudden changes
in species, appears to satisfy him of its truth. The theory of Pangenesis,
on the other hand, invented by Mr. Darwin for a different purpose, though
not inconsistent with the very slow generation of vital forces out of chemical
actions,–slow, that is, and insignificant compared to the normal actions
and productions of chemical forces,–is hardly compatible with the sudden
and conspicuous appearance of new life under the microscope of the observer. This theory was invented like other provisional theories,–like Newton's corpuscular theory of light, like the undulatory theory of light (though
this is no longer provisional), and like the chemical theory of atoms,–for
the purpose of giving a material or visual basis to the phenomena and
empirical laws of life in general, by embodying in such supposed properties
the phenomena of development, the laws of inheritance, and the various
modes of reproduction, just as the chemical theory of atoms embodies in
visual and tangible properties the laws of definite and multiple proportions,
and the relations of gaseous volumes in chemical unions, together with
the principle of isomerism and the relations of equivalent weights to specific
heats. The theory of Pangenesis presents life and vital forces in their ultimate and essential elements as perfectly continuous, and in great measure
isolated from other and coarser orders of forces, like the chemical and
mechanical, except so far as these are the necessary theatres of their actions.
Gemmules, or vital molecules, the smallest bodies which have separable
parts under the action of vital forces, and of the same order as the scope
of action in these forces,–these minute bodies, though probably as much
smaller than chemical molecules as these are smaller than rocks or pebbles,
may yet exist in unorganized materials as well as in the germs of eggs,
seeds, and spores, just as crystalline structures or chemical aggregations
may be present in bodies whose form and aggregation are mainly due
to mechanical forces. And, as in mechanical aggregations (like sedimentary
rocks), chemical actions and aggregations slowly supervene and give in
the metamorphosis of these rocks an irregular crystalline structure, so it
is supposable that finer orders of forces lying at the heart of fluid matter may
slowly produce imperfect and irregular vital aggregations. But *definite* vital
aggregations and *definite* actions of vital forces exist, for the most part, in a
world by themselves, as distinct from that of chemical forces, actions, and
aggregations as these are from the mechanical ones of dynamic surface-
geology, which produce and are embodied in visible and tangible masses
through forces the most directly apparent and best understood; or as distinct
as these are from the internal forces of geology and the masses of continents

and mountain formations with which they deal; or as distinct again as
these are from the actions of gravity and the masses in the solar system;
or, again, as these are from the unknown forces and conditions that regulate
sidereal aggregations and movements. These various orders of molar and
molecular sizes are limited in our powers of conception only by the needs
of hypothesis in the representation of actual phenomena under visual forms
and properties. Sir William Thomson has lately determined the probable
sizes of chemical molecules from the phenomena of light and experiments
relating to the law of the "conservation of force." According to these results,
these sizes are such that if a drop of water were to be magnified to the
size of the earth, its molecules, or parts dependent on the forces of chemical
physics, would be seen to range from the size of a pea to that of a billiard-
ball. But there is no reason to doubt that in every such molecule there
are still subordinate parts and structures; or that, even in these parts,
a still finer order of parts and structures exists, at least to the extent of
assimilated growth and *simple* division. Mr. Darwin supposes such growths
and divisions in the vital gemmules; but our author objects (p. 230) that,
"to admit the power of spontaneous division and multiplication in such
rudimentray structures seems a complete contradiction. The gemmules, by
the hypothesis of Pangenesis, are the ultimate organized components of
the body, the absolute organic atoms of which each body is composed;
how then *can* they be divisible? Any part of a gemmule would be an
impossible (because *less* than possible) quantity. If it is divisible into still
smaller organic wholes, as a germ-cell is, it must be made up, as the germ-
cell is, of subordinate component atoms, which are then the *true* gemmules."
But this is to suppose what is not implied in the theory (nor properly
even in the chemical theory of atoms), that the sizes of these bodies are
any more constant or determinate than those of visible bodies of any order.
It is the order only that is determinate; but within it there may be wide
ranges of sizes. A billiard-ball may be divided into parts as small as a
pea, or peas may be aggregated into masses as large as a billiard-ball,
without going beyond the order of forces that produce both sizes. Our
author himself says afterwards and in another connection (p. 290), "It
is possible that, in some minds, the notion may lurk that such powers
are simpler and easier to understand, because the bodies they affect are
so minute! This absurdity hardly bears stating. We can easily conceive
a being so small that a gemmule would be to it as large as St. Paul's
would be to us." This argument, however, is intended to discredit the theory
on the ground that it does not tend to simplify matters, and that we must

rest somewhere in "what the scholastics called 'substantial forms.' " But this criticism, to be just, ought to insist, not only that vital phenomena are due to "a special nature, a peculiar innate power and activity," but that chemical atoms only complicate the mysteries of science unnecessarily; that corpuscles and undulations only hide difficulties; and that we ought to explain very simply that crystalline bodies are produced by "polarity," and that the phenomena of light and vision are the effects of "luminosity." This kind of simplicity is not, however, the purpose which modern science has in view; and, consequently, our real knowledges, as well as our hypotheses, are much more complicated than were those of the schoolmen. It is not impossible that vital phenomena themselves include orders of forces as distinct as the lowest vital are from chemical phenomena. May not the contrast of merely vital or vegetative phenomena with those of *sensibility* be of such orders? But, in arriving at *sensibility*, we have reached the very elements out of which the conceptions of size and movement are constructed,–the elements of the tactual and visual constructions that are employed by such hypotheses. Can sensibility and the movements governed by it be derived directly by chemical synthesis from the forces of inorganic elements? It is probable, both from analogy and direct observation, that they cannot (though some of the believers in "spontaneous generation" think otherwise); or that they cannot, except by that great alchemic experiment which, employing all the influences of nature and all the ages of the world, has actually brought forth most if not all of the definite forms of life in the last and greatest work of creative power.

Comments on Mivart and Wright

St. George Jackson Mivart came from a well-to-do London family. His father had founded the Mivart Hotel, which was later to become the celebrated Claridge's of London. At a time when there were few Catholics in England and the Catholic Church was none too popular, Mivart then sixteen, converted to Catholicism. His subsequent conversion to evolutionism was just as sudden and just as unexpected. Mivart began his education in science under Owen and Huxley. His candidacy in 1862 as Lecturer at St. Mary's Hospital Medical School in London and later as Professor of Comparative Anatomy was supported by both men. If Mivart actually was the author of the anonymous review (1860) of the *Origin of Species* in which Owen was taken to task for his anonymous review (1860), then Mivart's penchant for defying authority also appeared early. Needless to say, Darwin and his friends were happy with this paper and found its author "a jolly good fellow." The harmony between Mivart and the Darwinians was to be short-lived.

Mivart was interested both in fostering the acceptance of evolutionary theory and in reconciling it with Catholic dogma. He wished to counter the materialistic and atheistic tendencies of the theory revealed so clearly in Haeckel's *General Morphology* (1866). Mivart agreed with Darwin that species evolved but disagreed with Darwin's mechanisms. In his book, *On the Genesis of Species* (1871), Mivart argued that species evolved by saltation, directed by some unknown "internal innate force," like a self-propelled spheroid tumbling from one facet to another. The surprising feature of Mivart's book was not his dissatisfaction with Darwin's mechanisms and his suggested alterations but his contention that evolutionary theory as he set it out was in full accord with the teachings of the Catholic Church. Mivart argued that evolution is restricted to the material world. Since mind and soul are immaterial, their origin and frame are outside the scope of evolutionary theory and in the proper province of theology.

At first there seemed to be some hope that the Church would accept Mivart's compromise. In 1876 Pius IX conferred on Mivart the degree of Doctor of Philosophy, implying tacit acceptance of his ideas, but time and circumstances were allied against him. In 1870 the Vatican Council had proclaimed the infallibility of the pope in matters of faith and morals. It also interposed papal authority in matters of science as well. Mivart was opposed to both claims. Eventually, Mivart was excommunicated from the Church, but not because of his views on evolution or his opposition to papal infallibility. Rather, the occasion for his fall from favor was an article he wrote entitled "Happiness in Hell" (Mivart, 1892), an attack on the traditional notion of hell as a place of eternal torture by a vindictive God. Mivart's total break from the Church came in 1899, a year before his death, over the Church's role in the persecution of Alfred Dreyfus in France. Mivart reasoned that just as the Galileo affair had proved the pope not to be infallible in matters of science, the Dreyfus affair cast doubt as well on his infallibility in matters of faith and morals. In 1900 Mivart died and was buried as an excommunicant in unhallowed ground. Four years later the ecclesiastical authorities were convinced by Mivart's family that he had been insane during the last years of his life, and his body was transferred to hallowed ground.

Mivart's compromise between theology and evolutionary theory was treated even more roughly by the Darwinians. His attack on the Darwinian mechanisms for evolution in his *Genesis of Species* (1871a) and his objections to Darwin's *Descent of Man* in an anonymous review in the *Quarterly Review* (1871b) could not go unanswered. Huxley, the man who invented the term "agnostic," took perverse pleasure in defending the true Catholic faith against Mivart's heresies in the *Contemporary Review* (1871). After thoroughly enjoying himself at Mivart's expense on theological matters, Huxley addressed himself to evolutionary theory and the origin of mind. Huxley could not very well object to Mivart's notion of saltative evolution, since he himself favored such a view. Mind was quite another matter. According to Huxley (1871), consciousness was "an expression of the molecular changes which take place in that nervous matter, which is the organ of consciousness." Huxley was also forced to reply, however reluctantly, to Wallace's opinions on the impossibility of the evolution of mind. Mivart's reply to Huxley (Mivart, 1872a) matched Huxley's review in sarcasm and disdain but lacked the light-hearted touch which tended to grace Huxley's polemics.

Darwin found Mivart's objections sufficiently important to warrant the insertion of several passages in the sixth edition of the *Origin* (1872) but he limits himself largely to factual matters: the putative correlation between dermal covering and teeth in mammals, the relation between vertebrate and cephalopod eyes, and the resemblance between the common mouse and an Australian marsupial (Darwin, 1872, pp. 140, 176, 395). However, he does deal at length with Mivart's charge that natural selection is inadequate to account for the incipient stages of useful structures. Darwin concludes:

> My judgment may not be trustworthy, but after reading with care Mr. Mivart's book, and comparing each section with what I have said on the same head, I never before felt so strongly convinced of the general truth of the conclusions here arrived at, subject, of course, in so intricate a subject, to much partial error (Darwin, 1872, p. 200).

Just as Huxley, a biologist, chose to contest Mivart's theology, it was left to Chauncey Wright, a mathematician and philosopher to challenge his biology. What little we know of Wright's life stems from his association with the Metaphysical Club at Cambridge, Massachusetts, and the founding of pragmatism (see Thayer, 1878; Wiener, 1949; and Madden, 1963). The Metaphysical club was an informal group whose meetings were attended by such men of letters as William James, Charles S. Peirce, and Oliver Wendell Holmes. Wright worked until 1870 as a computer for the

Cambridge office of the *Nautical Almanac,* after graduating from Harvard in 1852. He then lectured for a short while at Harvard in mechanics, but a new system had been instituted at the university which permitted students considerable freedom in electing their courses. Apparently Wright was not a very good lecturer, and his classes were poorly attended. Wright's contract was not renewed, perhaps because of his inadequacies as a teacher, perhaps because of his "irregular habits." He, like Peirce, "sought relief from recurrent physical suffering in stimulants."

Wright began his intellectual career as a disciple of William Hamilton and an admirer of Louis Agassiz's *Essay on Classification* (1857). Wright was teaching at Agassiz's school for girls in Cambridge when the *Origin of Species* appeared. Wright observed then that if Darwin's conclusions were true, Agassiz's essay became a "useless and mistaken speculation" (Wiener, 1949, p.33). Shortly thereafter he wrote a paper defending Darwin's views on the hive-making habits of bees (1860) and was promptly attacked by Francis Bowen. The most important event in Wright's conversion, however, was the publication of Mill's *Examination of Sir William Hamilton's Philosophy* (1865). Wright was converted to Mill's version of empiricism and nominalism. He saw in evolutionary theory a further enhancement of Mill's philosophy, in spite of Peirce's telling observation that "Mill's doctrine was nothing but a metaphysical point of view to which Darwin's . . . must be deadly." (Peirce, 1935, 6:297).

Wright's most fundamental conviction was that scientific method was metaphysically and ethically neutral. He argued on the one hand against what he termed the "German Darwinism" of Herbert Spencer, Ernst Haeckel, and John Fiske and, on the other, against the theological objections of such men as Mivart. Although Darwin did not think too highly of Wright's criticisms of Mivart, he had Wright's attack on Mivart's *Genesis of Species* reprinted separately at his own expense (1871). The first half of Wright's paper deals with issues already discussed at length in the Introduction to this volume—the role of hypotheses in science, the notion of accident and chance, and the causal role of such "abstractions" as gravity and polarity. When Wright turns to an evaluation of the more empirical aspects of the controversy between Darwin and Mivart, the caliber of his discussion declines sharply—as Darwin observed, from want of knowledge. For example, Wright fails to discern the crucial difference between the views of Darwin and Mivart on variation, and he completely misses the thrust of Jenkin's criticisms of evolutionary theory. Darwin believed that the raw material of evolution was to be found in the numerous slight

individual differences commonly observed among the offspring of any set
of parents. Mivart opted for large, abrupt changes occurring with comparative infrequency. Both Darwin and Mivart were partially correct. The variations which serve as the raw material for evolution are as slight as Darwin
believed, but they are as infrequent as Mivart believed.

Although Mivart's review of Darwin's *Descent of Man* (1871b) was
published anonymously, as were most reviews in the day, no one had any
doubts about its authorship. Nevertheless, all parties concerned maintained
the pretense of anonymity and continued to refer coyly to the "Quarterly
Reviewer." Like the *Descent of Man,* Mivart's review is divided into two
parts, one dealing with sexual selection, the other with the genesis of man.
Mivart asserts that in early editions of the *Origin of Species* natural selection
was supposed to be a sufficient cause of evolution but that in later editions
of the *Origin* and in the *Descent of Man,* Darwin had relegated it to
a subordinate position. Blatantly false as this contention clearly is, it has
passed into the folklore of science. Throughout his career, Darwin consistently maintained that natural selection was the major, though not the
sole, mechanism for evolution. In his later years, Darwin came to believe
that he had exaggerated the importance of natural selection slightly—but
only slightly. It was still the major mechanism.

Mivart tries this same gambit in his criticism of sexual selection. He
saddles Darwin with the view that "sexual selection is the universal cause
of sexual characters" and then proceeds to show numerous instances in
which sexual selection could not be the cause of differences between sexes.
Mivart argues, on the contrary, that sexual characters like all characters
are due to "internal spontaneous powers." Since Darwin had never claimed
that natural selection or sexual selection was sufficient even in its proper
sphere of influence, Mivart's criticisms are not nearly as devastating as
he believed them to be. But what of Mivart's alternative explanation?
Darwin agreed with Mivart that there was something about the internal
constitution of an organism which was passed on to its progeny to determine
ontogenetic development. Ontogenetic development was predictable. Once
a life cycle had been worked out, it was highly likely that successive life
cycles of individuals of that species would proceed along much the same
lines, even though the mechanism for this process was unknown. Darwin
disagreed with Mivart only on the role of this organization in phylogenetic
development. Mivart argued that this same unknown internal constitution also determined the phylogenetic development of the species. Darwin

thought not. Even when the mechanisms of heredity and variation became known, prediction would still be impossible. To use a modern metaphor, phylogenetic development was not programmed into the makeup of an organism, though the constitution of an organism might well limit its evolutionary possibilities. Instead, Darwin set out his theory of pangenesis. Mivart complained that Darwin's theory of pangenesis explained great difficulties only by presenting others no less great—"almost to be the explanation of *obscurum per obscuris*" (Mivart, 1871a, p. 39). In this instance Mivart's objection has some justification. The theory of pangenesis is much too loose in its construction. Darwin tells us a little about his gemmules and nascent cells, but not nearly enough. But by these standards, Mivart fails even more miserably, since he tells us nothing about his own theory of internal powers.

The second half of Mivart's paper is of greater interest than the first for our purposes because he finally brings to the forefront the issue which had been lurking behind so much of the hostility directed at Darwin's theory—the evolution of mind, soul, morality, and free will. Mivart's argument is fairly straightforward. It is the same argument which had been urged against the evolution of any species or any genuinely new character, not just man and his mind. Real differences are those of kind and not degree. If a character could change gradually through successive generations into another character, it could not have marked a genuine difference in kind. The difference between man as a rational, moral being and all other living creatures is one of kind, not degree. Man as a physical being may have evolved from lower creatures, but neither mind nor soul could have evolved. They are specially implanted in man by God. Mivart's position is typically that of the special creationist. It differs only in that he limits the special creation to man. No real differences exist between all other species of living creatures! (See Wright's discussion of natural kinds.)

Darwin approached the problem of man's evolution from the standpoint of a scientist. Man certainly could think and exhibit moral compunctions. Many, perhaps most, other creatures could not. What Darwin tried to show was that there were, nevertheless, intermediaries—his usual line of attack. If gradation occurred among contemporary forms in a particular character, then gradation in this character in time was at least possible. In the case of man, however, Darwin had his work cut out for him. The facts and arguments martialled in the *Descent of Man* for the evolution of man were not nearly so compelling as those set out in the *Origin of*

Species for evolution in general. The primary reason for accepting man's evolution remained the acceptance of evolutionary theory as such. If all species with their multifarious characters evolved, then why not man?

Mivart was right when he said that the difference between him and Darwin was primarily one of metaphysics. For Mivart, life and mind were fundamental ontological categories. This belief has been the source of the claims frequently made that life will never be created in a test tube or that machines will never be able to think. The conviction common among metaphysicians in Mivart's day was that ontological gaps were necessarily unbridgeable in time. Life could not evolve from non-life, nor mind from non-mind. Mivart was right when he said that Darwin failed to address himself to these metaphysical issues, and one suspects that Darwin's neglect was intentional.

In April of 1872 Mivart replied to Wright's paper (Mivart, 1872b) and in July Wright replied to Mivart (Wright, 1872). The exchange was none too productive. Mivart repeats the quotations which he had set out in his *Quarterly Review* paper which purport to show that Darwin had originally held natural selection to be *the* sufficient cause of evolution but had abandoned it later. Wright asserts once again that Darwin had never believed that natural selection was the one and only mechanism for the fixing of characters after they had arisen by the unknown laws of inheritance and that Darwin had modified this belief very little in his later writings. Mivart reasserts his faith in the existence of some unknown principle which guides the evolution of species in the same manner as ontogenetic development is guided, and once again Wright fails to accept, repudiate, or even discuss this contention. Nor does Wright address himself to the question of the evolution of mind. Instead they haggle over the analogy between crystals and organisms and the sense in which one "abstraction" can be said to "cause" another. A disproportionate amount of space is devoted to an analysis of Democritus and his atheistic tendencies. One positive contribution was the clarification of the term "accidental." In Mivart's sense it did not imply the absence of physical causation but the absence of design. Finally, the two authors reply to specific points (especially the reasons for a giraffe's having such a long neck) and give their views on the proper method of carrying on philosophic and scientific debates. It is unlikely that anyone reading this exchange between Mivart and Wright would have been persuaded one way or the other on any of the relevant issues.

Wright died quite suddenly in 1875 at the age of forty-five. The final break between Mivart and the Darwinians had occurred the previous year over a slur on the character of Darwin's son, George. Mivart had great difficulty in distinguishing between a man and his ideas. According to the mores of Victorian society, one could deal as harshly as one wished with a man's views but one had to be extremely careful to avoid personal slurs of any kind. Mivart stepped over this line once too often in alluding to the immoral implications of some of George Darwin's ideas on eugenics. Just as Mivart was being excommunicated from the Catholic Church because of his article on happiness in hell, he was being excluded from the scientific community by Darwin and his associates. In the end, Mivart's attempts to reconcile science and the Catholic Church led him to be excommunicated from both (see Gruber, 1960, for a well-balanced account of Mivart's life).

Karl Ernst von Baer (1792–1876)

I have to announce a new and great ally for you . . .
Von Baer writes to me thus: 'And besides, I find that you are still
writing reviews. You have written on Mr. Darwin's work a criticism of
which I have found only fragments in a German journal. I have forgotten
the terrible name of the English journal in which your review appeared.
Also in any case I cannot find the journal here. As I am much interested
in the ideas of Mr. Darwin, on which I have spoken publicly and on
which I shall perhaps have something put into print, you would oblige
me infinitely if you would be able to have forwarded to me what you
have written about these ideas. I have expressed the same ideas on the
transformation of types or origin of species as Mr. Darwin. But it is only
on zoological geography that I rely. You will find, in the last chapter
of the treatise "Über Papuas and Alfuren" that I speak of it positively
without knowing that Mr. Darwin was concerning himself with the subject.'
The treatise to which Von Bär refers he gave me when over there, but
I have not been able to lay hands on it since this letter reached me
two days ago. When I find it I will let you know what there is in it.—
T. H. Huxley to C. Darwin, August 6, 1860 (*Life and Letters,* 2:122–123),
trans. Jane Oppenheimer in *Forerunners of Darwin: 1745–1859,* ed.
B. Glass, O. Temkin, and W. L. Straus, Jr. (1959), p. 295.

Your note contained magnificent news, and thank you heartily for sending
me it. Von Baer weighs down with a vengeance all the virulence of Owen
and weak arguments of Agassiz. If you write to Von Baer, for heaven's
sake tell him that we should think one nod of approbation on our side,
of the greatest value; and if he does write anything, beg him to send us
a copy, for I would try and get it translated and published in the
Athenaeum and in 'Silliman' to touch up Agassiz.—C. Darwin to T. H.
Huxley, August 8, 1860 (*Life and Letters,* 2:123)

The Controversy over Darwinism*

K. E. VON BAER

Recently, in the Augsburger Allgemeine Zeitung the battle for and against
Darwinism has been waged with passion. Even though this battlefield does

* From *Augsburger Allgemeine Zeitung* (1873), no. 130, pp. 1986–1988. Translated
by D. L. H.

not seem to me to be the most appropriate place to settle such a question, I too must beg admittance to this arena. For I want to ask that my name not be included among those for or against Darwinism, since I have not previously expressed my views on the descent hypothesis and my work in the field of embryology was published prior to the rise of Darwinism. (In my work in embryology I took great pains to point out that the higher categories of animals differ greatly from each other in the course of their development; however, the essential characters of the higher groups are always formed first and completely, while the characters peculiar to the lower, less inclusive groups appear later, until finally those of the species category appear.)

As a matter of fact, a few years ago, when Darwinism was still fairly new, I began an essay on it. The printing of this work had already begun when Darwin announced that on his principles, he would demonstrate the genesis of the human species. I decided to wait for this book [*The Descent of Man*], since such a proof seemed impossible. I found it very commendable that Darwin had avoided passing judgment on the answer to this question in his earlier publications, though it certainly must have confronted him. While awaiting this book, I undertook other tasks which, I am sorry to say, have occupied me for a long time, less because of their inherent difficulty than because in the interim my eyesight has almost completely failed me. Even so, I still harbor the wish to make public my views on the scientific foundations of the Darwinian hypothesis of descent, whether before my death or after. The second volume of my *Reden und Aufsätze* [von Baer, 1876] which will appear after the third, is devoted to a discussion of Darwinism. However, here for the time being is a preliminary opinion.

Meanwhile Darwin's book on the descent of man has appeared, but it has not convinced me. I still cannot comprehend how man in the course of time could have developed from an ape-like animal. My doubts on this score are very simple. No matter how I view apes, they always seem to be organized for an arboreal life, whereas, in contrast, man is organized to walk upright on solid ground. Of course, it is said that both faculties have developed in the course of time through "adaptation." But then, for what environment should this problematic primate archetype have been organized, since, obviously, all animals require a specific environment? Were they perhaps climbing animals that had a few descendants so obsessed with the idea of progressing that they refrained from climbing trees for thousands and millions of years until their lower extremities became adapted for an upright gait? It would seem easier for me to imagine that this archetype was plantigrade. A few of their descendants were so glutinous that they refused to leave the trees which supplied their food, and from these evolved the "poor relations" which we sometimes call apes.

"But," someone might ask, "why bother yourself about the question of how these first primates may have lived? It is enough that they must have existed if the descent of man is to be explained." But to this I reply that no matter how necessary it may seem for such an explanation, I cannot believe that a creature has lived and reproduced itself which was not organized from the start as one of the possible forms of life that could inhabit this earth.

"Heaven help us! Destiny, goals, purposes[1] Don't you know that naturalists who adhere to similar superstitions are fast becoming extinct?"

Certainly I have read such claims, but they do not frighten me, since I have known for some time that I am nearing the end of my life. Besides, I console myself with the knowledge that I do not envy those who ascend the ladder of immortality with a step so much firmer than mine. I just wish that they would make the foundation upon which their ladder stands more substantial. These foundations must surely be inadequate, since only a short time ago a naturalist like Agassiz could say that Darwinism is a "mire of mere assertions." This is certainly very blunt; but the situation is aggravated by the fact that this bluntness comes from a naturalist who can hardly be reproached for having a mind closed to general ideas and who, moreover, possesses a most thorough knowledge of those branches of science concerned with the phylogenetic development of animal forms— palaeontology, embryology, and comparative anatomy.

I object not only to the foundations of the Darwinian system but also to the conclusions and embellishments which top it off. To be specific, the embellishments which I have in mind are the cynical attacks on religious conceptions which invariably ornament the pinnacle of the system. However, I will not pursue this matter further because I find it too offensive. But I still must say a word or two about other philosophical consequences!

In the anonymous essay, 'The Unconscious from the Viewpoint of Physiology and the Theory of Descent' (1872), we read: "Now, for the first time (i.e., since Darwin) it is possible to acknowledge purposiveness [*Zweckmässigkeit*] in nature but only as the result of precise, verifiable

1. Words like *Zweck* and *Zweckmässig* have a spectrum of meanings—especially in the German of a hundred years ago—which comparable words in modern English lack. They have accordingly been translated differently, depending on the context— D. L. H.

compensatory processes." More than anything else, it is very pleasing to see purposiveness given proper credit in the development of organic bodies, since, after all, this is obviously what is meant. But what about these demonstrable, mechanical compensatory relations which are supposed to explain purposiveness? Is it that the less fit [*Zweckmässige*] among the infinite variety of forms continually produced are annihilated in the "struggle for existence"? Somewhere in the back of my mind I dimly recall having actually read or heard about striving to attain fitness through accidental variability in the offspring and the elimination of the unfit. Since I am at this moment trying to coax this dim memory over the "threshold of consciousness," it comes to me with particular vividness! In the Academy of Lagado there was a philosopher who maintained, and rightly so, that all the truths which man can hope to know are expressible in words. He inscribed all of the words of his language in all their grammatical forms on dice and then devised a machine which would take these word-covered dice, shuffle them, and then combine them in various ways. After each turn of the machine, the words that were visible were read off. When three or four words went together to make sense, the resulting sentence fragment was jotted down. Thus, in this way, all possible truth was going to be obtained since, after all, truth can be expressed only in words. The elimination of those that did not go together was equally mechanical and was completed much more rapidly than occurs in the "struggle for existence." What was eventually accomplished? Unfortunately we don't know. [Lemuel] Gulliver is the only one who has recorded the history of the Academy of Lagado in his third voyage. When he was there, a number of volumes had been filled with individual sentences and the wish had been expressed that in the public interest and education, five hundred machines should be built exactly like the original machine and operated at the public expense! For a long time the author of these reports was taken to be joking, because it is self-evident that nothing useful and significant could ever result from chance events. On the contrary, order must emerge as a complete whole at the outset, even though there might well be room for considerable improvement. Now we must acknowledge this philosopher as a deep thinker since he foresaw the present triumphs of science!

Accidents!? But Professor Haeckel states quite categorically, "There is no such thing as accident!" Because everything has its sufficient cause, universal necessity or force prevails, and since these words did not appear sufficiently strong in German, he closes with a Greek phrase!

No thinking human being can doubt that every event has its sufficient cause. Nevertheless, accidents do exist and occur quite frequently. When I pass by a house—for whatever reason—and at that particular moment a stone falls off the roof, this is for me an accident—an unhappy one if the stone falls on my head and a very insignificant one if it drops in front of me or behind me. To be sure, the falling of the stone had its sufficient cause. It might have come loose from its moorings. The only accidental feature is that its fall coincided with my passing by, unless of course it was directed at me by someone. Then it ceases to be an accident. Accident, therefore, is the convergence of two or more chains of events whose efficient causes are different. This is reflected in the formation of words in most languages: *Zufall;* accident (English and French), *accidens* or *caus,* which means literally "fall." In general one event is always considered the more important, and the other is added as a coincident. In nature, no happening considered by itself is accidental, but it is an obvious misusage to call an event an accident just because we do not know its real cause.

If, as the descent hypothesis postulates, higher forms of animal life developed out of lower forms, then either the real causes for these transformations are causally connected with these lower forms of life or they are not—much as a cherry trees bears some buds that blossom and others that do not. In the latter case, later developments are accidental, and since this is true for the whole system we are discussing, the entire course of the development of the organic world is based on accidents or, for want of a better word, on chance. Since those who oppose the Darwinian system have brought this charge against it, the defenders of the system have thought it wise to eliminate chance entirely. But if the higher forms of animal life stand in a causal relationship to the lower, developing out of them, then how can we deny that nature has purposes or goals?

Apparently Darwinism has triumphed precisely because it denies purposes in nature and because it insists on explaining the appearance of purposiveness in nature by blind forces producing a host of life forms and the elimination of the less fit by natural selection. We see even philosophy bowing before these conclusions. Thus, when some of the proponents of this new theory—though definitely not Darwin himself, since he is much too cautious to do such a thing—can show that something developed necessarily, they exclaim with great delight, "necessity but no purpose!" Now that is really cute! Do purposes exclude the efficacy of necessity? Rather, don't they assure it? If I wanted to shoot a bird, perhaps with the intention of roasting or stuffing it, I would hardly attain this end, no matter how diligent and talented I might be, if I were simply to throw pellets at it. However, if I put gunpowder, a plug, and shot in the barrel of a gun, and then aim the gun accurately before I shoot, I will attain my end. The powder explodes, producing a large mass of gas which expands considerably and propels the shot with tremendous force out of the barrel of the gun in the direction toward which it was aimed. Isn't this the way that natural forces function in the attainment of ends? Need I remind you that a boy who wants to draw a circle will not get very far if he just draws it free-hand (no matter how talented he may be), but if he uses a compass (i.e., a mechanical device), he can easily attain his goal?

I am well aware that it is not considered polite to be too outspoken. But how am I to express my utter amazement at the frequency with which one meets the belief that all goals are completely absent in the workings of natural forces and material substances? I too am convinced that everything that exists and continues to exist in nature arose and will continue to arise through natural forces and material substances. But these natural forces must be coordinated or directed. Forces which are not directed—so-called blind forces—can never, as far as I can see, produce order. In the preceding examples the efficient causes and material substances were coordinated.

Now I must proceed more quickly to my original goal—space permitting. That goal is to defend teleology, after I attempt to show the readers of this newspaper how, in my opinion, purpose appears to have been spirited out of nature, much to the alarm of many and the joy of a few.

Over a century ago, when we had a much poorer understanding of organic beings and when the verification of the laws of nature (with the exception of the law of universal gravitation) seemed impossible, many people were so enthralled with the diversity and construction of organisms and their parts that they felt compelled to credit the Creator with every little detail. It was at this time that "Insectotheology," "Fishtheology," even "Rocktheology" and other books with equally comical titles, made their appearance. The advance of science proved that these views were much too superficial and that greater penetration could be obtained if self-regulating powers were recognized in nature. One considers these "forces" autonomous, because the mind operates more easily with this concept. Besides, comparative anatomy has shown that many of the things which had been admired so greatly before did not deserve this admiration in the least. For instance, serial homologies threw some observers into fits of admiration.

One need mention only Schäfer. However, it is precisely the lower organisms which exhibit serial homologies. In the more highly developed organisms one finds fewer organs, but these differ greatly from each other and are extremely diverse in their organization. Some worms, spiders, scorpions, and many insects have numerous eyes or eye spots. But some of the latter are so simple that there is no way of knowing whether they actually register images of the external world. Vertebrates never have more than two eyes, but these eyes are so marvelously constructed that they produce a sharp picture even of peripheral objects, while simultaneously being so mobile that they can focus on any given point. Some worms have numerous, sac-shaped hearts; a vertebrate has only one heart, but it is divided into several compartments, and each of these has a definite function. It seems that the admiration for serial homology arose through reasoning by analogy from the results of human effort. Because every little bead in a rosary has to be made separately, more work is required to make longer strand than a shorter one. However, the forces or powers of material substance through which nature works are incomparably broader in scope than those of man. One might say that the production of similar parts is frequently easier for nature, if this word were appropriate. Take, for example, a rainstorm in which millions of raindrops fall. What a task it would be if each drop had to be make separately. Nature does things much more simply. When air passes over a body of water, it absorbs the water in the form of an invisible vapor. The warmer the air, the more water it can absorb. But when this warm air is cooled upon meeting a cold mass of air, it can no longer hold so much water vapor. Part of it condenses into a fine mist. If this mist becomes sufficiently dense, droplets combine to form drops. Since these cannot continue to float in the air, they fall, and we have a shower of countless drops. However, on these same principles, a single drop falling from the sky would require such a peculiar combination of conditions that it probably has never happened.

The fact that the natural theologians of the last century used only human performance as the standard for their judgments is made even more obvious by their praise of the diversity and beauty of organic forms and by their tacit assumption that the Creator did all of this to show us his artistry.

The more one recognized various regularities and law-governed powers in the contruction and life of organisms, the more ludricous this naive view of nature appeared. Soon, only observations and their explanation were valued, that is, tracing back observations to various regularities and possibly even to general laws. It became customary, almost as if it were a tacit agreement among naturalists, to omit any mention of the Creator in their discourses, since no knowledge of him can be obtained in this manner anyway. They preferred to speak of nature—that is, all being and becoming which is independent of man. Even though it was completely justified, theologians used to find this custom of naturalists reprehensible. The naturalist, being what he is, wants to know what exists in nature and how it functions.

In doing so, however, naturalists frequently got out of the habit of referring to the goals of the processes and things which they observed. The main reason for this was that earlier investigators had been too rash in deciding what destiny [*Bestimmung*] a process or a thing might have, frequently without bothering to look into the situations themselves. They searched only for purposes, without searching further into the conditions of existence. Older physiology is full of it. Since their knowledge was so incomplete, it is easy to see why they were so often mistaken in their assignment of purposes. This, then, was how the teleological way of studying nature came to be completely discredited.

Surely it is wrong for a naturalist to attempt to assign a destiny to a thing before he understands it sufficiently. But today, when so many naturalists energetically reject the teleological viewpoint as being conducive to error, then to my mind it is they who err. I have already expressed my opinion that the answer to the question "What for?" must not be confused with the answer to the question "How come?"[2] The loud, fanatical rejection of all teleological conceptions of nature serves only to alarm and confuse the general public, who are primarily interested in the goals of things. Try as he may, even the most hard-nosed naturalist cannot avoid this question. He meets it at every turn. Darwinists lay tremendous stress on heredity, but what else is heredity if not the determination of something in the future? Is not this itself highly teleological? Indeed, is it not the object of all this transmissibility to initiate a new life? When one studies

2. *Bulletin de l'Academic de St. Petersburg* [1860], 9:126–127. With respect to what I have said above, I do not want to suppress the remark that many Darwinists completely misuse the concept of purposiveness [*Zweckmässigkeit*] when they mistakenly maintain that the meaning of the "struggle for existence" is equivalent to that of being fit [*zweckmässig*]. Persistence, not fitness, is promoted in the struggle for existence. The natural elements and the building which we inhabit clash in a struggle for existence, and in the end the former always win. Caves are our most durable dwelling places, but one would hardly term them the most fit. In this conception of nature, however, caves would be the most fit. On the contrary, human beings find the perishable fruits of the trees and the edible flesh of animals much more useful. In general, the notion of purpose has meaning and value only in a specific context.

the oviducts of a dissected insect and asks how they developed, the present-day naturalist can say only that through metabolism they came by necessity to be what they were in their ancestors. But what if one wants to know why they are that way? Is it wrong to say that they are constructed as they are so that, when the ripened eggs are fertilized, new individuals of the same kind can be produced?! The fervor displayed against narrow-minded teleology is justified, just as it is against all narrow-mindedness! But why should it be considered a sign of simple-mindedness if one enjoys the wonderful co-adaptions of natural processes or, starting with the empirical fact that man appeared quite late in the history of the earth, if I accept him as the highest goal of natural processes and recognize that the great diversity of plants and animals which appeared earlier were able to satisfy his material needs?

When the naturalist speaks of natural processes within the narrow confines which are set for research, he should perhaps avoid the two words "purpose" and "purposeful." The word "purpose" is used most properly in the context of man's ability to set himself a goal, to strive to attain it, to change it, to postpone its realization or to abandon it altogether. That which we observe in nature, especially in organisms, is of quite another kind. Every cause results in a process, which has an effect on some other object. But an individual natural process may not be credited with "purpose" because "purpose" presupposes conscious effort. Therefore, one cannot say that the egg has the purpose of becoming a chicken, because the egg has neither consciousness nor will. However, if the egg undergoes thousands of changes by the action of air and suitable warmth until finally a viable chick hatches from the egg, then it is all right to say that the chick is the result of all these processes. However, since every healthy egg under the same circumstances always produces the same results, we can safely say that the egg has the "goal" of developing into a chick and is organized accordingly. The material of which it is made is so organized, both quantitatively and qualitatively, that under pre-determined conditions a chick must develop. Therefore, the egg itself may be said to have the "goal," since it seems to me that the word "goal" need not imply a conscious will. (The center of the earth is the goal of a falling stone since that is where it would stop if it could ever reach that spot.). This is why in the previously mentioned essay, I have proposed to substitute the word "goal-directed" [zielstrebig] for the word "purposeful" [zweckmässig]. The goal for which the egg is organized is to become a chick. The organization of the egg is prepared inside the mother organism. Further development and eventual completion is brought about by fertilization and brooding, both of which are directed by instinct.

"Instincts are inherited habits," say the Darwinists. I must admit that at times it looks that way. Bu as far as our example is concerned the very first hen must have hatched, otherwise there would have been no subsequent hens. Rather it seems to me that instinct is the "goal-directedness of volitional processes," just as we can observe goal-directedness in physical processes and without which all orderly life processes are inconceivable.

Everywhere we find in individual organisms "goal-directedness" mediated by the forces of material substances. These operate according to prescribed principles, i.e., according to necessity. The highest level of goal-directedness in organisms is the life process itself. It cannot, however, continue without continuous interaction with the outside world. So throughout nature I find necessity mediating goal-directedness. It in turn can be traced back to material substances and their forces, but conversely, material substances and their forces must be directed if they are to lead to a goal.

Finally, this control could have been initiated only by a spiritual entity. This entity is of course the Creator. The precise determination of the essence of this entity and how He produces such a multiplicity of forms will never be decided by means of scientific investigation.

Those who do not want to acknowledge this original Spiritual Being must, therefore, assume that the world originated by means of an untold number of accidents. I find it impossible to embrace this conviction. These countless accidents would have to be in marvelous harmony if anything orderly were ever to result. Therefore, I call upon the teleophobes, i.e., those naturalists who are so afraid of any appeal to purpose, to substitute the word "goal-directed" or, if you like "goal-oriented," in the appropriate places for the word "purposeful," since this is the sense in which most naturalists use the word anyway.

Comments on von Baer

Karl Ernst von Baer was one of the most eminent of the old-guard biologists. He discovered the mammalian ovum in 1827, enunciated a germ layer theory of embryological development, and all but founded the science of embryology. One can imagine Darwin's elation when Huxley announced him as a "new and great ally." Prior to the appearance of the *Descent of Man,* von Baer was planning to publish a short piece on evolutionary theory. He may have intended to accept, at least tentatively, certain aspects

Darwin and His Critics

of it. However, Darwin's bold assertions about the evolution of man were more than von Baer could stomach: a man could never evolve into an ape or an ape into a man. His rejoinder is included in this volume. He followed up this short discussion three years later by a longer denunciation of evolutionary theory in his *Reden gehalten in wissenschaftlichen Versammlungen und kleinere Aufsätze vermischten Inhalts* (1876).

Even before the publication of the *Origin of Species* radically transformed biology, von Baer had found himself separated from his younger colleagues by what Kuhn has called a "paradigm switch." Natural processes might have goals, but it was no longer acceptable to talk of such goals in scientific publications. Von Baer explained this convention as an over-reaction to the naïve teleology of the previous century. In time, soberer minds would prevail, and scientists would come to realize once more the necessity of the teleological dimension to science. Von Baer simply could not see how undirected natural forces could produce organization. Darwin declared that natural selection acting on chance variation could do just that. And Darwin did not stop short at man. Even the organization present in man was the product of such undirected forces.

Von Baer's main objection to evolutionary theory was the role which "chance" or "accident" played in it, as opposed to his own teleological views. In his discussion of these issues, von Baer adheres to the basic terminology of Aristotle. In fact, he is so steeped in the metaphysics of Christian Aristotelianism that he was totally incapable of understanding Darwin's theory. According to von Baer, the only events which are truly caused are those that are intended by some conscious agent. Any event which occurs only through the agency of natural necessity is uncaused and accidental. For example, an eclipse of the sun by the moon as seen from the earth is not the result of any conscious intent on the part of these heavenly bodies. Certainly no human beings influence the event. If there were no creator, eclipses would be accidental. For von Baer, all purely physical causal chains are accidental—even the most regular and mechanical—in the absence of God. Introduce an all-powerful omniscient being, and all events are caused. From the human perspective, the coincidence of some causal chains is accidental; from the divine perspective, all events are totally caused. Using this terminology, von Baer was able to claim that anyone who does not acknowledge an original spiritual being must assume that the world emerged through countless accidents, an incredible assumption.

Darwin agreed with von Baer that all events were caused, including what Darwin termed "chance variations." Variations were certainly caused.

Darwin termed them "chance" to contrast his views with the travesties of Lamarck then prevailing. An organism would survive because of those traits which it happened to possess. If it possessed those traits that it needed, it survived; if not, it perished. But variation was not oriented toward the future. If a species of fish were living in water which was gradually becoming more alkaline, the modifications needed to survive in such an environment were no more likely to occur in the successive progeny of this species of fish than in any other species. In short, Darwin thought that there was no correlation between those variations which a species might need and those it might get. And Darwin saw no point in adding to his description of the organic world, "And whatever happens, God is responsible." Naïve teleology was false. Sophisticated teleology was redundant. Von Baer could not see how the movement of one billiard ball could be explained solely in terms of another billiard ball's hitting it. In increasing numbers, von Baer's fellow scientists failed to see how the presence of divine guidance added anything to their explanations.

Von Baer was especially irked because his views on the progressive development of embryos was being misconstrued by some as supporting evolutionary theory. On the contrary, von Baer's doctrine was no more evolutionary than that of Aristotle, who had anticipated him by two thousand years. According to both men, each organism has its place in a nested hierarchy of forms. During its embryological development, the essential characters of the organism emerged, first the essential characters of the highest category to which it belonged, then the essential characters of its next highest category, and so on down to the species level. "But," von Baer insisted, "there is no repetition of ancestral forms in any embryonic period. There is only a parallel development of different types up to a certain stage." For example, the embryological development of an octopus and a fish would diverge immediately after the appearance of those characters which made both of them animals. The embryological development of a fish and a frog would run in parallel much longer. They would diverge only after their vertebrate characters emerged. The embryological development of two different species of swans would parallel each other almost to the very end and then diverge. Darwin saw such parallels as hints as to the possible phylogenetic development of the organisms involved and as supporting his theory. Haeckel inflated this relation into his Biogenetic Law.*

* See also Haacke (1905), Holmes (1947), and Oppenheimer (1959).

Louis Agassiz (1807–1873)

It is delightful to hear all that he [Asa Gray] says on Agassiz: how very singular it is that so *eminently* clever a man, with such *immense* knowledge on many branches of Natural History, should write as he does. Lyell told me that he was so delighted with one of his (Agassiz) lectures on progressive development, &c., &c., that he went to him afterwards and told him, 'that it was so delightful, that he could not help all the time wishing it was true.'—C. Darwin to J. D. Hooker, Down, March 26, 1854 (*Life and Letters,* 1:403)

I have always suspected Agassiz of superficiality and wretched reasoning powers; but I think such men do immense good in their way. See how he stirred up all Europe about glaciers.—C. Darwin to T. H. Huxley, Down, September 26, 1857 (*More Letters,* 1:103–104)

Agassiz, when I saw him last, had read but a part of it [the *Origin*]. He says it is *poor—very poor!!* (*entre nous*). The fact [is] he is very much annoyed by it . . . and I do not wonder at it. To bring all *ideal* systems within the domain of science, and give good physical or natural explanations of all his capital points, is as bad as to have Forbes take the glacier materials . . . and give scientific explanation of all the phenomena.
Tell Darwin all this.—Asa Gray to J. D. Hooker, Cambridge, Massachusetts, January 5, 1860 (*Life and Letters,* 2:63)

I am amused by Asa Gray's account of the excitement my book has made amongst the naturalists in the U. States. Agassiz has denounced it in a newspaper, but yet in such terms that it is in fact a fine advertisement!—C. Darwin to J. Murray, Down, 1860 (*Life and Letters,* 2:64–65)

Have you seen Agassiz's weak metaphysical and theological attack on the 'Origin' in the last 'Silliman'?—C. Darwin to T. H. Huxley, Down, August 8, 1860 (*Life and Letters,* 2:123)

As you wish to hear what reviews have appeared, I may mention that Agassiz has fired off a shot in the last 'Silliman,' not good at all, denies variations and rests on the perfection of Geological evidence. Asa Gray tells me that a very clever friend has been almost converted to our side by this review of Agassiz's— To C. Lyell, August 11, 1860 (*Life and Letters,* 2:124)

I am surprised that Agassiz did not succeed in writing something better. How absurd that logical quibble—'if species do not exist, how can they vary?' As if anyone doubted their temporary existence. How coolly he assumes that there is some clearly defined distinction between individual differences and varieties. It is no wonder that a man who calls identical forms, when found in two countries, distinct species, cannot find variation in nature. Again, how unreasonable to suppose that domestic varieties selected by man for his own fancy should resemble natural varieties or species. The whole article seems to me poor; it seems to me hardly worth a detailed answer (even if I could do it, and I much doubt whether I possess your skill in picking out salient points and driving a nail into them), and indeed you have already answered several points. Agassiz's name, no doubt, is a heavy weight against us.— To Asa Gray, August 11, 1860 (*Life and Letters,* 2:124)

I am delighted at the manner in which you have bearded this lion [Agassiz] in his den. I agree most entirely with all that you have written. What I meant when I wrote to Agassiz to thank him for a bundle of his publications, was exactly what you suppose. I confess, however, I did not fully perceive how he had misstated my views; but I only skimmed through his *Methods of Study,* and thought it a very poor book. I am so much accustomed to be utterly misrepresented that it hardly excites my attention. But you really have hit the nail on the head capitally. All the younger good naturalists whom I know think of Agassiz as you do; but he did grand service about glaciers and fish. About the succession of forms, Pictet has given up his whole views, and no geologist now agrees with Agassiz. I am glad that you have attacked Dana's wild notions; [though] I have a great respect for Dana.—C. Darwin to B. D. Walsh, Down, December 4, 1864 (*More Letters,* 1:258)

You refer in your letters to the facts which Agassiz is collecting, against our views, on the Amazons. Though he has done so much for science, he seems to me so wild and paradoxical in all his views that I cannot regard his opinions as of any value.—C. Darwin to F. Müller, Down, January 11, 1866 (*More Letters,* 1:264)

Agassiz is back (I have not seen him), and he went at once down to the National Academy of Sciences, from which I sedulously keep away, and, I hear, proved to them that the Glacial period covered the whole continent of America with unbroken ice, and closed with a significant gesture and the remark: 'So here is the end of the Darwin theory.' How do you like that?

I said last winter that Agassiz was bent on covering the whole continent with ice, and that the motive of the discovery he was sure to make was to make sure that there should be no coming down of any terrestrial life from Tertiary or post-Tertiary period to ours. You cannot deny that he has done his work effectually in a truly imperial way.—Asa Gray to Charles Darwin, September 1866 (*More Letters*, 2:160)

Many thanks for the pamphlet [Agassiz's "Geology of the Amazons"], which was returned this morning. I was very glad to read it, though chiefly as a psychological curiosity. I quite follow you in thinking Agassiz glacier-mad.—C. Darwin to C. Lyell, Down, September 8, 1866 (*More Letters,* 2:159)

Alex Agassiz has just paid me a visit with his wife. He has been in England two or three months, and is now going to tour over the Continent to see all the zoologists. We liked him very much. He is a great admirer of yours, and he tells me that your correspondence and book first made him believe in evolution. This must have been a great blow to his father, who, as he tells me, is very well, and so vigorous that he can work twice as long as he (the son) can.—C. Darwin to F. Müller, Down, December 1, 1869 (*More Letters*, 2:357–358)

Evolution and Permanence of Type*

LOUIS AGASSIZ

In connection with modern views of science we hear so much of evolution and evolutionists that it is worth our while to ask if there is any such process as evolution in nature. Unquestionably, yes. But all that is actually known of this process we owe to the great embryologists of our century, Döllinger and his pupils K. E. von Baer, Pander, and others,–the men in short who have founded the science of Embryology. It is true there are younger men who have done since, and are doing now, noble work in this field of research; but the glory must, after all, be given to those who opened the way in which more recent students are pressing forward.

* From the *Atlantic Monthly,* January 1874, 33:92–101.

The pioneers in the science of Embryology, by a series of investigations which will challenge admiration as long as patience and accuracy of research are valued, have proved that all living beings produce eggs, and that these eggs contain a yolk-substance out of which new beings, identical with their parents, are evolved by a succession of gradual changes. These successive stages of growth constitute evolution, as understood by embryologists, and within these limits all naturalists who know anything of Zoölogy may be said to be evolutionists. The law of evolution, however, so far as its working is understood, is a law controlling development and keeping types within appointed cycles of growth, which revolve forever upon themselves, returning at appointed intervals to the same starting-point and repeating through a succession of phases the same course. These cycles have never been known to oscillate or to pass into each other; indeed, the only structural differences known between individuals of the same stock are monstrosities or peculiarities pertaining to sex, and the latter are as abiding and permanent as type itself. Taken together the relations of sex constitute one of the most obscure and wonderful features of the whole organic world, all the more impressive for its universality.

Under the recent and novel application of the terms "evolution" and "evolutionist," we are in danger of forgetting the only process of the kind in the growth of animals which has actually been demonstrated, as well as the men to whom we owe that demonstration. Indeed, the science of Zoölogy, including everything pertaining to the past and present life and history of animals, has furnished, since the beginning of the nineteenth century, an amount of startling and exciting information in which men have lost sight of the old landmarks. In the present ferment of theories respecting the relations of animals to one another, their origin, growth, and diversity, those broader principles of our science—upon which the whole animal kingdom has been divided into a few grand comprehensive types, each one a structural unit in itself—are completely overlooked.

It is not very long since, with the exception of Insects, all the lower animals were grouped together in one division as Worms, on account of their simple structure. A century ago this classification, established by Linnaeus, was still unquestioned. Cuvier was the first to introduce a classification based not merely upon a more or less complicated organization but upon ideas or plans of structure. He recognized four of these plans in the whole animal kingdom, neither more nor less. However, when this principle was first announced, the incompleteness of our knowledge made it impossible to apply it correctly in every case, and Cuvier himself placed certain animals of obscure or intricate structure under the wrong head. Nevertheless the law was sanctioned, and gave at once a new aim and impulse to investigation. This idea of structural plans, as the foundation of a natural classification, dates only from the year 1812, and was first presented by Cuvier in the Annals of the Museum in Paris.

About the same time another great investigator, Karl Ernest von Baer, then a young naturalist, Döllinger's favorite and most original pupil, was studying in Germany the growth of the chicken in the egg. In a different branch of research, though bearing equally on the structural relations of organized beings, he, without knowing of Cuvier's investigations, arrived at a like conclusion, namely, that there are four different modes of growth among animals. This result has only been confirmed by later investigators. Every living creature is formed in an egg and grows up according to a pattern and a mode of development common to its type, and of these embryonic norms there are but four. Here, then, was a double confirmation of the distinct circumscription of types, as based upon structure, announced almost simultaneously by two independent investigators, ignorant of each other's work, and arriving at the same result by different methods. The one, building up from the first dawn of life in the embryonic germs of various animals worked out the four great types of organic life from the beginning; while his co-worker reached the same end through a study of their perfected structure in adult forms. Starting from diametrically opposite points, they met at last on the higher ground to which they were both led by their respective studies.

For a quarter of a century following, the aim of all naturalists was to determine the relations of these groups to one another with greater precision, and to trace the affinities between the minor divisions of the whole animal kingdom. It was natural to suppose that all living beings were in some way or other connected; and, indeed, the discoveries in Geology, with its buried remains of extinct life, following fast upon those of Cuvier in structure and of Von Baer in Embryology, seemed to reveal, however dimly and in broken outlines, a consistent history carried on coherently through all times and extending gradually over the whole surface of the earth, until it culminated in the animal kingdom as it at present exists, with man at its head.

The next step, though a natural result of the flood of facts poured in upon us under the new stimulus to research, led men away from the simple and, as I believe, sound principles of classification established by the two great masters of zoölogical science. The announcement of four typical divisions in the animal kingdom stirred investigators to a closer comparison of their structure. The science of Comparative Anatomy made rapid strides; and since the ability of combining facts is a much rarer gift than that of discerning them, many students lost sight of the unity of structural design in the multiplicity of the structural detail. The natural result of this was a breaking up of the four great groups of Radiates, Mollusks, Articulates and Vertebrates into a larger number of primary divisions. Classifications were multiplied with astonishing rapidity, and each writer had his own system of nomenclature, until our science was perplexingly burdened with synonymes. I may mention, as a sample, one or two of the more prominent changes introduced at this time into the general classification of animals.

The Radiates had been divided by Cuvier into three classes, to which, on imperfect data, he erroneously added the Intestinal Worms and the Infusoria. These classes, as they now stand according to his classification, with some recent improvements, are Polyps (corals, sea-anemones, and the like), Acalephs (jellyfishes), and Echinoderms (star-fishes, sea-urchins, and holothurians, better known, perhaps, as Beche-de-mer). Of these three classes the two first, Polyps and Acalephs, were set apart by Leuckart and other naturalists as "coelenterata," while the Echinoderms by themselves were elevated into a primary division. There is, however, no valid ground for this. The plan of structure is the same in all three classes, the only difference being that various organs which in the Polyps and Acalephs are, as it were, simply hollowed out of the substance of the body, have in the Echinoderms walls of their own. This is a special complication of structural execution, but makes no difference in the structural plan. The organs and the whole structural combination are the same in the two divisions. In the same way Cephalopods, squids and cuttlefishes, which form the highest class among Mollusks, were separated from the Gasteropods and Acephala, and set apart as a distinct type, because their eggs undergo only a surface segmentation instead of being segmented through and through, as is the case with the members of the two other classes. But this surface segmentation leads ultimately to a structure which has the same essential features as that of the other Mollusks. Indeed, we find also in other branches of the animal kingdom, the Vertebrates, for instance, partial or total segmentation, in different classes; but it does not lead to any typical differences there, any more than among Mollusks. Another instance is that of the Bryozoa and Tunicata, which were separated from the Mollusks on account of the greater simplicity of their structure and associated with those simpler Worms in which articulated limbs are wanting. In short,

the numerous types admitted nowadays by most zoologists are founded
only upon structural complication, without special regard to the plan of
their structure; and the comprehensive principle of structural conception
or plan, as determining the primary types, so impressive when first announced, has gradually lost its hold upon naturalists through their very
familiarity with special complications of structure. But since we are still in
doubt as to the true nature of many organisms, such as the sponges and the
Protozoa so-called, it is too early to affirm positively that all the primary
divisions of the animal kingdom are included in Cuvier's four types. Yet
it is safe to say that no primary division will stand which does not bear
the test he applied to the four great groups, Radiates, Mollusks, Articulates,
and Vertebrates, namely, that of a distinct plan of structure for each.

The time has, perhaps, not come for an impartial appreciation of the
views of Darwin, and the task is the more difficult because it involves
an equally impartial review of the modifications his theory has undergone
at the hands of his followers. The aim of his first work on The Origin
of Species was to show that neither vegetable nor animal forms are so
distinct from one another or so independent in their origin and structural
relations as most naturalists believed. This idea was not new. Under
different aspects it had been urged repeatedly for more than a century
by DeMaillet, by Lamarck, by E. Geoffroy St. Hilaire and others; nor
was it wholly original even with them, for the study of the relations of
animals and plants has at all times been one of the principal aims of
all the more advanced students of Natural History; they have differed
only in their methods and appreciations. But Darwin has placed the subject
on a different basis from that of all his predecessors, and has brought
to the discussion a vast amount of well-arranged information, a convincing
cogency of argument, and a captivating charm of presentation. His doctrine
appealed the more powerfully to the scientific world because he maintained
it at first not upon metaphysical ground but upon observation. Indeed
it might be said that he treated his subject according to the best scientific
methods, had he not frequently overstepped the boundaries of actual knowledge and allowed his imagination to supply the links which science does
not furnish.

The excitment produced by the publication of The Origin of Species
may be fairly compared to that which followed the appearance of Oken's
Natur-Philosophie, over fifty years ago, in which it was claimed that the
key had been found to the whole system of organic life. According to
Oken, the animal kingdom, in all its diversity, is but the presentation in

detail of the organization of man. The Infusoria are the primordial material of life scattered broadcast everywhere, and man himself but a complex of such Infusoria. The Vertebrates represent what Oken calls flesh, that is, bones, muscles, nerves, and the senses, in various combinations; the Fishes are Bone-animals (Knochen-Thiere); the Reptiles, Muscle-animals (Muskel-Thiere); the Birds, Nerve-animals (Nerven Thiere); the Mammals—with man, combining in his higher structure the whole scheme of organic life, at their head—are Sense-animals (Sinnen-Thiere). The parallelism was drawn with admirable skill and carried into the secondary divisions, down to the families and even the genera. The Articulates were likened to the systems of respiration and circulation; the Mollusks to those of reproduction; the Radiates to those of digestion. The comprehensiveness and grandeur of these views, in which the scattered elements of organic life, serving distinct purposes in the lower animals, are gathered into one structural combination in the highest living being appealed powerfully to the imagination. In Germany they were welcomed with an enthusiasm such as is shown there for Darwinism. England was lukewarm, and France turned a cold shoulder, as she at present does to the theory of the great English naturalist. The influence of Cuvier and the Jussieux was deeply felt in Western Europe, and perhaps saved French naturalists from falling into a fanciful but attractive doctrine, numbered now among the exploded theories of the past.

Darwin's first work, though it did not immediately meet with the universal acceptance since accorded to it, excited, nevertheless, intense and general interest. The circumstance that almost identical views were simultaneously expressed by Wallace, and that several prominent investigators hailed them as the solution of the great problem, gave them double strength; for it seemed improbable that so many able students of nature should agree in their interpretation of facts, unless that interpretation were the true one. The Origin of Species was followed by a second work, The Variation of Animals and Plants under Domestication, to which a third soon succeeded, The Descent of Man. The last phase of the doctrine is its identification with metaphysics in Darwin's latest work on The Expression of the Emotions in Man and Animals. I can only rejoice that the discussion has taken this turn, much as I dissent from the treatment of the subject. It cannot be too soon understood that science is one, and that whether we investigate language, philosophy, theology, history, or physics, we are dealing with the same problem, culminating in the knowledge of ourselves. Speech is known only in connection with the organs of man, thought in connection with his brain, religion as the expression of his aspirations, history as the record of his deeds, and physical sciences as the laws under which he lives. Philosophers and theologians have yet to learn that a physical fact is as sacred as a moral principle. Our own nature demands from us this double allegiance.

It is hardly necessary to give here an analysis of the theory contained in these works of Darwin. Its watchwords, "natural selection," "struggle for existence," "survival of the fittest," are equally familiar to those who do and to those who do not understand them; as well known, indeed, to the amateur in science as to the professional naturalist. It is supported by a startling array of facts respecting the changes animals undergo under domestication, respecting the formation of breeds and varieties, respecting metamorphoses, respecting the dangers to life among all animals and the way in which nature meets them, respecting the influence of climate and external conditions upon superficial structural features, and respecting natural preferences and proclivities between animals as influencing the final results of interbreeding. In the Variation of Animals and Plants under Domestication all that experiments in breeding or fancy horticulture could teach, whether as recorded in the literature and traditions of the subject or gathered from the practical farmers, stock-breeders, and gardeners, was brought together and presented with equal erudition and clearness. No fact was omitted showing the pliability of plants and animals under the fostering care of man. The final conclusion of the author is summed up in his theory of Pangenesis. And yet this book does but prove more conclusively what was already known, namely, that all domesticated animals and cultivated plants are traceable to distinct species, and that the domesticated pigeons which furnish so large a portion of the illustration are, notwithstanding their great diversity under special treatment, no exception to this rule. The truth is, our domesticated animals, with all their breeds and varieties, have never been traced back to anything but their own species, nor have artificial varieties, so far as we know, failed to revert to the wild stock when left to themselves. Darwin's works and those of his followers have added nothing new to our previous knowledge concerning the origin of man and his associates in domestic life, the horse, the cow, the sheep, the dog, or, indeed, of any animal. The facts upon which Darwin, Wallace, Haeckel, and others base their views are in the possession of every well-educated naturalist. It is only a question of interpretation, not of discovery or of new and unlooked-for information.

Darwin's third book, The Descent of Man, treats a more difficult part of the subject. In this book the question of genealogy is the prominent topic. It has been treated already, it is true, in The Origin of Species, but with no special allusion to mankind. The structure was as yet a torso, a trunk without a head. In these two volumes the whole ground of heredity, of qualities transmitted to the new individual by his progenitors, and that of resemblance—whether physical, intellectual, or moral, between mankind and the higher mammalia, and especially between ourselves and our nearest relations, the anthropoid monkeys,—are brought out with the fulness of material and the skill of treatment so characteristic of the author. But here again the reader seeks in vain for any evidence of a transition between man and his fellow-creatures. Indeed, both with Darwin and his followers, a great part of the argument is purely negative. It rests partly upon the assumption that, in the succession of ages, just those transition types have dropped out from the geological record which would have proved the Darwinian conclusions had these types been preserved, and that in the living animal the process of transition is too subtle for detection. Darwin and his followers thus throw off the responsibility of proof with respect to the embryonic growth and geological succession.

Within the last three or four years, however, it has seemed as if new light were about to be thrown at least upon one of these problems. Two prominent naturalists announced that they had found indications of a direct structural connection between primary types; in the one case between Mollusks and Vertebrates, in the other between Radiates and Articulates. The first of these views was published by a Russian investigator of great skill and eminence, Kowalevsky. He stated that the Ascidians (the so-called soft-shelled clams) showed, in the course of their growth, a string of cells corresponding to the dorsal cord in Vertebrates. For the uninitiated I must explain that, at one stage of its development, in the upper layer of cells of which the Vertebrate germ consists, there arise two folds which, curving upward and inward, form first a longitudinal furrow and finally a cavity. for the nervous centres, the brain and spinal cord, while the lower layer of these cells folds downward to enclose the organs of digestion, circulation, and reproduction. Between these two fold, but on the dorsal side, that is, along the back, under the spinal marrow, arises a solid string of more condensed substance, which develops into the dorsal cord, the basis of the backbone. Kowalevsky describes, in the Ascidians, a formation of longitudinally arranged cells as representing an incipient backbone, running from the middle of the body into the tail, along a furrow of the germ of these animals in which the main nervous swelling is situated. This was hailed as a great discovery by the friends of the transmutation theory. At last the transition point was found between the lower and higher animals, and man himself was traced back to the Ascidians. One could hardly open a scientific journal or any popular essay on Natural History, without meeting some allusion to the Ascidians as our ancestors. Not only was it seized upon by the many amateur contributors to the literature of this subject, but Darwin himself, and his ardent followers, welcomed this first direct evidence of structural affinity between the Vertebrates and the lower animals.

The existence of these cells, though never thought of in this light before, was not unknown to naturalists. I have myself seen and examined them, and had intended to say something in this article of their nature and position; but while I was preparing it for the press the subject was taken from me and treated by the hand of a master whom all naturalists venerate. I have received very recently from the aged Nestor of the science of Embryology, K. E. von Baer, to whose early investigations I have already alluded, a pamphlet upon the development of the Ascidians as compared to that of the Vertebrates. There is something touching in the conditions under which he enters the lists with the younger men who have set aside the great laws of typical structure, to the interpretation of which his whole life has been given. He is now very feeble and nearly blind; but the keen, far-reaching, internal sight is undimmed by age. With the precision and ease which only a complete familiarity with all the facts can give, he shows that the actual development of the Ascidians has no true homology with that of the Vertebrates; that the string of cells in the former—compared to the dorsal cord of the latter—does not run along the back at all, but is placed on the ventral side of the body. To say that the first Vertebrates or their progenitors carried their backbones in this fashion is about as reasonable as to say that they walked on their heads. It is reversing their whole structure, and putting their vertebral column where the abdominal cavity should be. Von Baer closes his paper in these words: "It will readily be granted that I have written for zoologists and anatomists; but I may perhaps be blamed for being frequently very circumstantial where a brief allusion would have been sufficient. In so doing, I had the many dilettanti in view, who believe in complete transmutations, and who might be disposed to consider it mere conceit not to recognize the Ascidians as the ancestors of Man. I beg to apologize for some repetitions arising from this consideration for the dillentanti."

The other so-called discovery is that of Haeckel, that star-fishes are com-

pound animals, made up, as it were, or worm-like beings united like rays
in one organism. A similar opinion had already been entertained by
Duvernoy, and in a measure also by Oken, who described the Echinoderms
as Radiate-worms. This doctrine, if true, would at once establish a transition
from Radiates to Articulates. There is, in the first place, not the slightest
foundation for this assumption in the structure of the star-fish. The arms
of these animals are made up of the same parts as the vertical zones of
a sea-urchin and of all the Radiates, and have no resemblance whatever
to the structure of the Worms. Each ambulacral zone of a star-fish or
a sea-urchin is strictly homological to a structural segment of an Acaleph
or to a radiating chamber of a Polyp. Moreover, the homology between
a sea-urchin and a star-fish is complete; if one is an organic unit the
other must be so also, and no one ever suggested that the sea-urchin was
anything but a single organism. In comparing the Radiates with other
animals, it is essential to place them in the same attitude, so that we compare like with like; otherwise, we make the mistake of the Russian naturalist, and compare the front side of one animal with the dorsal side of another,
or the upper side of one with the lower side of another; thus taking mere
superficial resemblance between totally distinct parts for true homologies.
In all Mollusks, Articulates, and Vertebrates the parts are arranged along
a longitudinal axis; in Radiates alone they are disposed around a vertical
axis, like spherical wedges, comparable in some instances to the segments
of an orange. This organic formula, for so we may call it, is differently
expressed and more or less distinct in different Radiates. It may be built
up in a sphere, as in the sea-urchins, or opened out into a star, like the
five-finger; it may be in the form of a sac divided internally, as in the
sea-anemones, or in that of a disk, channelled or furrowed so as to divide
it into equal segments, like the jelly-fish; but upon comparison the same
structural elements are found in all. These structural elements bear an
identical relation to the vertical axis of the animals. To compare any
Radiate with any Articulate is therefore to compare the vertical axis of
one animal with the horizonatal axis of the other. The parallelism will
not bear examination any more than that between the Mollusks and Vertebrates. Even in those holothurians and sea-urchins in which one side of
the body is flattened, the structure exhibits the same plan and the parts
are arranged in the same way as in all other Radiates, whatever be their
natural attitude in the element in which they live; whether they stand
upright with the mouth turned upward, or hang down in the reverse position, or crawl about horizontally. In like manner the vertical position of

man in no way invalidates the homology of his organization with that of the fishes, reptiles, birds, and mammalia. These two cases are thus far the only instances which have been brought forward to prove actual structural affinity between distinct primary divisions of the animal kingdom.

It is not my intention to take up, categorically all the different points on which the modern theory of transmutation is based. Metamorphosis plays a large part in it, and is treated as an evidence of transition from one animal into another. The truth is that metamorphosis, like all embryonic growth, is a normal process of development, moving in regular cycles, returning always to the same starting-point, and leading always to the same end; such are the alternate generations in the lower animals and the metamorphoses in higher ones, as in the butterflies and other insects, or in certain reptiles, frogs, and toads, salamanders, and the like. In some of these types the development lasts for a long time and the stages of embryonic growth are often so distinct that, until the connection between them is traced, each phase may seem like a separate existence, whereas they are only chapters in one and the same life. I have myself watched carefully all the successive changes of development in the North American Axolotl, whose recently discovered metamorphoses have led to much discussion in connection with the modern doctrine of evolution. I can see no difference between this and other instances of metamorphosis. Certain organs, conspicuous in one phase of the animal's life, are resorbed and disappear in a succeeding phase. But this does not differ at all from like processes in the toads and frogs, for instance; nor does it even differ essentially from like processes in the ordinary growth of all animals. The higher Vertebrates, including man himself, breathe through gill-like organs in the early part of their life. These gills disappear and give place to lungs only in a later phase of their existence. Metamorphoses have all the constancy and invariability of other modes of embryonic growth, and have never been known to lead to any transition of one species into another.

Another fertile topic in connection with this theory is that of heredity. No one can deny that inheritance is a powerful factor in the maintenance of race and in the improvement of breeds and varieties. But it has never been known that acquired qualities, even though retained through successive generation, have led to the production of new species. Darwin's attractive style is never more alluring than in connection with this subject. His concise and effective phrases have the weight of aphorisms and pass current for principles, when they may be only unfounded assertions. Such is "the survival of the fittest." After reading some chapters of The Descent of Man, could any one doubt, unless indeed he happened to be familiar with the facts, that animals, possessing certain advantages over others, are necessarily winners in the race for life? And yet it is not true that, outside of the influence of man, there are, in nature, privileged individuals among animals capable of holding on to a positive gain, generation after generation, and of transmitting successfully their peculiarities until they become the starting point for another step; the descendants losing at last, through this cumulative process, all close resemblance to their progenitors. It is not true that a slight variation, among the successive offspring of the same stock, goes on increasing until the difference amounts to a specific distinction. On the contrary, it is a matter of fact that extreme variations finally degenerate or become sterile; like monstrosities they die out, or return to their type.

The whole subject of inheritance is exceedingly intricate, working often in a seemingly capricious and fitful way. Qualities, both good and bad, are dropped as well as acquired, and the process ends sometimes in the degradation of the type and the survival of the unfit rather than the fittest. The most trifling and fantastic tricks of inheritance are quoted in support of the transmutation theory; but little is said of the sudden apparition of powerful original qualities which almost always rise like pure creations and are gone with their day and generation. The noblest gifts are exceptional, and are rarely inherited; this very fact seems to me an evidence of something more and higher than mere evolution and transmission concerned in the problem of life.

In the same way, the matter of natural and sexual selection is susceptible of very various interpretations. No doubt, on the whole, Nature protects her best. But it would not be difficult to bring together an array of facts as striking as those produced by the evolutionists in favor of their theory, to show that sexual selection is by no means always favorable to the elimination of the chaff and the preservation of the wheat. A natural attraction, independent of strength or beauty, is an unquestionable element in this problem and its action is seen among animals as well as among men. The fact that fine progeny are not infrequently the offspring of weak parents and *vice versa* points perhaps to some innate power of redress by which the caprices of choice are counterbalanced. But there can be no doubt that types are as often endangered as protected by the so-called law of sexual selection.

As to the influence of climate and physical conditions, we all know their power for evil and for good upon living beings. But there is, nevertheless, nothing more striking in the whole book of nature than the power shown by types and species to resist physical conditions. Endless evidence may be brought from the whole expanse of land and air and water, showing that identical physical conditions will do nothing toward the merging of species into one another, neither will variety of conditions do anything toward their multiplication. One thing only we know absolutely, and in this treacherous, marshy ground of hypothesis and assumption, it is pleasant to plant one's foot occasionally upon a solid fact here and there. Whatever be the means of preserving and transmitting properties, the primitive types have remained permanent and unchanged—in the long succession of ages amid all the appearance and disappearance of kinds, the fading away of one species and the coming in of another—from the earliest geological periods to the present day. How these types were first introduced, how the species which have successively represented them have replaced one another,—these are the vital questions to which no answer has been given. We are as far from any satisfactory solution of this problem as if development theories had never been discussed.

This brings us to the geological side of the question. As a palaeontologist I have from the beginning stood aloof from this new theory of transmutation, now so widely admitted by the scientific world. Its doctrines, in fact, contradict what the animal forms buried in the rocky strata of our earth tell us of their own introduction and succession upon the surface of the globe. Let us therefore hear them;—for, after all, their testimony is that of the eye-witness and the actor in the scene. Take first the type to which we ourselves belong. If it be true that there has been a progressive transmutation of the whole type of Vertebrates, beginning with the lowest and culminating in the highest, the earlier should of course be structurally inferior to the later ones. What then is the lowest[1] living Vertebrate? Every zoologist will answer, The Amphioxus, that elongated, worm-like Vertebrate whose organization is nothing more than a dorsal cord, with a nervous thread above, and a respiratory and digestive cavity below, containing also the reproductive organs, the whole being clothed in flesh. Yet low as it is in the scale of life, the Amphioxus is, by virtue of its vertebral column, a member of the same type as ourselves. Next to the Amphioxus come the Myxinoids, structurally but little above them, and the Lamper-eels. These are the animals which Haeckel places at the base of his zoological tree, rooting the whole Vertebrate branch of the animal kingdom in the Amphioxus as the forefather (Stamm-Vater) of the type. Let us look now

1. I use the terms low and high throughout in the zoological sense; with reference to specialization of structure, as comparative anatomist understand it.

at the earliest Vertebrates, as known and recorded in geological surveys. They should of course, if there is any truth in the transmutation theory, correspond with the lowest in rank or standing. What then are the earliest known Vertebrates? They are Selachians (sharks and their allies) and Ganoids (garpikes and the like), the highest of all living fishes, structurally speaking. I shall be answered that these belong to the Silurian and Devonian periods, and that it is believed that Vertebrates may have existed before that time. It will also be argued that Myzonts, namely Amphioxus, Myxinoids, and Lamper-eels, have no hard parts and could not have been preserved on that account. I will grant both these points, though the fact is that Myzonts do possess solid parts, in the jaws, as capable of preservation as any bone, and that these solid parts, if ever found, even singly, would be as significant, for a zoologist, as the whole skeleton. Granting also that Amphioxus-like fishes may have lived and may have disappeared before the Silurian period; the Silurian deposits follow immediately upon those in which life first appeared, and should therefore contain not the highest fishes, but the fishes next in order to the Myzonts, and these are certainly neither the Ganoids nor the Selachians. The presence of the Selachians at the dawn of life upon earth is in direct contradiction to the idea of a gradual progressive development. They are nevertheless exceedingly abundant in the Palaeozoic beds, and these fossil forms are so similar to the living representatives of the same group that what is true of the organization and development of the latter is unquestionably equally true of the former. In all their features the Selachians, more than any other fishes, resemble the higher animals. They lay few eggs, the higher kinds giving birth only to three, four, or five at a brood, whereas the common fishes lay myriads of eggs, hundred of thousands in some instances, and these are for the greater part cast into the water to be developed at random. The limitation of the young is unquestionably a mark of superiority. The higher we rise in the scale of animal life the more restricted is the number of offspring. In proportion to this reduction in number, the connection of the offspring with the parent is drawn closer, organically and morally, till this relation becomes finally the foundation of all social organization, of all human civilization. In some Selachians there is an actual organic connection between parent and progeny, resembling the placental connection which marks the embryonic development of the higher Vertebrates. This feature is in harmony with the sexual relations among them; for it is of all facts in their organic history the most curious that, among Vertebrates, the Selachians are the only ones with whom the connection of the sexes recalls that of the human family. Now, these higher fishes being the first representatives of the Vertebrates on earth, or at least those next following their earliest representatives, where do we find the Myzonts, fishes which are structurally inferior to all others, and of which the Amphioxus is the lowest member? They come in during the latest period of our world's history, with what is called the present period, to which we ourselves belong. This certainly does not look like a connected series beginning with the lowest and ending with the highest, for the highest fishes come first and the lowest come last.

The companions of the Selachians in the earlier geological periods, the Ganoids, belong also to the higher representatives of the class of fishes. Some of them have the ball-and-socket vertebral joint of the reptiles and birds, enabling the head to move upon the neck with greater freedom than in the lower fishes. I am aware that these synthetic and prophetic types, which I have myself been the first to point out, and in which features of higher and later groups are combined or hinted at in lower and earlier ones, have been interpreted as transition types. It has even been said that I have myself furnished the strongest evidence of the transmutation theory. This might perhaps be so, did these types follow, instead of preceding, the lower fishes. But the whole history of geological succession shows us that the lowest in structure is by no means necessarily the earliest in time, either in the Vertebrate type or any other. Synthetic and prophetic types have accompanied the introduction of all the primary divisions of the animal kingdom. With these may be found what I have called embryonic types, which never rise, even in their adult state, above those conditions which in higher structures are but the prelude to the adult state. It may, therefore, truly be said that a great diversity of types has existed from the beginning.

The most advanced Darwinians seem reluctant to acknowledge the intervention of an intellectual power in the diversity which obtains in nature, under the plea that such an admission implies distinct creative acts for every species. What of it, if it were true? Have those who object to repeated acts of creation ever considered that no progress can be made in knowledge without repeated acts of thinking? And what are thoughts but specific acts of the mind? Why should it then be unscientific to infer that the facts of nature are the result of a similar process, since there is no evidence of any other cause? The world has arisen in some way or other. How it originated is the great question, and Darwin's theory, like all other attempts to explain the origin of life, is thus far merely conjectural. I believe he has not even made the best conjecture possible in the present state of our knowledge.

The more I look at the great complex of the animal world, the more sure do I feel that we have not yet reached its hidden meaning, and the more do I regret that the young and ardent spirits of our day give themselves to speculation rather than to close and accurate investigation.

I hope in future articles to show, first, that, however broken the geological record may be, there is a complete sequence in many parts of it, from which the character of the succession may be ascertained; secondly, that, since the most exquisitely delicate structures, as well as embryonic phases of growth of the most perishable nature, have been preserved from very early deposits, we have no right to infer the disappearance of types because their absence disproves some favorite theory; and, lastly, that there is no evidence of a direct descent of later from earlier species in the geological succession of animals.

Comments on Agassiz

Louis Agassiz successfully pursued a career in science which took him from Paris, where he studied under Cuvier, to Neuchatel, Switzerland, where he was resident professor between 1832 and 1842, and finally to the United States and Harvard University, where he established what came to be known as the Agassiz Museum of Comparative Zoology. His major scientific contributions were his studies in extant and fossil fishes, completed in Europe, his glacier theory published in 1840, and the three volumes of his massive *Contributions to the Natural History of the United States.* The first two volumes of this work were published in 1857 and included an introductory "Essay on Classification." In this essay Agassiz thought he had supplied the definitive explanation of the order in the organic world: species were thoughts in the mind of the creator. Species came into existence and were extinguished as God thought of them or ceased to think of them. The order to be found in the organic world was none other than the order of God's mind. Thus Agassiz could not begin to understand why Darwin's *Origin of Species,* should receive so much attention—in fact, a good deal more attention than his own classic exposition published two years before.

Anticipating the publication of the *Origin* by two months, Owen, Buckland and Sedgwick persuaded Agassiz to publish his "Essay on Classification" in a separate volume. Agassiz made several changes in the new edition,

but he barely recognized the existence of Darwin and his theory. He had
already disproved such nonsense for all time in his refutations of Lamarck,
DeMaillet, Saint-Hilaire, and Chambers. Darwin's theory warranted a single
footnote:

> I hope the day is not far distant when zoologists and botanists will
> equally disclaim having shared in the physical doctrines more or less
> prevailing now, respecting the origin and existence of organic beings.
> Should the time come when my present efforts may appear like fighting
> against windmills, I shall not regret having spent so much labor in urging
> my fellow-laborers in a right direction; but at the same time I must
> protest now and forever against the bigotry spreading in some quarters,
> which would press upon science doctrines not immediately flowing from
> scientific premises and check its free progress (Agassiz, 1859, pp. 71–72).

Agassiz viewed any doctrines which differed from his utterly dogmatic
pronouncements as sheer bigotry. His own premises concerning divine
thoughts were scientific. Those of Darwin concerning natural selection were
not. Agassiz was totally out of touch with the naturalistic tendency of
science in his day. Even those naturalists who rejected Darwin's theory
found it difficult to swallow Agassiz's bald theology. Pious allusions were
one thing; species as ideas in God's mind were quite another. As Lyell
observed in a letter to P. Dawson of May 15, 1860, the license which
"[Agassiz] indulges in multiplying the miracle of creation whenever he has
the slightest difficulty of making out how a bird or a fish could have mi-
grated to some distant point from its first or other habitat, prepared many
to embrace Darwin's or Lamarck's hypothesis" (Lyell, 1881, 2:322). Agassiz
was apparently aware that some naturalists found his talk of divine thoughts
unsatisfactory, too metaphysical. Agassiz in turn found their reservations
inexplicable. No one viewed the initiation and cessation of thought in
human beings especially metaphysical or mysterious. Why should they find
God doing the same any more mysterious?

Since the third volume of his *Natural History* was still in preparation
in 1859, Agassiz was able to include in it a chapter containing a fuller
discussion of Darwin's peculiar brand of evolutionism.* His decision to
introduce the topic in his chapter on "Individuality and Specific Differences
among Acalephs" at first might seem incongruous, but, given his conceptual
outlook, the connection was direct. Before settling in Neuchatel, Agassiz
had studied under Cuvier in Paris. If Owen was the English Cuvier, Agassiz

* This chapter was published separately in the *American Journal of Science*
(formerly *Silliman's Journal*), 1860, 30:142–155, and in *Annals and Magazine of
Natural History*, 1860, 6:219–232.

was to become the American Cuvier. For Cuvier, evolution was impossible because each organism had a definite, finely balanced organization which imposed on it a very strict set of conditions of existence. If an individual strayed too far from its inherent organization, it could not survive. If the conditions of existence for a species changed to any extent, the species became extinct. Individuals were in no sense labile or plastic. They could not adapt to new conditions beyond a certain and definite limit. Thus, species were static.

The Acalephs (as they were then recognized) presented Cuvier's beliefs about the rigidity of organization in the individual with some of its most difficult counter-examples because of the prevalence of asexual modes of reproduction, extreme polymorphism, and colonial forms. The only "evolution" which Agassiz would countenance was embryological development. In certain polymorphic species, ontogeny took the individuals through decidedly different stages—stages which differed from each other at least as much as many species differed from each other. Their constitutions permitted them at least this much plasticity. But embryological development was cyclical and predetermined. Evolutionary development, on Darwin's theory, was neither. The great mystery for Agassiz was not the nature of species but the nature of individuality: What is it in the individual which guides its steady progressive and cyclical development?

Agassiz combined the Cuvierian notion of the individual with the Platonic conception of plans or archetypes. Cuvier recognized four basic plans in the animal kingdom, neither more nor less. Between these four types there could be no community of descent. Within each of these four basic plans were additional plans of more restricted scope; within each of these additional plans, more limited plans, and so on until the species level was reached. But no matter how restricted it might be, each plan was unique and distinct. No intermediaries could exist between any of these groups. In his earliest writings Agassiz chose birds as being the group whose boundaries were the sharpest and most indubitable. In 1868 the *Archaeopteryx* was discovered, and we hear no more about the absence of intermediaries between birds and reptiles. Aves was a class within one of Agassiz's four basic plans. It was only a matter of time before the boundaries between these groups were to be questioned. Since certain Ascidians seemed to have a notochord, one of the defining characteristics of the vertebrates, Kowalevsky argued that they formed an intermediary between the vertebrates and invertebrates. Von Baer, who had first described the notochord and its universal distribution among the vertebrates, disagreed. Whatever structure the Ascidians might have, it could not be a notochord. Haeckel further claimed that starfish were geunine intermediaries between the Radiates and Articulates. Kowalevsky's contention turned out to have considerable merit; Haeckel's did not. It should be kept in mind, however, that in neither the case of the Ascidians nor the case of the starfish could these animals have been the stem species from which their respective groups evolved, since both were contemporary. The finding of extant intermediaries between groups guaranteed neither that these groups had evolved from these intermediaries nor that they had a common ancestor. It served only as evidence that such intermediary common ancestors might exist.

In his "Essay on Classification" and in his discussion of evolution in Volume III of his *Contributions to the Natural History of the United States* (1860), Agassiz expressed himself with complete assurance that this evolutionary nonsense would soon be forgotten. By the time he came to write his article for the *Atlantic Monthly* (1874), he had been forced to admit that Darwin's theory was not just a temporary aberration. Even his own son Alex had become a convinced evolutionist. Agassiz acknowledged that Darwin had placed the subject of evolution on a different basis from that of all his predecessors: "Indeed it might be said that he treated his subject according to the best scientific methods, had he not frequently overstepped the boundaries of actual knowledge and allowed his imagination to supply the links which science does not provide." But no sooner did Agassiz make this concession than he withdrew it, comparing the excitement produced by the *Origin* with that elicited by Oken's *Natur-Philosophie* in Germany.

Lorenz Oken (1779–1851) was the most influential champion of a highly romantic view of science in Germany. Although Oken himself was a prolific writer in the field of anatomy, many of his works read like pure metaphysics to the empirically oriented scientists of England. Richard Owen was one of the few English scientists who received Oken's ideas with any enthusiasm. On the one hand, Agassiz implicitly condemned such metaphysical speculation, commenting that France had been spared such fanciful doctrines through the influence of Cuvier and the Jussieux. On the other hand, he complimented Darwin on the "metaphysical" turn which his work had taken in his discussion of mental attributes. Darwin had treated the evolution of mind, and was not mind precisely the ordering principle which he, Agassiz, had long postulated as the cause of order in the universe?

By the time his *Natural History* appeared, Agassiz had ceased to make significant contributions to science. His Amazon glacier theory was received with derision by some and in embarrassed silence by others. Although

Agassiz continued to raise money for his collections and for the museum which was to bear his name, his opinions came to count for less and less among scientists. Prior to the publication of the *Origin*, Agassiz was considered to be slightly behind the times; after its appearance, he became a living fossil. It is significant that his final word on evolutionary theory appeared posthumously, not in a scientific journal but in the literary pages of the *Atlantic Monthly*. The future papers on evolutionary theory which he promised were never forthcoming.*

* See also L. Agassiz (1840, 1863); E. C. Agassiz (1885); Lurie (1959); Mayr (1959a).

Conclusion

Philosophers of science have provided rational reconstructions of the logic of confirmation, rational criteria for deciding between competing theories, and such. In many respects, these reconstructions do not have a great deal to do with the actual course of events which lead up to the acceptance of a new scientific theory. One of the purposes of this anthology is to indicate some of the salient features of the reception of evolutionary theory among the scientific community. Too often, all the opponents of evolutionary theory are lumped together and their resistance to it explained away as religious bigotry. For example, John Dalton Hooker, Darwin's closest friend, looking back over the nearly thirty years that had elapsed since the publication of the *Origin,* complained:

> It is not for us, who repeat *ad nauseum* our contempt for the persecutors of Galileo and the sneers at Franklin, to conceal the fact that our own great discoverers met the same fate at the hands of the highest in the land of history and science, as represented by its most exalted organ, the *Quarterly Review.*[1]

Hooker was referring to such men as Bishop Wilberforce and Adam Sedgwick who rejected evolutionary theory primarily for theological reasons and would not have been able to accept it even if all the evidence had been overwhelmingly in its favor—which it was not. Modern readers are likely to concur in Hooker's judgment. Evolutionary theory has turned out to be basically correct. Hence, anyone who opposed it must have been an obscurantist. Furthermore, incompatibility with theology is now looked upon as an illicit justification for opposing a scientific theory. Hence, such opponents of evolutionary theory were doubly obscurantists. However, a little thought should make one realize how ill-considered these conclusions are, since they would require us to judge Lyell an obscuranist. For years

1. J. D. Hooker to Francis Darwin, October 22, 1886, *Life and Letters of Joseph Dalton Hooker,* 1918, 2:302–303.

he refused to accept evolutionary theory, both for evidential and theological reasons (Wilson, 1970).

One of the most fundamental principles of both deductive and inductive logic is that the goodness of a line of reasoning is independent of the truth of the resulting conclusion. It is possible for a logically impeccable argument to lead to a false conclusion and an incredibly bad argument to lead to a true conclusion. The premises supporting evolutionary theory in Darwin's day were riddled with errors. Many of these errors have since been corrected. Similarly, most of the factual objections have since been met and the theological objections discounted. Nevertheless, it does not follow that those men who opposed evolutionary theory on mistaken factual grounds were narrow-minded bigots. One should also be cautious in condemning those who opposed evolutionary theory for extra-scientific reasons. Although the scientific community is now in general agreement that theological considerations are irrelevant to the acceptance of a scientific theory, other extra-scientific considerations still do enter into science.

The book which has addressed itself most squarely to the problems raised in this volume concerning the acceptance of scientific theories by the scientific community is Thomas Kuhn's *The Structure of Scientific Revolutions* (1962). In spite of the fact that no two readers can agree on exactly what Kuhn intended to say in this extremely popular book, the ideas expressed there have been influential. In the most popular interpretation, Kuhn's pronouncements concern the logic of justification and are highly controversial. Apparently Kuhn (1970a and 1970b) now disowns this position. With one possible exception, he now sets out a position on the philosophy of science which differs hardly at all from the received doctrines in that field which have been prevalent for two decades or more. The exception concerns the role of scientific communites in defining what counts as a new scientific "paradigm" and in its promulgation.

The controversial interpretation of Kuhn has been most popular among humanists, social scientists, and -historians. Humanists seem attracted to Kuhn because they have long chafed under the arrogance of scientists in claiming that they have sole access to truth. The Kuhnian analysis of science, as humanists interpret it, seems to put science on the same level as ethics, aesthetics, and literary criticism. Social scientists also object to the traditional analysis of science as cumulative knowledge attained by the scientific method—a method derived from a study of physics. The Kuhnian analysis of science seems more closely attuned to developments in the social sciences. Finally, historians are attracted to Kuhn's work be-

cause he claims a greater role for history of science in the philosophy
of science.

Traditional analyses of science and the scientific method have been set
out in the Introduction to this anthology and need not be repeated here.
The popular interpretation of Kuhn's view has not. The reader should
be warned, however that Kuhn himself now disowns the following interpretation. According to the popular (mis?) interpretation of Kuhn's analysis
of science, a pre-paradigm period precedes the emergence of theoretical
science. After much investigation of natural phenomena from widely different points of view, a paradigm emerges, usually explicated in such books
as Aristotle's *Physics,* Newton's *Principia,* Linnaeus's *System of Nature,*
and Darwin's *Origin of Species.* Such paradigms exemplify the proper way
of "doing science," the proper techniques of investigation, acceptable modes
of explanations, and the like. As science proceeds under one of these accepted paradigms, anomalous facts gradually accumulate, producing a period
of crisis. Such a crisis is dissipated only by the emergence of a new paradigm
which can account for these anomalies. It is at this point that the Kuhnian
analysis becomes most original and most controversial. Kuhn was interpreted
as arguing that each paradigm determines not only its own rules of evidence
but also to some extent the evidence itself. Hence, two different paradigms
cannot actually be brought into direct conflict since they are never about
the same phenomena. The observation languages of the two theories are
incommensurable. Instead, one theory is preferred to another for quite
different reasons. Objective truth and progress in science are myths. Perhaps
evidence, experiments, and rational arguments play a major role in science
during its periods of normal development, but they are never decisive factors
in the transition from one paradigm to another. Scientific revolutions are
fundamentally arational affairs. To quote Kuhn (1962:93):

> Like the choice between competing political institutions, that between
> competing paradigms proves to be a choice between incompatible modes
> of community life. Because it has that character, the choice is not and
> cannot be determined merely by the evaluation procedures characteristic
> of normal science, for these depend in part upon a particular paradigm,
> and that paradigm is at issue. When paradigms enter, as they must,
> into a debate about paradigm choice, their role is necessarily circular.
> Each group uses its own paradigm to argue in that paradigm's defense.
>
> Yet whatever its force, the status of the circular argument is only
> that of persuasion. It cannot be made logically or even probabilistically
> compelling for those who refuse to step into the circle. The premises
> and values shared by the two parties are not sufficiently extensive for

that. As in political revolutions, so in paradigm choice—there is no standard higher than the assent of the relevant community.

If the preceding comments are taken literally and other comments made by Kuhn in other places are ignored, Kuhn is arguing for a relative notion of truth. Truth is decided in science—as many have argued it is in theology, politics, and the arts—by propaganda, coercion, and attrition. Whether or not a theory is true in any sort of objective sense is irrelevant, and should be irrelevant, to the acceptance of this theory. All that matters is that members of the scientific community be brought to accept it as true. Old men die off, and young men jump on the currently fashionable bandwagon. Among scientists, evidence and rational argument do play a role in convincing them to adhere to a new doctrine, but if they do not, in the most extreme cases one could always settle the matter by killing recalcitrant members of the scientific community. Of course, Kuhn does not advocate such extreme measures, but on a literal interpretation of what he says, they would be justified. If conviction and not approximation to objective truth is all that matters in science, there is no reason to reject any of the effective means of bringing about consensus (see Peirce, 1877, for a discussion of precisely this issue).

Until recently the arguments for and against this extreme interpretation of Kuhn have been conducted in the context of examples drawn from the physical sciences. Several authors in the past two years have attempted to apply the Kuhnian analysis of science to biology, specifically to the introduction of evolutionary theory and the discovery of the molecular basis of heredity (Ruse, 1970, 1971; Greene, 1971; and Ghiselin, 1971). As these authors have discovered, the major stumbling block in their enterprise is deciding exactly what the Kuhnian analysis of science actually is. The main difficulty is trying to keep straight the wide variety of things which Kuhn might mean by "paradigm." His examples range all the way from something as specific as the introduction of the telescope to something as amorphous as conflicting world-views. These various meanings of "paradigm" must be kept separate if any sense is to be made of Kuhn. For example, one can trace the development of scientific theories from Aristotle through Newton to Einstein, concentrating on them as scientific theories, as the notion is now understood. But numerous additional developments were taking place concurrently. For example, Aristotle was completely cognizant of the metaphysical foundations of his various scientific theories and the relations which existed between his theories and his metaphysics. His metaphysics, however, was not "theological" in the modern sense of the word. In Newton's day metaphysics had a definite theological dimension. For Einstein, the notion of "God" functions once again in the innocuous way it did in Aristotle.

The Kuhnian analysis of science in terms of "leaps of faith" seems to apply to the Darwinian revolution, but not for the generally assumed reasons. There was more to the comparison of special creationism and Darwin's theory than evidence and rational argument, since one of the main issues was the relevance of evidence and rational argument as such (Ruse, 1970, 1971). When "paradigm" is interpreted as referring to the metaphysical foundations of science, and the change that one is following is the replacement of one scientific theory by another, the Kuhnian analysis seems relevant. When "paradigm" is interpreted more cautiously to refer to scientific theories as such, then the applicability seems less obvious. As we pointed out in the Introduction, the conflict between special creationism and evolutionary theory was not simply a matter of two conflicting scientific theories but of two conflicting world-views as well. Special creationism had been a partially supernatural theory and remained so long after naturalism had become firmly established in the physical sciences. Pious allusions to God were made by physicists, but God was no longer an operative factor in the phenomena treated by physical science. Before special creationists were able to devise a natural explanation for the sudden appearance of new species, Darwin published his completely naturalistic theory of the gradual evolution of species. Darwin's naturalism was a major factor in certain scientists' acceptance of evolutionary theory, others' rejection of it, and still others' desire to accept or reject it, even though they could not bring themselves to do so.

By and large, philosophers have been much less impressed by Kuhn's work than have scientists and historians of science. Just as scientists demand clarity and precision in the elucidation of scientific ideas, and historians require it in the study of the past, philosophers have the right to expect a certain level of competence in the explication of notions in their field. They find Kuhn seriously deficient on this count. For example, Dudley Shapere (1971) concludes his careful critique of Kuhn as follows:

> The point I have tried to make is not merely that Kuhn's is a view that denies the objectivity and rationality of the scientific enterprise; I have tried to show that the arguments by which Kuhn arrives at his conclusion are unclear and unsatisfactory.

Kuhn's defenders reply that if his *Structure of Scientific Revolutions* is anything, it is an exemplification of a new paradigm of science, a paradigm in which arguments and clarity are irrelevant. All that matters is propaganda, and, as any demagogue knows, clear, careful arguments are extremely ineffective methods of persuasion. According to these evangelists of the new idea of science, it does not matter what Kuhn meant to say in his seminal work. It was good propaganda, and has set loose a powerful new force in science and the philosophy of science. As much as I disagree with the extreme, popular interpretation of Kuhn set out above, I must admit that Kuhn's disciples are being consistent. Just as J. B. Watson in psychology with his radical behaviorism and P. W. Bridgman in physics with his operationism, Kuhn has given birth to a Frankenstein monster in science which no amount of reformulation, disclaimers, and reconsideration can call back.

But I believe that there is also a kernel of an original truth in Kuhn's work—the role of scientific communities in the reception of scientific theories. I do not know whether this truth belongs in the philosophy of science or in the sociology of science and do not much care. It is an interesting feature of science which needs investigating. New ideas, like mutations, stand a better chance of surviving and being promulgated if they develop for a time in an isolated community. Then, after they have been established firmly in it, they are injected more massively into the larger community as a whole. One of the difficulties for this particular Kuhnian thesis is defining "scientific community" independent of the members' accepting a common paradigm. So far, suggestions for delineating emerging subpopulations of scientists have been on the level of counting references in the bibliographies of their published works! A more plausible suggestion is defining a scientific community in terms of the interaction of its members, whether at meetings, in publications, or in private correspondence, whether they agree completely or not. An examination of Darwin's correspondence provides an amazingly complete record of the effective population in the development and eventual acceptance of evolutionary theory. In addition, the boundaries are surprisingly sharp. In any case, the readers of this anthology will have to decide for themselves how appropriate these musings are in the dispute between Darwin and his critics.

Bibliography   Index

Bibliography

Agassiz, Elizabeth C. 1885. *Agassiz: Louis Agassiz, His Life and Correspondence.* Boston: Houghton, Mifflin & Co.

Agassiz, Louis. 1840. *Studies on Glaciers.* Translated by Albert V. Carozzi (1967). New York: Hafner.

—— 1857–62. *Contributions to the Natural History of the United States.* vols. I–III. Boston: Little, Brown, and Company.

—— 1859. *Essay on Classification.* Edited by Edward Lurie (1962). Cambridge, Massachusetts: Belknap Press of Harvard University. An extract from volume I of Agassiz's *Contributions to the Natural History of the United States* (1857).

—— 1860. Prof. Agassiz on the Origin of Species. *American Journal of Science and Arts* (formerly *Silliman's Journal*), 30:142–154; also in *Annals and Magazine of Natural History,* 6:219-232; this paper was extracted from volume III of Agassiz's *Contributions to the Natural History of the United States* (1862).

—— 1863. *Methods of Study in Natural History.* Boston: Ticknor and Fields.

—— 1874. "Evolution and Permanence of Type," *Atlantic Monthly* 33:92–101.

Allan, W. 1967. *The Hookers of Kew, 1785–1911.* London: Joseph.

Anonymous. 1860. Professor Owen on the Origin of Species. *Saturday Review* 9:573-574. [Probably authored by St. George Jackson Mivart.]

Anonymous. 1866. "Review of the Origin of Species," *Popular Science Review* 5:215.

Anschutz, Richard P. 1953. *The Philosophy of John Stuart Mill.* London: Oxford University Press.

Appleman, Philip. 1959. "The Logic of Evolution: Some Reconsiderations," *Victorian Studies* 3:115–125.

Aristotle. 1928. *The Works of Aristotle.* Translated under the editorship of W. D. Ross. Oxford: Clarendon Press.

Babbage, Charles. 1837. *The Ninth Bridgewater Treatise. A Fragment.* London: Murray.

Bacon, Francis. 1620. *The New Organon and Related Writings.* Edited by F. H. Anderson (1960). New York: Liberal Arts Press.

—— 1626. *Sylva Sylvarum.* London: John Haviland for William Lee.

Baer, Karl Ernst von. 1873. "Zum Streit über den Darwinismus," *Augsburger Allgemeine Zeitung* No. 130, pp. 1986–1988.

—— 1876. *Reden gehalten in wissenschaftlichen Versammlungen und kleinere Aufsätze vermischten Inhalts.* St. Petersburg: H. Schnitzdorf.

Baker, J. J., and G. Allen. 1968. *Hypothesis, Prediction, and Implication in Biology* Boston: Addison-Wesley Publishing Co.

Basalla, George, William Coleman, and Robert H. Kargon. 1970. *Victorian Science*. Garden City, N.Y.: Doubleday and Company.

Bastian, H. C. 1872. *The Beginnings of Life*. London: Macmillan.

Bentley, Richard. 1693. *A Confutation of Atheism from the Origin and Frame of the World*. Reprinted in I. B. Cohen, *Isaac Newton's Papers and Letters on Natural Philosophy and Related Documents* (1958). Cambridge, Mass.: Harvard University Press.

Bertalanffy, Ludwig von. 1952. *Problems of Life*. London: Watts & Co.

Blunt, H. W. 1903. "Bacon's Method of Science." *Proceedings of the Aristotelian Society*, n.s., 4:16–31.

Bowen, Francis. 1860. "Darwin on the Origin of Species," *North American Review* 90:474–506.

Boyd, Thomas. 1859. "On the Tendency of Species to Form Varieties," *Zoologist* 17:6357–6359.

Bronn, H. G. 1858. *Untersuchungen über die Entwickelungsgesetze der organischen Welt während der Bildungs Zeit unserer Oberfläche*. [Investigations into the developmental laws of the organic world during the formation of the crust of the earth.] Stuttgart: Heinrich Georg.

—— 1859. "On the Laws of Evolution of the Organic World during the Formation of the Crust of the Earth," *Annals and Magazine of Natural History* 4:81–184. [A translation of the last chapter of Bronn (1858).]

—— 1860. Review of the *Origin of Species*. *Neues Jahrbuch für Mineralogie*. 112–116.

Buchdahl, Gerd. 1969 *Metaphysics and the Philosophy of Science*. Oxford: Basil Blackwell.

Buckland, William E. 1836. *Geology and Mineralogy. Considered with Reference to Natural Theology*. Volume VI of the Bridgewater Treatises. London: Pickering.

Burckhardt, Richard W. 1970. "Lamarck, Evolution, and the Politics of Science," *Journal of the History of Biology*. 3:275–298.

Butts, Robert E. 1968. *William Whewell's Theory of Scientific Method*. Pittsburgh: University of Pennsylvania Press.

Canfield, J. V. 1966. *Purpose in Nature*. Englewood Cliffs; N.J.: Prentice-Hall.

Cannon, Walter F. 1961. "John Herschel and the Idea of Science," *Journal of the History of Ideas* 22:215–239.

Carpenter, W. B. 1839. *The Principles of Physiology, General and Comparative*. London: J. Churchill.

—— 1860a. "The Theory of Development in Nature," *British and Foreign Medico-Chirurgical Review* 25:367–404.

—— 1860b. "Darwin and the Origin of Species," *National Review* 10:188–214.

—— 1862. *Introduction to the Study of Foraminifera*. London: The Ray Society.

—— 1872a. Man the Interpreter of Nature. Presidential Address to the British Association for the Advancement of Science. Reprinted in *Victorian Science*, Edited by G. Basalla, W. Coleman, and R. Kargon (1970). New York: Doubleday.

—— 1872b. "On Mind and Will in Nature," *Contemporary Review* 20:738–762.

Chambers, Robert. 1845. *Vestiges of the Natural History of Creation*, 3rd ed. New York: Wiley and Putnam.

/

Clark, John Willis and Thomas McKenney Hughes. 1890. *The Life and Letters of the Reverend Adam Sedgwick*. Cambridge, [England] Universtiy Press.

Colvin, S., and J. A. Ewing. 1887. *Papers, Literary, Scientific, etc. by the Late Fleeming Jenkin*. London: Longmans, Green.

Cressey, D. R. 1950. "Criminal Violation of Financial Trust," *American Sociological Review* 15:738–743.

Dana, J. D. 1857. "Thoughts on Species," *Annals and Magazine of Natural History* 20:485–497.

Daniels, George. 1968. *Darwin Comes to America*. Walthem, Mass.: Blaisdell Publishing Co.

Darwin, Charles. 1839. "Observations on the Parallel Roads of Glen Roy, and of Other Parts of Lochaber in Scotland, with an Attempt to Prove that They are of Marine Origin," *Philosophical Transactions, Royal Society of London* 129:39–82.

—— 1842. *The Structure and Distribution of Coral Reefs*. Reprinted with Forward by H. W. Menard (1962). Berkeley: University of California Press.

—— 1858. An Extract from an Unpublished Work on Species by C. Darwin, and an Abstract of a Letter from C. Darwin to Prof. Asa Gray, Read July 1, 1858 before the Linnaean Society. *Journal of the Proceedings of Linnaean Society* (1858), 3:45–62; *Zoologist* (1859), 17:6293–6308. Reprinted in *The Darwin Reader*. Edited by M. Bates and P. Humphrey (1956). New York: Charles Scribner's Sons.

—— 1859. *On the Origin of Species, A Facsimile of the First Edition*. Reprinted with Introduction by E. Mayr (1966). Cambridge, Mass.: Harvard University Press.

—— 1868. *Variation of Animals and Plants under Domestication*. London: Murray.

—— 1871. *The Descent of Man and Selection in Relation to Sex*. London: Murray.

—— 1872. *On the Origin of Species by Means of Natural Selection or the Preservation of Favoured Races in the Struggle for Life*, 6th ed. London: Murray.

—— 1958. *The Autobiography of Charles Darwin, 1809–1882*. Edited by Nora Barlow. London: Collins.

Darwin, Francis. 1887. *The Life and Letters of Charles Darwin*. London: Murray.

—— 1903. *More Letters of Charles Darwin*. London: Murray.

De Beer, Gavin, ed. 1960. "Darwin's Notebooks on Transmutation of Species," *Bulletin of the British Museum*, 2:1–83.

—— 1964. *Charles Darwin, A Scientific Biography*. Garden City, N.Y.: Doubleday.

Dewey, John. 1910. *The Influence of Darwin on Philosophy and Other Essays*. New York: Henry Holt.

Dobzhansky, Theodosius. 1937. *Genetics and the Origin of Species*. New York: Columbia University Press.

Douglas, J. M. 1881. *Life and Selected Correspondence of William Whewell*. London: Kegan Paul & Co.

Ducasse, C. J. 1951. Whewell's Philosophy of Scientific Discovery. *The Philosophical Review* 60:56–69.

—— 1960. John F. W. Herschel's Methods of Experimental Inquiry. *Theories of Scientific Method: The Renaissance Through the Nineteenth Century*. Edited by R. M. Blake, C. J. Ducasse, and E. H. Madden. Seattle: University of Washington Press.

Eiseley, Loren. 1958. *Darwin's Century*. New York: Doubleday.

Ellegård, Alvar. 1957. "The Darwinian Revolution and Nineteenth-Century Philosophies of Science," *Journal of the History of Ideas* 18:362–393.

Ellegård, Alvar. 1958. *Darwin and the General Reader: The Reception of Darwin's Theory of Evolution in the British Periodical Press, 1859–1872*. Göthenburg: Goteborgs Universitets Årsskrift.

Elliot, H. S. R. 1910. *The Letters of John Stuart Mill*. London: Longmans, Green.

Fawcett, Henry. 1860. "A Popular Exposition of Mr. Darwin on the Origin of Species," *Macmillan's Magazine* 3:81–92.

Flourens, Marie Jean Pierre. 1850. *Histoire des travaux et des idées de Buffon*. Paris: L. Hachette et cie.

—— 1864. *Examen du livre de M. Darwin sur l'origine des especis*. Paris: Garnier.

Forbes, E. 1854. "On the Manifestation of Polarity in the Distribution of Organized Beings in Time," *Proceedings of the Royal Institution of Great Britain* 1:428–433.

Ghiselin, Michael. 1969. *The Triumph of the Darwinian Method*. Berkeley: University of California Press.

—— 1971. "The Individual in the Darwinian Revolution," *New Literary History*, 3:113–134.

Gillispie, Charles. 1951. *Genesis and Geology: The Impact of Scientific Discoveries upon Religious Beliefs in the Decades before Darwin*. Cambridge, Mass.: Harvard University Press.

Glass, Bentley, Owsei Temkin, and William L. Straus, Jr. 1959. *Forerunners of Darwin: 1745–1859*. Baltimore: The Johns Hopkins Press.

Gray, Asa. 1876. *Darwiniana*. Reprinted with Foreword by A. H. Dupree (1963). Cambridge, Mass.: Harvard University Press.

Greene, John C. 1959. *The Death of Adam: Evolution and Its Impact on Western Thought*. Ames, Iowa: Iowa State University Press.

—— 1963. *Darwin and the Modern World View*. New York: Mentor.

—— 1971. The Kuhnian Paradigm and the Darwinian Revolution. In *Perspectives in the History of Science and Technology*. Edited by Duane H. D. Roller. Norman: University of Oklahoma Press.

Grene, Marjorie. 1958. "Two Evolutionary Theories," *The British Journal for the Philosophy of Science* 9:110–127; 185–193.

—— 1963. *A Portrait of Aristotle*. Chicago, Ill.: University of Chicago Press.

Gruber, Jacob. 1960. *A Conscience in Conflict: The Life of St. George Mivart*. New York: Columbia University Press.

Haacke, W. 1905. *Karl Ernst von Baer*. Leipzig: T. Thomas.

Haeckel, Ernst. 1866. *Generelle Morphologie der Organismen*. Berlin: Reimer.

Hall, Thomas S. 1968. "On Biological Analogs of Newtonian Paradigms," *Philosophy of Science* 35:6–27.

Haughton, Samuel. 1846. "On the Laws of Equilibrium and Motion of Solid and Fluid Bodies," *Cambridge and Dublin Mathematical Journal* 1:173–182.

—— 1860. "*Bioyeveois,*" *Natural History Review* 7:23–32; also *Annals and Magazine of Natural History* 11:420–427.

—— 1873. *Principles of Animal Mechanics*. London: Longmans, Green.

Heathcote, A. W. 1953. "William Whewell's Philosophy of Science," *The British Journal for the Philosophy of Science* 4:302–314.

Hempel, Carl G. 1965. *Aspects of Scientific Explanation*. New York: The Free Press.

Herschel, John. 1830. *Preliminary Discourse on the Study of Natural Philosophy*. Reprinted with an Introduction by M. Partridge (1966). New York: Johnson Reprint Corporation.

—— 1833. *A Treatise On Astronomy*. London: Longmans, Green.

Herschel, John. 1841. "William Whewell's Philosophy of the Inductive Sciences," *Quarterly Review* 68:177–238.
—— 1849. *Outlines of Astronomy*. London: Longmans, Green.
—— 1850. "Review of Quetelet on Probabilities," *Edinburgh Review* 42:1–57.
—— 1857. *Essays from the Edinburgh and Quarterly Reviews with Addresses and Other Pieces*. London: Longmans, Green.
—— 1861. *Physical Geography of the Globe*. London: Longmans, Green.
—— 1864. Resolution on Scripture and Science. *Athenaeum*, September 17, p. 375.
—— 1912. *Scientific Papers of Sir John William Herschel*. London: The Royal Society.
Himmelfarb, Gertrude. 1959. *Darwin and the Darwinian Revolution*. Garden City: Doubleday.
Hochberg, H. 1953. The Empirical Philosophy of Roger and Francis Bacon. *Philosophy of Science* 20:313–326.
Hofstadter, Richard. 1944. *Social Darwinism in American Thought*. Philadelphia: University of Pennsylvania Press.
Holmes, S. J. 1947. "K. E. von Baer's Perplexities over Evolution," *Isis* 37:7–14.
Hooker, J. D. 1859. *On the Flora of Australia, Its Origins, Affinities and Distribution: Being an Introductory Essay to the Flora of Tasmania*. London: Lovell Reeve. Reprinted in *American Journal of Science and Arts*, 2nd ser. 29:1–25; 305–326.
—— 1860. "Review of the *Origin of Species*," *Gardeners' Chronicle*, December 31, 1859, pp. 1051–1053.
—— 1868. Address of the President. *Report of the Thirty-Eighth Meeting of the British Association for the Advancement of Science held at Norwich*. 68:lviii–lxxv. London: Murray.
Hopkins, William. 1839–40. "Researches in Physical Geology," *Philosophical Transactions of the Royal Society of London* 129:381–423.
—— 1860. "Physical Theories of the Phenomena of Life," *Fraser's Magazine* 61:739–752; 62:74–90.
Hull, David L. 1965. The Effect of Essentialism on Taxonomy," *The British Journal for the Philosophy of Science* 15:314–326; 16:1–18.
—— 1967. "The Metaphysics of Evolution," *The British Journal for the History of Science* 3:309–337.
—— 1968. "The Operational Imperative: Sense and Nonsense in Operationism," *Systematic Zoology* 17:438–457.
—— 1974. *The Philosophy of the Biological Sciences*. Englewood Cliffs, N.J.: Prentice-Hall.
Humboldt, Alexander von. 1818. *Personal Narrative of Travels to the Equinoctial Regions of the New Continent during the Years 1799 to 1804*. Translated by H. N. Williams. London: Longmans, Green.
Hussey, Arthur. 1859. "The Tendency of Species to Form Varieties," *Zoologist* 17:6474–6475.
Hutton, F. W. 1860. Review of the *Origin of Species*. *The Geologist* 3:464–472.
—— 1861. "Some Remarks on Mr. Darwin's Theory," *The Geologist* 4:132–136; 183–188.
—— 1899. *Darwinism and Lamarckism, Old and New*. London: Macmillan.
Huxley, Leonard. 1902. *Life and Letters of Thomas Henry Huxley*. London: Macmillan.
—— 1918. *Life and Letters of Sir Joseph Dalton Hooker*. London: Murray.

Huxley, T. H. 1845. Review of the *Vestiges of Creation*. *British and Foreign Medico-Chirurgical Review* 19:425–439.
—— 1860a. On Species and Races and Their Origin. *Notices of the Proceedings of the Royal Institution of Great Britain* 3:195–200.
—— 1860b. "The Origin of Species," *Westminster Review* 17:541–570; reprinted in *Darwiniana* (1896).
—— 1863. *On Our Knowledge of the Causes of the Phenomena of Organic Nature.* London: Hardwicke. Reprinted as *On the Origin of Species or, The Causes of the Phenomena of Organic Nature* (1968). Ann Arbor: The University of Michigan Press.
—— 1869. Anniversary Presidential Address. *Quarterly Journal of the Geological Society of London* 25:xxxviii–liii.
—— 1871. "Mr. Darwin's Critics," *Contemporary Review* 18:443–476.
—— 1896. *Darwiniana*. New York: Appleton & Co.
Irvine, William. 1955. *Apes, Angels and Victorians*. New York: McGraw Hill.
Jenkin, Fleeming. 1867. "The Origin of Species," *North British Review* 46:149–171.
Kneale, William. 1949. *Probability and Induction*. Oxford: Clarendon Press.
Kölliker, Rudolf Albert von. 1861. *Entwickelungsgeschichte des Menschen und höheren Thiere*. [Developmental history of man and higher animals]. Leipzig: W. Engelmann.
—— 1864. Ueber die Darwinische Schöpfungstheorie; ein Vortrag [An examination of Darwin's theory of creation]. Leipzig: W. Engelmann.
Kuhn, Thomas. 1962. *The Structure of Scientific Revolutions*. Chicago, Ill.: University of Chicago Press.
—— 1970a. Logic of Discovery, a Psychology of Research. *Criticism and the Growth of Knowledge*. Edited by I. Lakatos and A. Musgrave. London: Cambridge University Press.
—— 1970b. Reflections of My Critics. *Criticism and the Growth of Knowledge*. Edited by I Lakatos and A. Musgrave. London: Cambridge University Press.
Lamarck, J. B. 1809. *Zoological Philosophy*. Translated by Hugh Elliot (1963). New York: Hafner.
Lewontin, Richard C. 1969. "The Bases of Conflict in Biological Explanation," *Journal of the History of Biology* 2:35–45.
Lurie, Edward. 1959. "Louis Agassiz and the Idea of Evolution," *Victorian Studies* 3:87–108.
—— 1960. *Louis Agassiz: A Life in Science*. Chicago, Ill.: University of Chicago Press.
Lyell, Charles. 1851. "The Theory of the Successive Geological Development of Plants, from the Earliest Periods to Our Own, as Deduced from Geological Evidence," *Edinburgh New Philosophy Journal* 51:1–31; 213–226.
—— 1873. *The Geological Evidence of the Antiquity of Man*. 4th ed. London: Murray.
—— 1889. *Principles of Geology*. 11th ed. New York: D. Appleton.
Lyell, K. M. 1881. *Life, Letters and Journals of Sir Charles Lyell*. 2 vols. London: Murray.
MacArthur, R. H., and E. O. Wilson. 1967. *The Theory of Island Biogeography* Princeton: Princeton University Press.
Madden, E. H. 1958. *The Philosophical Writings of Chauncey Wright*. New York: Liberal Arts Press.
—— 1963. *Chauncey Wright and the Foundations of Pragmatism*. Seattle: University of Washington Press.

Manser, A. R. 1965. "The Concept of Evolution," *Philosophy* 40:18–34.

Maxwell, J. C. 1890. *The Scientific Papers of James Clerk Maxwell.* Edited by W. D. Niven. Cambridge [England] University Press.

Mayo, Herbert. 1838. *The Philosophy of the Living.* London: J. W. Parker.

Mayr, Ernst. 1942. *Systematics and the Origin of Species.* New York: Columbia University Press.

—— 1959a. "Agassiz, Darwin and Evolution," *Harvard Library Bulletin* 13:165–194.

—— 1959b. Review of *Darwin and the Darwinian Revolution. Scientific American* 201:209–216.

—— 1960. The Emergence of Evolutionary Novelties. *Evolution After Darwin.* Edited by Sol Tax. Chicago, Ill: University of Chicago Press.

—— 1961. Cause and Effect in Biology. *Science* 134:1501–1506.

—— 1963. *Animal Species and Evolution.* Cambridge, Mass.: Belknap Press of Harvard University.

—— 1969. *Principles of Systematic Zoology.* New York: McGraw-Hill.

Mendel, Gregor. 1865. "Experiment in Plant-Hybridization." *Classic Papers in Genetics.* Edited by J. A. Peters (1959). Englewood Cliffs, N.J.: Prentice-Hall.

—— 1867. "Letter to Carl Nägeli." *Great Experiments in Biology.* Edited by M. L. Gabriel and S. Fogel (1955). Englewood Cliffs, N.J.: Prentice-Hall.

Mill, John Stuart. 1843. *A System of Logic, Ratiocinative and Inductive, Being a Connected View of the Principles of Evidence, and the Methods of Scientific Investigation.* London: Longmans, Green.

—— 1865. *An Examination of Sir William Hamilton's Philosophy, and of the Principal Philosophies and Questions discussed in his Writings.* Boston: W. V. Spencer.

—— 1874. *A System of Logic, etc.* 8th ed. London: Longmans, Green.

—— 1910. *The Letters of John Stuart Mill.* Edited by H. S. R. Elliot. London: Longmans, Green.

—— 1924. *The Autobiography of John Stuart Mill.* New York. Columbia. University Press.

—— 1969. *Three Essays on Religion.* New York: Greenwood Press.

Mivart, St. George Jackson. 1860. "Professor Owen on the Origin of Species." *The Saturday Review* 9:573–574.

—— 1871a. *On the Genesis of Species.* London: Macmillan.

—— 1871b. "Darwin's Descent of Man," *Quarterly Review* 131:47–90.

—— 1872a. "Evolution and Its Consequences: A Reply to Professor Huxley," *Contemporary Review* 19:168–197.

—— 1872b. "Specific Genesis," *North American Review* 114:451–468.

—— 1892. "Happiness in Hell," *Nineteenth Century* 32:899–919.

Moorhead, P. S., and M. M. Kaplan. 1967. *Mathematical Challenges to the Neo-Darwinian Interpretation of Evolution.* Philadelphia: The Wistar Institute Press.

Newton, Isaac. 1958. *Isaac Newton's Papers & Letters on Natural Philosophy and Related Documents.* Edited by I. B. Cohen. Cambridge, Mass.: Harvard University Press.

—— 1960. *Sir Isaac Newton's Mathematical Principles of Natural Philosophy and his System of the World.* Translated by Andrew Motte (1729) and Florian Cajori (1960). Berkeley: University of California Press.

Oppenheimer, Jane. 1959. An Embryological Enigma in the Origin of Species.

Forerunners of Darwin: 1745–1859. Edited by Bentley Glass, Owsei Temkin, and W. L. Straus, Jr. Baltimore, Md.: The Johns Hopkins Press.

Owen, Richard. 1843. *Lectures on the Comparative Anatomy and Physiology of the Invertebrate Animals.* London: Longmans, Green.

—— 1847. Report on the Archetype and Homologies of the Vertebrate Skeleton. *Report of the Sixteenth Meeting of the British Association for the Advancement of Science* 16:169–340. London: Murray.

—— 1849. *On the Nature of Limbs.* London: John van Voorst.

—— 1855. *Recherches sur l'archetype et les homologies du squelette vertebre.* Paris. Translation of Owen (1847).

—— 1860a. "Darwin on the Origin of Species," *The Edinburgh Review* 111:487–532 (English ed.); 225:251–275 (American ed.).

—— 1860b. *Paleontology; or a Systematic Summary of Extinct Animals and Their Geological Relation.* Edinburgh: Adam and Charles Block.

—— 1862. "Characters of the Skull of the Male *Pithecus Morio* with Remarks on the Varieties of the Male *Pithecus Satyrus.*" *Zoological Transactions of the Zoological Society of London*" 4:165–178.

—— 1868. *The Anatomy of Vertebrates.* Vol. III. London: Longmans, Green.

—— 1869. Address of the President. *Report of the Twenty-Seventh Meeting of the British Association for the Advancement of Science.* Reprinted in *Victorian Science.* Edited by George Basalla, William Coleman, and R. H. Kargon (1970). New York: Doubleday.

Owen, Richard S. 1894. *The Life of Richard Owen.* London: Murray.

Passmore, John. 1959. "Darwin's Impact on British Metaphysics," *Victorian Studies* 3:41–54.

Peirce, Charles Sanders. 1877. "The Fixation of Belief," *Popular Science Monthly* 12:1–15.

—— 1931. *Principles of Philosophy.* Vol. 1 of *Collected Papers of Charles Sanders Peirce.* Edited by Charles Hartshorne and Paul Weiss. Cambridge, Mass.: Belknap Press of Harvard University.

—— 1935. *Scientific Metaphysics.* Vol. VI of *Collected Papers of Charles Sanders Peirce.* Edited by Charles Hartshorne and Paul Weiss. Cambridge, Mass. Belknap Press of Harvard University.

Pictet, Francois Jules. 1860. *On the Origin of Species* by Charles Darwin. *Archives des Sciences de la Bibliotheque Universelle* 3:231–247.

—— and A. Humbert. 1866. "Recent Researches on the Fossil Fishes of Mount Lebanon," *Annals and Magazine of Natural History* 18:237–247.

Plato. 1952. *Plato's Phaedo.* Translated by R. Hackforth. New York: The Liberal Arts Press.

Popper, Karl R. 1959. *The Logic of Scientific Discovery.* New York: Basic Books, Inc.

Pouchet, F. A. 1859. *Hétérogénie; ou traité de la génération spontanée.* Paris: J. B. Bailliere et fils.

Powell, Baden. 1855. *Essays on the Spirit of the Inductive Philosophy, the Unity of Worlds and the Philosophy of Creation.* London: Longmans, Green.

Priestley, Joseph. 1779. *Experiments and Observations Relating to Various Branches of Natural Philosophy.* London: J. Johnson.

Provine, William B. 1971. *The Orgins of Theoretical Population Genetics.* Chicago, Ill.: University of Chicago Press.

Quetelet, M. A. 1846. *Letters on the Application of the Theory of Probabilities to the Moral and Political Sciences.* Translated by O. G. Downes (1849). London: C. and E. Layton.

Randall, John Herman, Jr. 1961. "The Changing Impact of Darwin on Philosophy," *Journal of the History of Ideas* 22:435–462.

Robinson, W. B. 1951. "The Logical Structure of Analytic Induction," *American Sociological Review* 16:812–818.

Ruse, Michael. 1970. The Revolution in Biology. *Theoria* 36:1–22.

—— 1971. "Two Biological Revolutions," *Dialectica* 25:17–38.

Schaffner, Kenneth F. 1969. "The Watson-Crick Model and Reductionism," *British Journal for the Philosophy of Science* 20:325–348.

Schonland, Basil. 1968. *The Atomists (1805–1933).* Oxford: Clarendon Press.

Sedgwick, Adam. 1834. *Discourse on the Studies of Cambridge.* London: Parker.

—— 1845. Review of the Vestiges of Creation. *Edinburgh Review* 82:1–45.

—— 1860. "Objections to Mr. Darwin's Theory of the Origin of Species," *Spectator,* March 24, pp. 285–286; April 7, pp. 334–335.

—— 1890. *The Life and Letters of the Reverend Adam Sedgwick.* 2 vols. Edited by John Willis Clark and Thomas McKenny Hughes. Cambridge [England] University Press.

Shapere, Dudley. 1971. "The Paradigm Concept," *Science* 172:706–709.

Simpson, G. G. 1961. *Principles of Animal Taxonomy.* New York: Columbia University Press.

Smart, J. J. C. 1963. *Philosophy and Scientific Realism.* London: Routledge & Kegan Paul.

Sokal, R. R., and P. H. A. Sneath. 1963. *The Principles of Numerical Taxonomy.* San Francisco: Freeman.

Stephen, Leslie. 1886. *Life of Henry Fawcett.* 5th ed. London: Smith, Elder, & Co.

Stevenson, R. L. 1912. *Memoir of Fleeming Jenkin.* London: Longmans, Green.

Strong, W. W. 1955. "William Whewell and John Stuart Mill: Their Controversy about Scientific Knowledge," *Journal of the History of Ideas* 16:209–231.

Tait, Peter. 1869. "Geological Time," *North British Review* 50:215–233.

Thayer, James Bradley. 1878. *Letters of Chauncey Wright, with Some Account of His Life.* Cambridge, Mass.: Harvard University Press.

Thomson, William, Lord Kelvin. 1865. The Doctrine of Uniformity in Geology Briefly Refuted. *Proceedings, Royal Society of Edinburgh* 5:512.

—— 1899. "The Age of the Earth as an Abode Fitted for Life," *Philosophical Magazine,* 5th ser. 47:66–90.

—— and Peter Guthrie Tait. 1867. *Treatise on Natural Philosophy.* Cambridge, England: The University Press.

—— and Peter Guthrie Tait. 1890. *Treatise on Natural Philosophy.* 2nd ed. Edited by George Howard Darwin. Cambridge, England: The University Press.

Todhunter, Isaac. 1876. *William Whewell, D. D., Master of Trinity College, Cambridge, An Account of his Writings with Selections from his Literary and Scientific Correspondence.* London: Macmillan.

Turrill, W. B. 1953. *Pioneer Plant Geography: the Phytogeographical Researches of Sir Joseph Dalton Hooker.* The Hague: Nijoff.

—— 1963. *Joseph Dalton Hooker.* London: Nelson.

Vorzimmer, Peter. 1963. "Charles Darwin and Blending Inheritance," *Isis* 54:371–390.

—— 1970. *Charles Darwin, The Years of Controversy.* Philadelphia: Temple University Press.

Wallace, A. R. 1858. "On the Tendency of Varieties to Depart Indefinitely from the Original Type, Read July 1, 1858 before the Linnaean Society," *Journal of the Proceedings of the Linnaean Society* (1858), 3:45–62; *Zoologist* (1859), 17:6293–6308. Reprinted in *The Darwin Reader*. Edited by M. Bates and P. Humphrey (1956). New York: Charles Scribner's Sons.

—— 1863. "Remarks on the Rev. S. Haughton's Paper on the Bee's Cell, and on the Origin of Species," *Annals and Magazine of Natural History* 12:303–309.

—— 1870. *Contributions to the Theory of Natural Selection*. London: Macmillan.

—— 1871. Review of *The Descent of Man*. *Academy* 2:177–182.

—— 1872. Review of Bastian's *The Beginnings of Life*. *Nature* 6:284–287; 299–303.

—— 1889. *Darwinism*. London: Macmillan.

Walsh, H. T. 1962. "Whewell and Mill on Induction," *Philosophy of Science* 29:279–284.

Wells, William Chambers. 1818. *Two Essays*. London: Archibald Constable.

Westwood, J. O. 1860. "Mr. Darwin's Theory of Development," *Gardeners' Chronicle and Agricultural Gazette,* p. 122.

Whewell, William. 1833. *Astronomy and General Physics Considered with Reference to Natural Theology*. Vol. III of the Bridgewater Treatises. London: Pickering.

—— 1837. *History of the Inductive Sciences*. London: Parker.

—— 1840. *The Philosophy of the Inductive Sciences, Founded upon Their History*. London: Parker.

—— 1849. *Of Induction, with Especial Reference to Mr. J. Stuart Mill's System of Logic*. London: Parker.

Wiener, Philip P. 1949. *Evolution and the Founders of Pragmatism*. Cambridge, Mass.: Harvard University Press.

Wilberforce, Samuel. 1860. "Darwin's *Origin of Species*," *Quarterly Review* 108:225–264.

Williams, Mary B. 1970. "Deducing the Consequences of Evolution: A Mathematical Model," *Journal of Theoretical Biology* 29:343–385.

Wilson, Leonard G. 1970. *Sir Charles Lyell's Scientific Journals on the Species Question*. New Haven Ct.: Yale University Press.

Wollaston, T. V. 1856. *On the Variation of Species: with Special Reference to the Insecta, followed by an Inquiry into the Nature of Genera*. London: Van Voost.

—— 1860. Review of the *Origin of Species*. *Annals and Magazine of Natural History* 5:132–143.

Woodger, J. H. 1929. *Biological Principles*. London: Routledge & Kegan Paul.

Wright, Chauncey. 1860. "On the Architecture of Bees," *Proceedings of the American Academy of Arts and Science* 4:432.

—— 1871. "The Genesis of Species." *The North American Review* 113:63–103; reprinted as *Darwinism: Being and Examination of Mr. St. George Mivart's 'Genesis of Species'* (*1871*), with an Appendix on Final Causes. London: Murray.

—— 1872. "Evolution by Natural Selection," *The North American Review* 115:1–30.

—— 1877. *Philosophical Discussions*. Edited by C. E. Norton. New York: Henry Holt.

Index